Graduate Texts in Mathematics 180

S.M. Srivastava

A Course on Borel Sets

With 11 Illustrations

S.M. Srivastava
Stat-Math Unit
Indian Statistical Institute
203 B.T. Road
Calcutta, 700 035
India

Mathematics Subject Classification (1991): 04-01, 04A15, 28A05, 54H05

Library of Congress Cataloging-in-Publication Data
Srivastava, S.M. (Sashi Mohan)
A course on Borel sets / S.M. Srivastava.
p. cm. — (Graduate texts in mathematics ; 180)
Includes index.
ISBN 0-387-98412-7 (hard : alk. paper)
1. Borel sets. I. Title. II. Series.
QA248.S74 1998
511,3'2—dc21 97-43726

ISBN 0-387-98412-7 Springer-Verlag New York Berlin Heidelberg SPIN 10660569

Srivastava: A Course on Borel Sets

First Indian Reprint 2010

ISBN 978-81-8489-672-5

This edition is published by Springer (India) Private Limited, (a part of Springer Science+Business Media), Registered Office: 3rd Floor, Gandharva Mahavidyalaya, 212, Deen Dayal Upadhyaya Marg, New Delhi – 110 002, India.

Printed in India by Rashtriya Printers, Delhi.

This book is dedicated to the memory of
my beloved wife, Kiran
who passed away soon after this book was completed.

Acknowledgments

I am grateful to many people who have suggested improvements in the original manuscript for this book. In particular I would like to thank S. C. Bagchi, R. Barua, S. Gangopadhyay (née Bhattacharya), J. K. Ghosh, M. G. Nadkarni, and B. V. Rao. My deepest feelings of gratitude and appreciation are reserved for H. Sarbadhikari who very patiently read several versions of this book and helped in all possible ways to bring the book to its present form. It is a pleasure to record my appreciation for A. Maitra who showed the beauty and power of Borel sets to a generation of Indian mathematicians including me. I also thank him for his suggestions during the planning stage of the book.

I thank P. Bandyopadhyay who helped me immensely to sort out all the LaTeX problems. Thanks are also due to R. Kar for preparing the LaTeX files for the illustrations in the book.

I am indebted to S. B. Rao, Director of the Indian Statistical Institute for extending excellent moral and material support. All my colleagues in the Stat – Math Unit also lent a much needed and invaluable moral support during the long and difficult period that the book was written. I thank them all.

I take this opportunity to express my sincere feelings of gratitude to my children, Rosy and Ravi, for their great understanding of the task I took onto myself. What they missed during the period the book was written will be known to only the three of us. Finally, I pay homage to my late wife, Kiran who really understood what mathematics meant to me.

S. M. Srivastava

Contents

Acknowledgments vii

Introduction xi

About This Book xv

1 Cardinal and Ordinal Numbers 1

1.1 Countable Sets 1
1.2 Order of Infinity 4
1.3 The Axiom of Choice 7
1.4 More on Equinumerosity 11
1.5 Arithmetic of Cardinal Numbers 13
1.6 Well-Ordered Sets 15
1.7 Transfinite Induction 18
1.8 Ordinal Numbers 21
1.9 Alephs 24
1.10 Trees 26
1.11 Induction on Trees 29
1.12 The Souslin Operation 31
1.13 Idempotence of the Souslin Operation 34

2 Topological Preliminaries 39

2.1 Metric Spaces 39
2.2 Polish Spaces 52
2.3 Compact Metric Spaces 57
2.4 More Examples 63

2.5 The Baire Category Theorem 69
2.6 Transfer Theorems 74

3 Standard Borel Spaces **81**
3.1 Measurable Sets and Functions 81
3.2 Borel-Generated Topologies 91
3.3 The Borel Isomorphism Theorem 94
3.4 Measures 100
3.5 Category 107
3.6 Borel Pointclasses 115

4 Analytic and Coanalytic Sets **127**
4.1 Projective Sets 127
4.2 $\boldsymbol{\Sigma}_1^1$ and $\boldsymbol{\Pi}_1^1$ Complete Sets 135
4.3 Regularity Properties 141
4.4 The First Separation Theorem 147
4.5 One-to-One Borel Functions 150
4.6 The Generalized First Separation Theorem 155
4.7 Borel Sets with Compact Sections 157
4.8 Polish Groups 160
4.9 Reduction Theorems 164
4.10 Choquet Capacitability Theorem 172
4.11 The Second Separation Theorem 175
4.12 Countable-to-One Borel Functions 178

5 Selection and Uniformization Theorems **183**
5.1 Preliminaries 184
5.2 Kuratowski and Ryll-Nardzewski's Theorem 189
5.3 Dubins – Savage Selection Theorems 194
5.4 Partitions into Closed Sets 195
5.5 Von Neumann's Theorem 198
5.6 A Selection Theorem for Group Actions 200
5.7 Borel Sets with Small Sections 204
5.8 Borel Sets with Large Sections 206
5.9 Partitions into G_δ Sets 212
5.10 Reflection Phenomenon 216
5.11 Complementation in Borel Structures 218
5.12 Borel Sets with σ-Compact Sections 219
5.13 Topological Vaught Conjecture 227
5.14 Uniformizing Coanalytic Sets 236

References **241**

Glossary **251**

Index **253**

Introduction

The roots of Borel sets go back to the work of Baire [8]. He was trying to come to grips with the abstract notion of a function introduced by Dirichlet and Riemann. According to them, a function was to be an arbitrary correspondence between objects without giving any method or procedure by which the correspondence could be established. Since all the specific functions that one studied were determined by simple analytic expressions, Baire delineated those functions that *can be constructed starting from continuous functions and iterating the operation of pointwise limit on a sequence of functions.* These functions are now known as **Baire functions**. Lebesgue [65] and Borel [19] continued this work. In [19], Borel sets were defined for the first time. In his paper, Lebesgue made a systematic study of Baire functions and introduced many tools and techniques that are used even today. Among other results, he showed that Borel functions coincide with Baire functions. The study of Borel sets got an impetus from an error in Lebesgue's paper, which was spotted by Souslin. Lebesgue was trying to prove the following:

Suppose $f : \mathbb{R}^2 \longrightarrow \mathbb{R}$ is a Baire function such that for every x, the equation

$$f(x, y) = 0$$

has a unique solution. Then y as a function of x defined by the above equation is Baire.

The wrong step in the proof was hidden in a lemma stating that a set of real numbers that is the projection of a Borel set in the plane is Borel. (Lebesgue left this as a trivial fact!) Souslin called the projection of a Borel set **analytic** because such a set can be constructed using analytical operations of union and intersection on intervals. He showed that there are analytic sets that are not Borel. Immediately after this, Souslin [111] and Lusin [67] made a deep study of analytic sets and established most of the basic results about them. Their results showed that analytic sets are of fundamental importance to the theory of Borel sets and give it its power. For instance, Souslin proved that *Borel sets are precisely those analytic sets whose complements are also analytic*. Lusin showed that *the image of a Borel set under a one-to-one Borel map is Borel*. It follows that Lebesgue's thoerem—though not the proof—was indeed true.

Around the same time Alexandrov was working on the continuum hypothesis of Cantor: *Every uncountable set of real numbers is in one-to-one correspondence with the real line*. Alexandrov showed that *every uncountable Borel set of reals is in one-to-one correspondence with the real line* [2]. In other words, a Borel set cannot be a counterexample to the continuum hypothesis.

Unfortunately, Souslin died in 1919. The work on this new-found topic was continued by Lusin and his students in Moscow and by Sierpiński and his collaborators in Warsaw.

The next important step was the introduction of **projective sets** by Lusin [68], [69], [70] and Sierpiński [105] in 1925: *A set is called projective if it can be constructed starting with Borel sets and iterating the operations of projection and complementation*. Since Borel sets as well as projective sets are sets that can be described using simple sets like intervals and simple set operations, their theory came to be known as **descriptive set theory**. It was clear from the beginning that the theory of projective sets was riddled with problems that did not seem to admit simple solutions. As it turned out, logicians did show later that most of the regularity properties of projective sets, e.g., whether they satisfy the continuum hypothesis or not or whether they are Lebesgue measurable and have the property of Baire or not, are independent of the axioms of classical set theory.

Just as Alexandrov was trying to determine the status of the continuum hypothesis within Borel sets, Lusin [71] considered the status of the axiom of choice within "Borel families." He raised a very fundamental and difficult question on Borel sets that enriched its theory significantly. Let B be a subset of the plane. A subset C of B **uniformizes** B if it is the graph of a function such that its projection on the line is the same as that of B. (See Figure 1.)

Lusin asked, When does a Borel set B in the plane admit a Borel uniformization? By Lusin's theorem stated earlier, if B admits a Borel uniformization, its projection to the line must be Borel. In [16] Blackwell [16] showed that this condition is not sufficient. Several authors considered this problem and gave sufficient conditions under which Lusin's question has a positive answer. For instance, *a Borel set admits a Borel uniformization if the sections of B are countable (Lusin [71]) or compact (Novikov [90]) or σ-compact (Arsenin [3] and Kunugui [60]) or nonmeager (Kechris [52] and Sarbadhikari* [100]). Even today these results are ranked among the

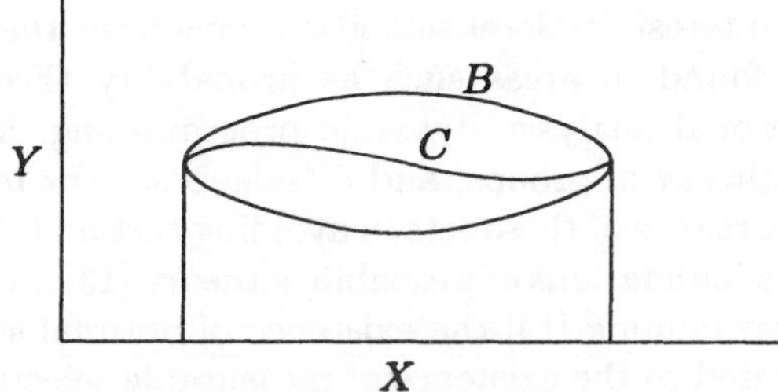

Figure 1. Uniformization

finest results on Borel sets. For the uniformization of Borel sets in general, the most important result proved before the war is due to Von Neumann [124]: *For every Borel subset B of the square* $[0,1] \times [0,1]$, *there is a null set N and a Borel function* $f : [0,1] \setminus N \longrightarrow [0,1]$ *whose graph is contained in B*. As expected, this result has found important applications in several branches of mathematics.

So far we have mainly been giving an account of the theory developed before the war; i.e., up to 1940. Then for some time there was a lull, not only in the theory of Borel sets, but in the whole of descriptive set theory. This was mainly because most of the mathematicians working in this area at that time were trying to extend the theory to higher projective classes, which, as we know now, is not possible within Zermelo – Fraenkel set theory. Fortunately, around the same time significant developments were taking place in logic that brought about a great revival of descriptive set theory that benefited the theory of Borel sets too. The fundamental work of Gödel on the incompleteness of formal systems [44] ultimately gave rise to a rich and powerful theory of recursive functions. Addison [1] established a strong connection between descriptive set theory and recursive function theory. This led to the development of a more general theory called **effective descriptive set theory**. (The theory as developed by Lusin and others has become known as **classical descriptive set theory**.)

From the beginning it was apparent that the effective theory is more powerful than the classical theory. However, the first concrete evidence of this came in the late seventies when Louveau [66] proved a beautiful theorem on Borel sets in product spaces. Since then several classical results have been proved using effective methods for which no classical proof is known yet; see, e.g., [47]. Forcing, a powerful set-theoretic technique (invented by Cohen to show the independence of the continuum hypothesis and the axiom of choice from other axioms of set theory [31]), and other set-theoretic tools such as determinacy and constructibility, have been very effectively used to make the theory of Borel sets a very powerful theory. (See Bartoszyński and Judah [9], Jech [49], Kechris [53], and Moschovakis [88].)

Much of the interest in Borel sets also stems from the applications that its theory has found in areas such as probability theory, mathematical statistics, functional analysis, dynamic programming, harmonic analysis, representation theory of groups, and C^*-algebras. For instance, Blackwell showed the importance of these sets in avoiding certain inherent pathologies in Kolmogorov's foundations of probability theory [13]; in Blackwell's model of dynamic programming [14] the existence of optimal strategies has been shown to be related to the existence of measurable selections (Maitra [74]); Mackey made use of these sets in problems regarding group representations, and in particular in defining topologies on measurable groups [72]; Choquet [30], [34] used these sets in potential theory; and so on. The theory of Borel sets has found uses in diverse applied areas such as optimization, control theory, mathematical economics, and mathematical statistics [5], [10], [32], [42], [91], [55]. These applications, in turn, have enriched the theory of Borel sets itself considerably. For example, most of the measurable selection theorems arose in various applications, and now there is a rich supply of them. Some of these, such as the cross-section theorems for Borel partitions of Polish spaces due to Mackey, Effros, and Srivastava are basic results on Borel sets.

Thus, today the theory of Borel sets stands on its own as a powerful, deep, and beautiful theory. This book is an introduction to this theory.

About This Book

This book can be used in various ways. It can be used as a stepping stone to descriptive set theory. From this point of view, our audience can be undergraduate or beginning graduate students who are still exploring areas of mathematics for their research. In this book they will get a reasonably thorough introduction to Borel sets and measurable selections. They will also find the kind of questions that a descriptive set theorist asks. Though we stick to Borel sets only, we present quite a few important techniques, such as universal sets, prewellordering, and scales, used in descriptive set theory. We hope that students will find the mathematics presented in this book solid and exciting.

Secondly, this book is addressed to mathematicians requiring Borel sets, measurable selections, etc., in their work. Therefore, we have tried our best to make it a convenient reference book. Some applications are also given just to show the way that the results presented here are used.

Finally, we desire that the book be accessible to all mathematicians. Hence the book has been made self-contained and has been written in an easygoing style. We have refrained from displaying various advanced techniques such as games, recursive functions, and forcing. We use only naive set theory, general topology, some analysis, and some algebra, which are commonly known.

The book is divided into five chapters. In the first chapter we give the set-theoretic preliminaries. In the first part of this chapter we present cardinal arithmetic, methods of transfinite induction, and ordinal numbers. Then we introduce trees and the Souslin operation. Topological preliminaries are presented in Chapter 2. We later develop the theory of Borel sets in the general context of Polish spaces. Hence we give a fairly complete account of Polish spaces in this chapter. In the last section of this chapter we prove several theorems that help in transferring many problems from general Polish spaces to the space of sequences $\mathbb{N}^{\mathbb{N}}$ or the Cantor space $2^{\mathbb{N}}$. We introduce Borel sets in Chapter 3. Here we develop the theory of Borel sets as much as possible without using analytic sets. In the last section of this chapter we introduce the usual hierarchy of Borel sets. For the first time, readers will see some of the standard methods of descriptive set theory, such as universal sets, reduction, and separation principles. Chapter 4 is central to this book, and the results proved here bring out the inherent power of Borel sets. In this chapter we introduce analytic and coanalytic sets and prove most of their basic properties. That these concepts are of fundamental importance to Borel sets is amply demonstrated in this chapter. In Chapter 5 we present most of the major measurable selection and uniformization theorems. These results are particularly important for applications. We close this chapter with a discussion on Vaught's conjecture—an outstanding open problem in descriptive set theory, and with a proof of Kondô's uniformization of coanalytic sets.

The exercises given in this book are an integral part of the theory, and readers are advised not to skip them. Many exercises are later treated as proved theorems.

Since this book is intended to be introductory only, many results on Borel sets that we would have much liked to include have been omitted. For instance, Martin's determinacy of Borel games [80], Silver's theorem on counting the number of equivalence classes of a Borel equivalence relation [106], and Louveau's theorem on Borel sets in the product [66] have not been included. Similarly, other results requiring such set-theoretic techniques as constructibility, large cardinals, and forcing are not given here. In our insistence on sticking to Borel sets, we have made only a passing mention of higher projective classes. We are sure that this will leave many descriptive set theorists dissatisfied.

We have not been able to give many applications, to do justice to which we would have had to enter many areas of mathematics, sometimes even delving deep into the theories. Clearly, this would have increased the size of the book enormously and made it unwieldy. We hope that users will find the passing remarks and references given helpful enough to see how results proved here are used in their respective disciplines.

1
Cardinal and Ordinal Numbers

In this chapter we present some basic set-theoretical notions. The first five sections[1] are devoted to cardinal numbers. We use Zorn's lemma to develop cardinal arithmetic. Ordinal numbers and the methods of transfinite induction on well-ordered sets are presented in the next four sections. Finally, we introduce trees and the Souslin operation. Trees are also used in several other branches of mathematics such as infinitary combinatorics, logic, computer science, and topology. The Souslin operation is of special importance to descriptive set theory, and perhaps it will be new to some readers.

1.1 Countable Sets

Two sets A and B are called **equinumerous or of the same cardinality**, written $A \equiv B$, if there exists a one-to-one map f from A onto B. Such an f is called a **bijection**. For sets A, B, and C we can easily check the following.

$$A \equiv A,$$

$$A \equiv B \Longrightarrow B \equiv A, \text{ and}$$

$$(A \equiv B \;\&\; B \equiv C) \Longrightarrow A \equiv C.$$

[1]These are produced here from my article [117] with the permission of the Indian Academy of Sciences.

A set A is called **finite** if there is a bijection from $\{0, 1, \ldots, n-1\}$ (n a natural number) onto A. (For $n = 0$ we take the set $\{0, 1, \ldots, n-1\}$ to be the empty set $\emptyset$.) If A is not finite, we call it **infinite**. The set A is called **countable** if it is finite or if there is a bijection from the set $\mathbf{N}$ of natural numbers $\{0, 1, 2, \ldots\}$ onto A. If a set is not countable, we call it **uncountable**.

Exercise 1.1.1 Show that a set is countable if and only if its elements can be enumerated as $a_0, a_1, a_2, \ldots$, (perhaps by repeating some of its elements); i.e., A is countable if and only if there is a map f from $\mathbf{N}$ onto A.

Exercise 1.1.2 Show that every subset of a countable set is countable.

Example 1.1.3 We can enumerate $\mathbf{N} \times \mathbf{N}$, the set of ordered pairs of natural numbers, by the diagonal method as shown in the following diagram

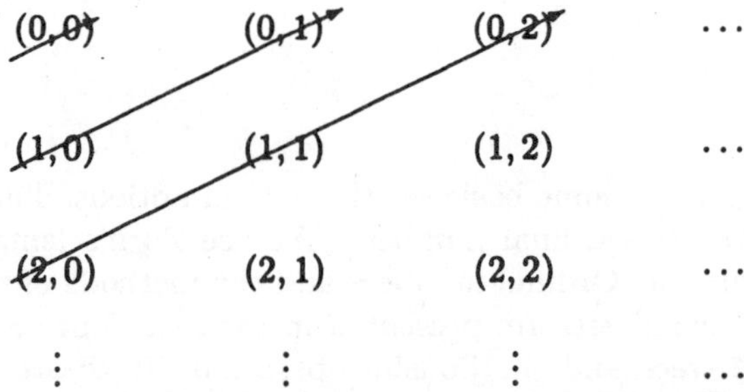

That is, we enumerate the elements of $\mathbf{N} \times \mathbf{N}$ as $(0,0), (1,0), (0,1), (2,0), (1,1), (0,2), \ldots$. By induction on k, k a positive integer, we see that $\mathbf{N}^k$, the set of all k-tuples of natural numbers, is also countable.

Theorem 1.1.4 *Let $A_0, A_1, A_2, \ldots$ be countable sets. Then their union $A = \bigcup_0^\infty A_n$ is countable.*

Proof. For each n, choose an enumeration $a_{n0}, a_{n1}, a_{n2}, \ldots$ of A_n. We enumerate $A = \bigcup_n A_n$ following the above diagonal method.

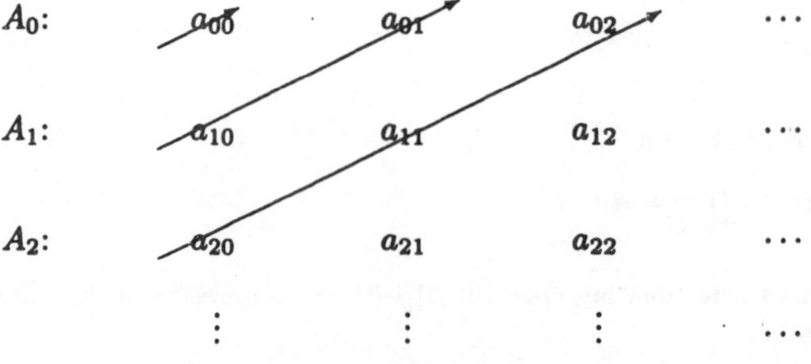

■

Example 1.1.5 Let $\mathbb{Q}$ be the set of all rational numbers. We have

$$\mathbb{Q} = \bigcup_{n>0} \{m/n : m \text{ an integer}\}.$$

By 1.1.4, $\mathbb{Q}$ is countable.

Exercise 1.1.6 Let X be a countable set. Show that $X \times \{0,1\}$, the set X^k of all k-tuples of elements of X, and $X^{<\mathbb{N}}$, the set of all finite sequences of elements of X including the empty sequence e, are all countable.

A real number is called **algebraic** if it is a root of a polynomial with integer coefficients.

Exercise 1.1.7 Show that the set $\mathbb{K}$ of algebraic numbers is countable.

The most natural question that arises now is; Are there uncountable sets? The answer is yes, as we see below.

Theorem 1.1.8 *(Cantor) For any two real numbers a, b with $a < b$, the interval $[a, b]$ is uncountable.*

Proof. (Cantor) Let (a_n) be a sequence in $[a, b]$. Define an increasing sequence (b_n) and a decreasing sequence (c_n) in $[a, b]$ inductively as follows: Put $b_0 = a$ and $c_0 = b$. For some $n \in \mathbb{N}$, suppose

$$b_0 < b_1 < \cdots < b_n < c_n < \cdots < c_1 < c_0$$

have been defined. Let i_n be the first integer i such that $b_n < a_i < c_n$ and j_n the first integer j such that $a_{i_n} < a_j < c_n$. Since $[a, b]$ is infinite i_n, j_n exist. Put $b_{n+1} = a_{i_n}$ and $c_{n+1} = a_{j_n}$.

Let $x = \sup\{b_n : n \in \mathbb{N}\}$. Clearly, $x \in [a, b]$. Suppose $x = a_k$ for some k. Clearly, $x \le c_m$ for all m. So, by the definition of the sequence (b_n) there is an integer i such that $b_i > a_k = x$. This contradiction shows that the range of the sequence (a_n) is not the whole of $[a, b]$. Since (a_n) was an arbitrary sequence, the result follows. ∎

Let X and Y be sets. The collection of all subsets of a set X is itself a set, called the **power set** of X and denoted by $\mathcal{P}(X)$. Similarly, the collection of all functions from Y to X forms a set, which we denote by X^Y.

Theorem 1.1.9 *The set $\{0,1\}^{\mathbb{N}}$, consisting of all sequences of 0's and 1's, is uncountable.*

Proof. Let (α_n) be a sequence in $\{0,1\}^{\mathbb{N}}$. Define $\alpha \in \{0,1\}^{\mathbb{N}}$ by

$$\alpha(n) = 1 - \alpha_n(n), \; n \in \mathbb{N}.$$

Then $\alpha \neq \alpha_i$ for all i. Since (α_n) was arbitrary, our result is proved. ∎

Exercise 1.1.10 (a) Show that the intervals $(0, 1)$ and $(0, 1]$ are of the same cardinality.

(b) Show that any two nondegenerate intervals (which may be bounded or unbounded and may or may not include endpoints) have the same cardinality. Hence, any such interval is uncountable.

A number is called **transcendental** if it is not algebraic.

Exercise 1.1.11 Show that the set of all transcendental numbers in any nondegenerate interval is uncountable.

1.2 Order of Infinity

So far we have seen only two different "orders of infinity"—that of $\mathbb{N}$ and that of $\{0,1\}^{\mathbb{N}}$. Are there any more? In this section we show that there are many.

We say that **the cardinality of a set A is less than or equal to the cardinality of a set B**, written $A \leq_c B$, if there is a one-to-one function f from A to B. Note that $\emptyset \leq_c A$ for all A (Why?), and for sets A, B, C,

$$(A \leq_c B \,\&\, B \leq_c C) \Longrightarrow A \leq_c C.$$

If $A \leq_c B$ but $A \not\equiv B$, then we say that **the cardinality of A is less than the cardinality of B** and symbolically write $A <_c B$. Notice that $\mathbb{N} <_c \mathbb{R}$.

Theorem 1.2.1 *(Cantor) For any set X, $X <_c \mathcal{P}(X)$.*

Proof. First assume that $X = \emptyset$. Then $\mathcal{P}(X) = \{\emptyset\}$. The only function on X is the empty function $\emptyset$, which is not onto $\{\emptyset\}$. This observation proves the result when $X = \emptyset$.

Now assume that X is nonempty. The map $x \longrightarrow \{x\}$ from X to $\mathcal{P}(X)$ is one-to-one. Therefore, $X \leq_c \mathcal{P}(X)$. Let $f : X \longrightarrow \mathcal{P}(X)$ be any map. We show that f cannot be onto $\mathcal{P}(X)$. This will complete the proof.

Consider the set

$$A = \{x \in X | x \notin f(x)\}.$$

Suppose $A = f(x_0)$ for some $x_0 \in X$. Then

$$x_0 \in A \Longleftrightarrow x_0 \notin A.$$

This contradiction proves our claim. ■

Remark 1.2.2 This proof is an imitation of the proof of 1.1.9. To see this, note the following. If A is a subset of a set X, then its **characteristic**

function is the map $\chi_A : X \longrightarrow \{0,1\}$, where

$$\chi_A(x) = \begin{cases} 1 & \text{if } x \in A, \\ 0 & \text{otherwise.} \end{cases}$$

We can easily verify that $A \longrightarrow \chi_A$ defines a one-to-one map from $\mathcal{P}(X)$ onto $\{0,1\}^X$. We have shown that there is no map f from X onto $\mathcal{P}(X)$ in exactly the same way as we showed that $\{0,1\}^{\mathbb{N}}$ is uncountable.

Now we see that

$$\mathbb{N} <_c \mathcal{P}(\mathbb{N}) <_c \mathcal{P}(\mathcal{P}(\mathbb{N})) <_c \cdots.$$

Let T be the union of all the sets $\mathbb{N}$, $\mathcal{P}(\mathbb{N})$, $\mathcal{P}(\mathcal{P}(\mathbb{N}))$, ... Then T is of cardinality larger than each of the sets described above. We can now similarly proceed with T and get a never-ending class of sets of higher and higher cardinalities! A very interesting question arises now: Is there an infinite set whose cardinality is different from the cardinalities of each of the sets so obtained? In particular, is there an uncountable set of real numbers of cardinality less than that of $\mathbb{R}$? These turned out to be among the most fundamental problems not only in set theory but in the whole of mathematics. We shall briefly discuss these later in this chapter.

The following result is very useful in proving the equinumerosity of two sets. It was first stated and proved (using the axiom of choice) by Cantor.

Theorem 1.2.3 *(Schröder – Bernstein Theorem) For any two sets X and Y,*

$$(X \leq_c Y \;\&\; Y \leq_c X) \Longrightarrow X \equiv Y.$$

Proof. (Dedekind) Let $X \leq_c Y$ and $Y \leq_c X$. Fix one-to-one maps $f : X \longrightarrow Y$ and $g : Y \longrightarrow X$. We have to show that X and Y have the same cardinality; i.e., that there is a bijection h from X onto Y.

We first show that there is a set $E \subseteq X$ such that

$$g^{-1}(X \setminus E) = Y \setminus f(E). \qquad (\star)$$

(See Figure 1.1.) Assuming that such a set E exists, we complete the proof as follows. Define $h : X \longrightarrow Y$ by

$$h(x) = \begin{cases} f(x) & \text{if } x \in E, \\ g^{-1}(x) & \text{otherwise.} \end{cases}$$

The map $h : X \longrightarrow Y$ is clearly seen to be one-to-one and onto.

We now show the existence of a set $E \subseteq X$ satisfying $(\star)$. Consider the map $\mathcal{H} : \mathcal{P}(X) \longrightarrow \mathcal{P}(X)$ defined by

$$\mathcal{H}(A) = X \setminus g(Y \setminus f(A)), \quad A \subseteq X.$$

It is easy to check that

(i) $A \subseteq B \subseteq X \Longrightarrow \mathcal{H}(A) \subseteq \mathcal{H}(B)$, and

(ii) $\mathcal{H}(\bigcup_n A_n) = \bigcup_n \mathcal{H}(A_n)$.

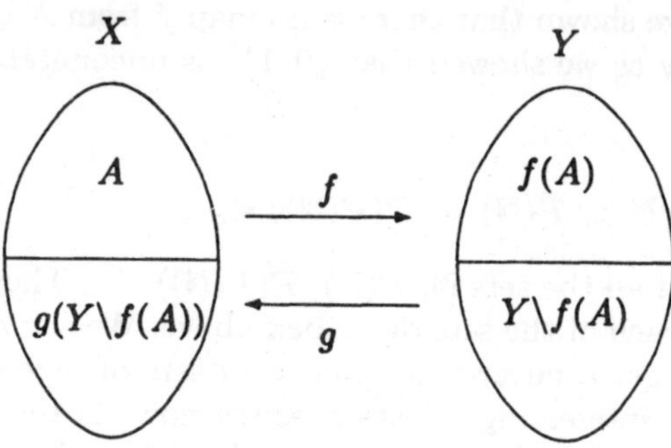

Figure 1.1

Now define a sequence (A_n) of subsets of X inductively as follows:

$A_0 = \emptyset$, and

$A_{n+1} = \mathcal{H}(A_n)$, $n = 0, 1, 2, \ldots$.

Let $E = \bigcup_n A_n$. Then, $\mathcal{H}(E) = E$. The set E clearly satisfies $(\star)$. ■

Corollary 1.2.4 *For sets A and B,*

$$A <_c B \Longleftrightarrow A \leq_c B \ \&\ B \not\leq_c A.$$

Here are some applications of the Schröder – Bernstein theorem.

Example 1.2.5 Define $f : \mathcal{P}(\mathbb{N}) \longrightarrow \mathbb{R}$, the set of all real numbers, by

$$f(A) = \sum_{n \in A} \frac{2}{3^{n+1}},\ A \subseteq \mathbb{N}.$$

Then f is one-to-one. Therefore, $\mathcal{P}(\mathbb{N}) \leq_c \mathbb{R}$. Now consider the map $g : \mathbb{R} \longrightarrow \mathcal{P}(\mathbb{Q})$ by

$$g(x) = \{r \in \mathbb{Q} | r < x\},\ x \in \mathbb{R}.$$

Clearly, g is one-to-one and so $\mathbb{R} \leq_c \mathcal{P}(\mathbb{Q})$. As $\mathbb{Q} \equiv \mathbb{N}$, $\mathcal{P}(\mathbb{Q}) \equiv \mathcal{P}(\mathbb{N})$. Therefore, $\mathbb{R} \leq_c \mathcal{P}(\mathbb{N})$. By the Schröder – Bernstein theorem, $\mathbb{R} \equiv \mathcal{P}(\mathbb{N})$. Since $\mathcal{P}(\mathbb{N}) \equiv \{0,1\}^{\mathbb{N}}$, $\mathbb{R} \equiv \{0,1\}^{\mathbb{N}}$.

Example 1.2.6 Fix a one-to-one map $x \longrightarrow (x_0, x_1, x_2, \ldots)$ from $\mathbb{R}$ onto $\{0,1\}^{\mathbb{N}}$, the set of sequences of 0's and 1's. Then the function $(x, y) \longrightarrow (x_0, y_0, x_1, y_1, \ldots)$ from $\mathbb{R}^2$ to $\{0,1\}^{\mathbb{N}}$ is one-to-one and onto. So, $\mathbb{R}^2 \equiv \{0,1\}^{\mathbb{N}} \equiv \mathbb{R}$. By induction on the positive integers k, we can now show that $\mathbb{R}^k$ and $\mathbb{R}$ are equinumerous.

Exercise 1.2.7 Show that $\mathbb{R}$ and $\mathbb{R}^{\mathbb{N}}$ are equinumerous, where $\mathbb{R}^{\mathbb{N}}$ is the set of all sequences of real numbers.

(Hint: Use $\mathbb{N} \times \mathbb{N} \equiv \mathbb{N}$.)

Exercise 1.2.8 Show that the set of points on a line and the set of lines in a plane are equinumerous.

Exercise 1.2.9 Show that there is a family $\mathcal{A}$ of infinite subsets of $\mathbb{N}$ such that

(i) $\mathcal{A} \equiv \mathbb{R}$, and

(ii) for any two distinct sets A and B in $\mathcal{A}$, $A \bigcap B$ is finite.

1.3 The Axiom of Choice

Are the sizes of any two sets necessarily comparable? That is, for any two sets X and Y, is it true that at least one of the relations $X \leq_c Y$ or $Y \leq_c X$ holds? To answer this question, we need a hypothesis on sets known as the axiom of choice.

The Axiom of Choice (AC) *If $\{A_i\}_{i \in I}$ is a family of nonempty sets, then there is a function $f : I \longrightarrow \bigcup_i A_i$ such that $f(i) \in A_i$ for every $i \in I$.* ■

Such a function f is called a **choice function** for $\{A_i : i \in I\}$. Note that if I is finite, then by induction on the number of elements in I we can show that a choice function exists. If I is infinite, then we do not know how to prove the existence of such a map. The problem can be explained by the following example of Russell. Let $A_0, A_1, A_2, \ldots$ be a sequence of pairs of shoes. Let $f(n)$ be the left shoe in the nth pair A_n, and so the choice function in this case certainly exists. Instead, let $A_0, A_1, A_2, \ldots$ be a sequence of pairs of socks. Now we are unable to give a rule to define a choice function for the sequence $A_0, A_1, A_2, \ldots$! **AC** asserts the existence of such a function without giving any rule or any construction for defining it. Because of its nonconstructive nature, **AC** met with serious criticism at first. However, **AC** is indispensable, not only for the theory of cardinal numbers, but for most branches of mathematics.

From now on, we shall be assuming AC.

Note that we used **AC** to prove that the union of a sequence of countable sets $A_0, A_1, \ldots$ is countable. For each n, we chose an enumeration of A_n.

But usually there are infinitely many such enumerations, and we did not specify any rule to choose one. It should, however, be noted that for some important specific instances of this result **AC** is not needed. For instance, we did not use **AC** to prove the countability of the set of rational numbers (1.1.5) or to prove the countability of $X^{<\mathbb{N}}$, X countable (1.1.6).

The next result shows that every infinite set X has a proper subset Y of the same cardinality as X. We use **AC** to prove this.

Theorem 1.3.1 *If X is infinite and $A \subseteq X$ finite, then $X \setminus A$ and X have the same cardinality.*

Proof. Let $A = \{a_0, a_1, \ldots, a_n\}$ with the a_i's distinct. By **AC**, there exist distinct elements $a_{n+1}, a_{n+2}, \ldots$ in $X \setminus A$. To see this, fix a choice function $f : \mathcal{P}(X) \setminus \{\emptyset\} \longrightarrow X$ such that $f(E) \in E$ for every nonempty subset E of X. Such a function exists by **AC**. Now inductively define $a_{n+1}, a_{n+2}, \ldots$ such that

$$a_{n+k+1} = f(X \setminus \{a_0, a_1, \ldots, a_{n+k}\}),$$

$k = 0, 1, \ldots$ Define $h : X \longrightarrow X \setminus A$ by

$$h(x) = \begin{cases} a_{n+k+1} & \text{if } x = a_k, \\ x & \text{otherwise.} \end{cases}$$

Clearly, $h : X \longrightarrow X \setminus A$ is one-to-one and onto. ∎

Corollary 1.3.2 *Show that for any infinite set X, $\mathbb{N} \leq_c X$; i.e., every infinite set X has a countable infinite subset.*

Exercise 1.3.3 Let X, Y be sets such that there is a map from X onto Y. Show that $Y \leq_c X$.

There are many equivalent forms of **AC**. One such is called **Zorn's lemma**, of which there are many natural applications in several branches of mathematics. In this chapter we shall give several applications of Zorn's lemma to the theory of cardinal numbers. We explain Zorn's lemma now.

A **partial order** on a set P is a binary relation R such that for any x, y, z in P,

xRx (reflexive),

$(xRy \;\&\; yRz) \Longrightarrow xRz$ (transitive), and

$(xRy \;\&\; yRx) \Longrightarrow x = y$ (anti-symmetric).

A set P with a partial order is called a **partially ordered set** or simply a **poset**. A **linear order** on a set X is a partial order R on X such that any two elements of X are comparable; i.e., for any $x, y \in X$, at least one of xRy or yRx holds. If X is a set with more than one element, then the inclusion relation $\subseteq$ on $\mathcal{P}(X)$ is a partial order that is not a linear order. Here are a few more examples of partial orders that are not linear orders.

Example 1.3.4 Let X and Y be any two sets. A **partial function** $f : X \longrightarrow Y$ is a function with domain a subset of X and range contained in Y. Let $f : X \longrightarrow Y$ and $g : X \longrightarrow Y$ be partial functions. We say that g **extends** f, or f is a **restriction** of g, written $g \succeq f$ or $f \preceq g$, if domain(f) is contained in domain(g) and $f(x) = g(x)$ for all $x \in$ domain(f). If f is a restriction of g and domain(f) $= A$, we write $f = g|A$. Let

$$Fn(X,Y) = \{f : f \text{ a one-to-one partial function from } X \text{ to } Y\}.$$

Suppose Y has more than one element and $X \neq \emptyset$. Then $(Fn(X,Y), \preceq)$ is a poset that is not linearly ordered.

Example 1.3.5 Let V be a vector space over any field F and P the set of all independent subsets of V ordered by the inclusion $\subseteq$. Then P is a poset that is not a linearly ordered set.

Fix a poset (P, R). A **chain in** P is a subset C of P such that R restricted to C is a linear order; i.e., for any two elements x, y of C at least one of the relations xRy or yRx must be satisfied. Let $A \subseteq P$. An **upper bound** for A is an $x \in P$ such that yRx for all $y \in A$. An $x \in P$ is called a **maximal element** of P if for no $y \in P$ different from x, xRy holds. In 1.3.4, a chain C in $Fn(X,Y)$ is a consistent family of partial functions, their common extension $\bigcup C$ an upper bound for C, and any partial function f with domain X or range Y a maximal element. So, there may be more than one maximal element in a poset that is not linearly ordered.

In 1.3.5, Let C be a chain in P. Then for any two elements E and F of P, either $E \subseteq F$ or $F \subseteq E$. It follows that $\bigcup C$ itself is an independent set and so is an upper bound of C.

Let $(L, \leq)$ be a linearly ordered set. An element x of L is called the **first (last) element** of L if $x \leq y$ (respectively $y \leq x$) for every $y \in L$. A linearly ordered set L is called **order dense** if for every $x < y$ there is a z such that $x < z < y$. Two linearly ordered sets are called **order isomorphic** or simply **isomorphic** if there is a one-to-one, order-preserving map from one onto the other.

Exercise 1.3.6 **(i)** Let L be a countable linearly ordered set. Show that there is a one-to-one, order-preserving map $f : L \longrightarrow \mathbb{Q}$, where $\mathbb{Q}$ has the usual order.

(ii) Let L be a countable linearly ordered set that is order dense and that has no first and no last element. Show that L is order isomorphic to $\mathbb{Q}$.

Zorn's Lemma *If P is a nonempty partially ordered set such that every chain in P has an upper bound in P, then P has a maximal element.* ■

As mentioned earlier, Zorn's lemma is equivalent to **AC**. We can easily prove **AC** from Zorn's lemma. To see this, fix a family $\{A_i : i \in I\}$ of nonempty subsets of a set X. A **partial choice function** for $\{A_i : i \in I\}$ is a choice function for a subfamily $\{A_i : i \in J\}$, $J \subseteq I$. Let P be the set of all partial choice functions for $\{A_i : i \in I\}$. As before, for f, g in P, we put $f \preceq g$ if g extends f. Then the poset $(P, \preceq)$ satisfies the hypothesis of Zorn's lemma. To see this, let $C = \{f_a : a \in A\}$ be a chain in P. Let $D = \bigcup_{a \in A} \text{domain}(f_a)$. Define $f : D \longrightarrow X$ by

$$f(x) = f_a(x) \quad \text{if} \quad x \in \text{domain}(f_a).$$

Since the f_a's are consistent, f is well defined. Clearly, f is an upper bound of C. By Zorn's lemma, let g be a maximal element of P. Suppose g is not a choice function for the family $\{A_i : i \in I\}$. Then domain$(g) \neq I$. Choose $i_0 \in I \setminus$ domain(g) and $x_0 \in A_{i_0}$. Let

$$h : \text{domain}(g) \bigcup \{i_0\} \longrightarrow \bigcup_i A_i$$

be the extension of g such that $h(i_0) = x_0$. Clearly, $h \in P$, $g \preceq h$, and $g \neq h$. This contradicts the maximality of g.

We refer the reader to [62] (Theorem 7, p. 256) for a proof of Zorn's lemma from **AC**.

Here is an application of Zorn's lemma to linear algebra.

Proposition 1.3.7 *Every vector space V has a basis.*

Proof. Let P be the poset defined in 1.3.5; i.e., P is the set of all independent subsets of V. Since every singleton set $\{v\}$, $v \neq 0$, is an independent set, $P \neq \emptyset$. As shown earlier, every chain in P has an upper bound. Therefore, by Zorn's lemma, P has a maximal element, say B. Suppose B does not span V. Take $v \in V \setminus$ span(B). Then $B \bigcup \{v\}$ is an independent set properly containing B. This contradicts the maximality of B. Thus B is a basis of V. ■

Exercise 1.3.8 Let F be any field and V an infinite dimensional vector space over F. Suppose V^* is the space of all linear functionals on V. It is well known that V^* is a vector space over F. Show that there exists an independent set B in V^* such that $B \equiv \mathbb{R}$.

Exercise 1.3.9 Let (A, R) be a poset. Show that there exists a linear order R' on A that extends R; i.e., for every $a, b \in A$,

$$aRb \Longrightarrow aR'b.$$

Exercise 1.3.10 Show that every set can be linearly ordered.

1.4 More on Equinumerosity

In this section we use Zorn's lemma to prove several general results on equinumerosity. These will be used to develop cardinal arithmetic in the next section.

Theorem 1.4.1 *For any two sets X and Y, at least one of*

$$X \leq_c Y \text{ or } Y \leq_c X$$

holds.

Proof. Without loss of generality we can assume that both X and Y are nonempty. We need to show that either there exists a one-to-one map $f : X \longrightarrow Y$ or there exists a one-to-one map $g : Y \longrightarrow X$. To show this, consider the poset $Fn(X, Y)$ of all one-to-one partial functions from X to Y as defined in 1.3.4. It is clearly nonempty. As shown earlier, every chain in $Fn(X, Y)$ has an upper bound. Therefore, by Zorn's lemma, P has a maximal element, say f_0. Then, either domain$(f_0) = X$ or range$(f_0) = Y$. If domain$(f_0) = X$, then f_0 is a one-to-one map from X to Y. So, in this case, $X \leq_c Y$. If range$(f_0) = Y$, then f_0^{-1} is a one-to-one map from Y to X, and so $Y \leq_c X$. ∎

As a corollary to the above theorem and the Schröder – Bernstein theorem, we get the following trichotomy theorem.

Corollary 1.4.2 *Let A and B be any two sets. Then exactly one of*

$$A <_c B, \quad A \equiv B, \quad \text{and } B <_c A$$

holds.

Theorem 1.4.3 *For every infinite set X,*

$$X \times \{0, 1\} \equiv X.$$

Proof. Let

$$P = \{(A, f) : A \subseteq X \text{ and } f : A \times \{0, 1\} \longrightarrow A \text{ a bijection}\}.$$

Since X is infinite, it contains a countably infinite set, say D. By 1.1.3, $D \times \{0, 1\} \equiv D$. Therefore, P is nonempty. Consider the partial order $\propto$ on P defined by

$$(A, f) \propto (B, g) \Longleftrightarrow A \subseteq B \; \& \; f \preceq g.$$

Following the argument contained in the proof of 1.4.1, we see that the hypothesis of Zorn's lemma is satisfied by P. So, P has a maximal element, say (A, f).

To complete the proof we show that $A \equiv X$. Since X is infinite, by 1.3.1, it will be sufficient to show that $X \setminus A$ is finite. Suppose not. By 1.3.2, there is a $B \subseteq X \setminus A$ such that $B \equiv \mathbb{N}$. So there is a one-to-one map g from $B \times \{0,1\}$ onto B. Combining f and g we get a bijection

$$h : (A \bigcup B) \times \{0,1\} \longrightarrow A \bigcup B$$

that extends f. This contradicts the maximality of (A, f). Hence, $X \setminus A$ is finite. Therefore, $A \equiv X$. The proof is complete. ∎

Corollary 1.4.4 *Every infinite set can be written as the union of k pairwise disjoint equinumerous sets, where k is any positive integer.*

Theorem 1.4.5 *For every infinite set X,*

$$X \times X \equiv X.$$

Proof. Let

$$P = \{(A, f) : A \subseteq X \text{ and } f : A \times A \longrightarrow A \text{ a bijection}\}.$$

Note that P is nonempty.

Consider the partial order $\propto$ on P defined by

$$(A, f) \propto (B, g) \Longleftrightarrow A \subseteq B \;\&\; f \preceq g.$$

By Zorn's lemma, take a maximal element (A, f) of P as in the proof of 1.4.3. Note that A must be infinite. To complete the proof, we shall show that $A \equiv X$. Suppose not. Then $A <_c X$. We first show that $X \setminus A \equiv X$.

Suppose $X \setminus A <_c X$. By 1.4.1, either $A \leq_c X \setminus A$ or $X \setminus A \leq_c A$. Assume first $X \setminus A \leq_c A$. Using 1.4.3, take two disjoint sets A_1, A_2 of the same cardinality as A and $A_1 \bigcup A_2 = A$. Now,

$$X = A \bigcup (X \setminus A) \leq_c A_1 \bigcup A_2 \equiv A <_c X.$$

This is a contradiction. Similarly we arrive at a contradiction from the other inequality. Thus, by 1.4.2, $X \setminus A \equiv X$.

Now choose $B \subseteq X \setminus A$ such that $B \equiv A$. By 1.4.4, write B as the union of three disjoint sets, say B_1, B_2, and B_3, each of the same cardinality as A. Since there is a one-to-one map from $A \times A$ onto A, there exist bijections $f_1 : B \times A \longrightarrow B_1$, $f_2 : B \times B \longrightarrow B_2$, and $f_3 : A \times B \longrightarrow B_3$. Let $C = A \bigcup B$. Combining these four bijections, we get a bijection $g : C \times C \longrightarrow C$ that is a proper extension of f. This contradicts the maximality of (A, f). Thus, $A \equiv X$. The proof is now complete. ∎

Exercise 1.4.6 Let X be an infinite set. Show that X, $X^{<\mathbb{N}}$, and the set of all finite sequences of X are equinumerous.

A **Hamel basis** is a basis of $\mathbb{R}$ considered as a vector space over the field of rationals $\mathbb{Q}$. Since every vector space has a basis, a Hamel basis exists.

Exercise 1.4.7 Show that if B is a Hamel basis, then $B \equiv \mathbb{R}$.

The next proposition, though technical, has important applications to cardinal arithmetic, as we shall see in the next section.

Proposition 1.4.8 *(J. König, [58]) Let $\{X_i : i \in I\}$ and $\{Y_i : i \in I\}$ be families of sets such that $X_i <_c Y_i$ for each $i \in I$. Then there is no map f from $\bigcup_i X_i$ onto $\Pi_i Y_i$.*

Proof. Let $f : \bigcup_i X_i \longrightarrow \Pi_i Y_i$ be any map. For any $i \in I$, let

$$A_i = Y_i \setminus \pi_i(f(X_i)),$$

where $\pi_i : \prod_j Y_j \longrightarrow Y_i$ is the projection map. Since for evry i, $X_i <_c Y_i$, each A_i is nonempty. By **AC**, $\Pi_i A_i \neq \emptyset$. But

$$\Pi_i A_i \bigcap \text{range}(f) = \emptyset.$$

It follows that f is not onto. ■

1.5 Arithmetic of Cardinal Numbers

For sets X, Y, and Z, we know the following.

$$X \equiv X,$$

$$X \equiv Y \Longrightarrow Y \equiv X, \text{ and}$$

$$(X \equiv Y \;\&\; Y \equiv Z) \Longrightarrow X \equiv Z.$$

So, to each set X we can assign a symbol, say $|X|$, called its **cardinal number**, such that

$$X \equiv Y \iff |X| \text{ and } |Y| \text{ are the same.}$$

In general, cardinal numbers are denoted by Greek letters κ, λ, μ with or without suffixes. However, some specific cardinals are denoted by special symbols. For example, we put

$$\begin{aligned} |\{0, 1, \ldots, n-1\}| &= n \qquad (n \text{ a natural number}), \\ |\mathbb{N}| &= \aleph_0, \text{ and} \\ |\mathbb{R}| &= \mathfrak{c}. \end{aligned}$$

As in the case of natural numbers, we can add, multiply and compare cardinal numbers. We define these notions now. Let λ and μ be two cardinal numbers. Fix sets X and Y such that $|X| = \lambda$ and $|Y| = \mu$. We define

$$\begin{array}{rcl} \lambda + \mu & = & |(X \times \{0\}) \bigcup (Y \times \{1\})|, \\ \lambda \cdot \mu & = & |X \times Y|, \\ \lambda^{\mu} & = & |X^Y|, \\ \lambda \leq \mu & \text{if} & X \leq_c Y, \text{ and} \\ \lambda < \mu & \text{if} & X <_c Y. \end{array}$$

The above definitions are easily seen to be independent of the choices of X and Y. Further, these extend the corresponding notions for natural numbers. Note that $2^{\lambda} = |\mathcal{P}(X)|$ if $|X| = \lambda$. We can define the sum and the product of infinitely many cardinals too. Let $\{\lambda_i : i \in I\}$ be a set of cardinal numbers. Fix a family $\{X_i : i \in I\}$ of sets such that $|X_i| = \lambda_i$, $i \in I$. We define

$$\Pi_i \lambda_i = |\Pi_i X_i|.$$

To define $\sum_i \lambda_i$, first note that there is a family $\{X_i : i \in I\}$ of pairwise disjoint sets such that $|X_i| = \lambda_i$; simply take a family $\{Y_i : i \in I\}$ of sets such that $|Y_i| = \lambda_i$ and put $X_i = Y_i \times \{i\}$. We define

$$\sum_i \lambda_i = |\bigcup_i X_i|.$$

With these notations, note that

$$\aleph_0 < 2^{\aleph_0} = \mathfrak{c},$$

$$\aleph_0 + \aleph_0 = \aleph_0 \cdot \aleph_0 = \aleph_0,$$

$$\mathfrak{c}^n = \mathfrak{c}^{\aleph_0} = \mathfrak{c} (n > 1), \text{ etc.}$$

Whatever we have proved about equinumerosity of sets; i.e., the results concerning union, product, $\leq_c$, etc., translate into corresponding results about cardinal numbers. For instance, by 1.2.1,

$$\forall \lambda (\lambda < 2^{\lambda}).$$

The Schröder – Bernstein theorem translates as follows:

$$\lambda \leq \mu \;\&\; \mu \leq \lambda \Longrightarrow \lambda = \mu.$$

The result on comparabilty of cardinals (1.4.1) becomes; For cardinals λ and μ at least one of

$$\lambda \leq \mu \text{ and } \mu \leq \lambda$$

holds. If λ is infinite, then

$$\lambda = \lambda + \lambda = \lambda \cdot \lambda.$$

Exercise 1.5.1 Let $\lambda \leq \mu$. Show that for any κ,

$$\lambda + \kappa \leq \mu + \kappa,\ \lambda \cdot \kappa \leq \mu \cdot \kappa, \lambda^\kappa \leq \mu^\kappa, \text{ and } \kappa^\lambda \leq \kappa^\mu.$$

Example 1.5.2

$$\begin{array}{lll} 2^{\mathfrak{c}} & \leq \aleph_0^{\mathfrak{c}} & (\text{since } 2 \leq \aleph_0) \\ & \leq \mathfrak{c}^{\mathfrak{c}} & (\text{since } \aleph_0 \leq \mathfrak{c}) \\ & = (2^{\aleph_0})^{\mathfrak{c}} & (\text{since } \mathfrak{c} = 2^{\aleph_0}) \\ & = 2^{\aleph_0 \cdot \mathfrak{c}} & (\text{since for nonempty sets } X, Y, Z, (X^Y)^Z \equiv X^{Y \times Z}) \\ & \leq 2^{\mathfrak{c} \cdot \mathfrak{c}} & (\text{since } \aleph_0 < \mathfrak{c}) \\ & = 2^{\mathfrak{c}} & (\text{since } \mathfrak{c} \cdot \mathfrak{c} = \mathfrak{c}). \end{array}$$

So, by the Schröder – Bernstein theorem, $2^{\mathfrak{c}} = \aleph_0^{\mathfrak{c}} = \mathfrak{c}^{\mathfrak{c}}$. It follows that

$$\{0,1\}^{\mathbb{R}} \equiv \mathbb{N}^{\mathbb{R}} \equiv \mathbb{R}^{\mathbb{R}}.$$

Exercise 1.5.3 (König's theorem, [58]) Let $\{\lambda_i : i \in I\}$ and $\{\mu_i : i \in I\}$ be nonempty sets of cardinal numbers such that $\lambda_i < \mu_i$ for each i. Show that

$$\sum_i \lambda_i < \Pi_i \mu_i.$$

(Hint: Use 1.4.8.)

1.6 Well-Ordered Sets

A **well-order** on a set W is a linear order $\leq$ on W such that every nonempty subset A of W has a least (first) element; i.e., A has an element x such that $x \leq y$ for all $y \in A$. If $\leq$ is a well-order on W then $(W, \leq)$, or simply W, will be called a **well-ordered set**. For $w, w' \in W$, we write $w < w'$ if $w \leq w'$ and $w \neq w'$. The usual order on $\mathbb{R}$ or that on $\mathbb{Q}$ is a linear order that is not a well-order.

Exercise 1.6.1 Show that every linear order on a finite set is a well-order.

If n is a natural number, then the well-ordered set $\{0, 1, \ldots, n-1\}$ with the usual order will be denoted by n itself. The usual order on the set of natural numbers $\mathbb{N} = \{0, 1, 2, \ldots\}$ is a well-order. We denote this well-ordered set by ω_0.

Proposition 1.6.2 *A linearly ordered set $(W, \leq)$ is well-ordered if and only if there is no descending sequence $w_0 > w_1 > w_2 > \cdots$ in W.*

Proof. Let W be not well-ordered. Then there is a nonempty subset A of W not having a least element. Choose any $w_0 \in A$. Since w_0 is not the first element of A, there is a $w_1 \in A$ such that $w_1 < w_0$. Since w_1 is not the first element of A, we get $w_2 < w_1$ in A. Proceeding similarly, we get a descending sequence $\{w_n : n \geq 0\}$ in W. This completes the proof of the "if" part of the result. For the converse, note that if $w_0 > w_1 > w_2 > \cdots$ is a descending sequence in W, then the set $A = \{w_n : n \geq 0\}$ has no least element. ■

Let W_1 and W_2 be two well-ordered sets. If there is an order-preserving bijection $f : W_1 \longrightarrow W_2$, then we call W_1 and W_2 **order isomorphic** or simply **isomorphic**. Such a map f is called an **order isomorphism**. If two well-ordered sets W_1, W_2 are order isomorphic, we write $W_1 \sim W_2$. Note that if W_1 and W_2 are isomorphic, they have the same cardinality.

Example 1.6.3 Let $W = \mathbb{N}\bigcup\{\infty\}$. Let $\leq$ be defined in the usual way on $\mathbb{N}$ and let $i < \infty$ for $i \in \mathbb{N}$. Clearly, W is a well-ordered set. Since W has a last element and ω_0 does not, $(W, \leq)$ is not isomorphic to ω_0. Thus there exist nonisomorphic well-ordered sets of the same cardinality.

Let W be a well-ordered set and $w \in W$. Suppose there is an element w^- of W such that $w^- < w$ and there is no $v \in W$ satisfying $w^- < v < w$. Clearly such an element, if it exists, is unique. We call w^- the **immediate predecessor** of w, and w the **successor** of w^-. An element of W that has an immediate predecessor is called a **successor element**. A well-ordered set W may have an element w other than the first element with no immediate predecessor. Such an element is called a **limit element** of W. Let W be as in 1.6.3. Then ∞ is a limit element of W, and each n, $n > 0$, is a successor element.

Let W be a well-ordered set and $w \in W$. Set

$$W(w) = \{u \in W : u < w\}.$$

Sets of the form $W(w)$ are called **initial segments** of W.

Exercise 1.6.4 Let W be a well-ordered set and $w \in W$. Show that

$$\bigcup_{u<w} W(u) = \begin{cases} W(w) & \text{if } w \text{ is a limit element,} \\ W(w^-) & \text{if } w \text{ is a successor.} \end{cases}$$

Proposition 1.6.5 *No well-ordered set W is order isomorphic to an initial segment $W(u)$ of itself.*

Proof. Let W be a well-ordered set and $u \in W$. Suppose W and $W(u)$ are isomorphic. Let $f : W \longrightarrow W(u)$ be an order isomorphism. For $n \in \mathbb{N}$, let $w_n = f^n(u)$. Note that

$$w_0 = f^0(u) = u > f^1(u) = f(u) = w_1.$$

By induction on n, we see that $w_n > w_{n+1}$ for all n, i.e., (w_n) is a descending sequence in W. By 1.6.2, W is not well-ordered. This contradiction proves our result. ■

Exercise 1.6.6 Let $(W_1, \leq_1)$ and $(W_2, \leq_2)$ be well-ordered sets. Define an order $\leq$ on $W_1 \times W_2$ as follows. For $(w_1, w_2), (w_1', w_2') \in W_1 \times W_2$,

$$(w_1, w_2) \leq (w_1', w_2') \iff w_2 <_2 w_2' \text{ or } (w_2 = w_2' \ \& \ w_1 \leq_1 w_1').$$

Show that $\leq$ is a well-order on $W_1 \times W_2$. The ordering $\leq$ on $W_1 \times W_2$ is called the **antilexicographical ordering.**

Exercise 1.6.7 Let $(W, \leq)$ be a well-ordered set and $\{(W_\alpha, \leq_\alpha) : \alpha \in W\}$ a family of well-ordered sets such that the W_α's are pairwise disjoint. Put $W' = \bigcup_\alpha W_\alpha$ and define an order $\leq'$ on W' as follows. For $w, w' \in W'$, put $w \leq' w'$ if

(i) there exists an $\alpha \in W$ such that $w, w' \in W_\alpha$ and $w \leq_\alpha w'$, or

(ii) there exist $\alpha, \beta \in W$ such that $\alpha < \beta$, $w \in W_\alpha$, and $w' \in W_\beta$.

Show that $\leq'$ is a well-order on W'.

If W' is as in 1.6.7, then we write $W' = \sum_{\alpha \in W} W_\alpha$. In the special case where W consists of two elements a and b with $a \leq b$, we simply write $W_a + W_b$ for $\sum_{\alpha \in W} W_\alpha$.

Remark 1.6.8 Let $(W_1, \leq_1)$ and $(W_2, \leq_2)$ be as in 1.6.6. For each $w \in W_1$, let $(W_w, \leq_w)$ be a well-ordered set isomorphic to $(W_2, \leq_2)$. Further, assume that $W_w \bigcap W_v = \emptyset$ for all pairs of distinct elements v, w of W_1. Then $W_1 \times W_2 \sim \sum_{w \in W_1} W_w$, where $W_1 \times W_2$ has the antilexicographical ordering.

Exercise 1.6.9 Give an example of a pair of well-ordered sets W_1, W_2 such that $W_1 + W_2$ and $W_2 + W_1$ are not isomorphic.

Exercise 1.6.10 Show that

$$\omega_0 \sim A_n + \omega_0 \sim n \times \omega_0,$$

where A_n is a well-ordered set of cardinality n disjoint from ω_0.

Using the operations on well-ordered sets described in 1.6.6 and 1.6.7 we can now give more examples of nonisomorphic well-ordered sets.

Exercise 1.6.11 For each $n \geq 0$, fix a well-ordered set A_n of cardinality n disjoint from ω_0. Also take a well-ordered set $\omega_0' \sim \omega_0$ disjoint from ω_0. Show that the well-ordered sets

$$\omega_0 + A_n (n \geq 0),\ \omega_0 + \omega_0',\ \omega_0 \times n (n > 2),\ \omega_0 \times \omega_0$$

are pairwise nonisomorphic.

Proceeding similarly, we can give more and more examples of well-ordered sets. However, note that all well-ordered sets thus obtained are countable. So, the following question arises: Is there an uncountable well-ordered set? There are many. But we shall have to wait to see an example of an uncountable well-ordered set. Another very natural question is the following: Can every set be well-ordered? In particular, can $\mathbb{R}$ be well-ordered? Recall that (using **AC**) every set can be linearly ordered and every countable set can be well-ordered. This brings us to another very useful and equivalent form of **AC**.

Well-Ordering Principle (WOP) *Every set can be well-ordered.* ■

Let $\{A_i : i \in I\}$ be a family of nonempty sets and $A = \bigcup_i A_i$. By **WOP**, there is a well-order, say $\leq$, on A. For $i \in I$, let $f(i)$ be the least element of A_i. Clearly, f is a choice function for $\{A_i\}$. Thus we see that **WOP** implies **AC**.

Exercise 1.6.12 Prove **WOP** using Zorn's lemma.

We refer the reader to [62] (Theorem 1, p. 254) for a proof of **WOP** from **AC**.

1.7 Transfinite Induction

In this section we extend the method of induction on natural numbers to general well-ordered sets. To some readers some of the results in this section may look unmotivated and unpleasantly complicated. However, these are preparatory results that will be used to develop the theory of ordinal numbers in the next section.

It will be convenient to recall the principles of induction on natural numbers.

Proposition 1.7.1 *(Proof by induction) For each $n \in \mathbb{N}$, let P_n be a mathematical proposition. Suppose P_0 is true and for every n, P_{n+1} is true whenever P_n is true. Then for every n, P_n is true. Symbolically, we can express this as follows.*

$$(P_0 \ \& \ \forall n(P_n \Longrightarrow P_{n+1})) \Longrightarrow \forall n P_n.$$

The proof of this proposition uses two basic properties of the set of natural numbers. First, it is well-ordered by the usual order, and second, every nonzero element in it is a successor. A repeated application of 1.7.1 gives us the following.

Proposition 1.7.2 *(Definition by induction) Let X be any nonempty set. Suppose x_0 is a fixed point of X and $g : X \longrightarrow X$ any map. Then there is a unique map $f : \mathbb{N} \longrightarrow X$ such that $f(0) = x_0$ and $f(n+1) = g(f(n))$ for all n.*

We wish to extend these two results to general well-ordered sets. Since a well-ordered set may have limit elements, we only have the so-called complete induction on well-ordered sets.

Theorem 1.7.3 *(Proof by transfinite induction) Let $(W, \leq)$ be a well-ordered set, and for every $w \in W$, let P_w be a mathematical proposition. Suppose that for each $w \in W$, if P_v is true for each $v < w$, then P_w is true. Then for every $w \in W$, P_w is true. Symbolically, we express this as*

$$(\forall w \in W)(((\forall v < w)P_v) \Longrightarrow P_w) \Longrightarrow (\forall w \in W)P_w.$$

Proof. Let

$$(\forall w \in W)(((\forall v < w)P_v) \Longrightarrow P_w). \qquad (*)$$

Suppose P_w is false for some $w \in W$. Consider

$$A = \{w \in W : P_w \text{ does not hold}\}.$$

By our assumptions, $A \neq \emptyset$. Let w_0 be the least element of A. Then for every $v < w_0$, P_v holds. However, P_{w_0} does not hold. This contradicts $(\star)$. Therefore, for every $w \in W$, P_w holds. ∎

Theorem 1.7.4 *(Definition by transfinite induction) Let $(W, \leq)$ be a well-ordered set, X a set, and $\mathcal{F}$ the set of all maps with domain an initial segment of W and range contained in X. If $G : \mathcal{F} \longrightarrow X$ is any map, then there is a unique map $f : W \longrightarrow X$ such that for every $u \in W$,*

$$f(u) = G(f|W(u)). \qquad (*)$$

Proof. For each $w \in W$, let P_w be the proposition "there is a unique map $g_w : W(w) \longrightarrow X$ such that $(\star)$ is satisfied for $f = g_w$ and $u \in W(w)$." Let $w \in W$ be such that P_v holds for each $v < w$. For each $v < w$, choose the function $g_v : W(v) \longrightarrow X$ satisfying $(\star)$ on $W(v)$. If $v' < v < w$, then $g_v|W(v')$ also satisfies $(\star)$ on $W(v')$. Therefore, by the uniqueness of $g_{v'}$,

$$g_v|W(v') = g_{v'};$$

i.e., $\{g_v : v < w\}$ is a consistent set of functions. So, there is a common extension $h : \bigcup_{v<w} W(v) \longrightarrow X$ of the functions g_v, $v < w$. If w is a limit element, then $W(w) = \bigcup_{w'<w} W(w')$ and we take $g_w = h$. If w is a successor, then we extend h on $W(w)$ to a function g_w by putting $g(w^-) = G(h)$. The uniqueness of g_w easily follows from the fact that $\{g_v : v < w\}$ are unique. Thus by 1.7.3, P_w holds for all w.

Now take

$$h : \bigcup_{w \in W} W(w) \longrightarrow X$$

to be the common extension of the functions $\{g_w : w \in W\}$. If W has no last element, then take $f = h$; Suppose W has a last element, say w. Take

f to be the extension of h to W such that $f(w) = G(h)$. As before, we see that f is unique. ■

Let W and W' be well-ordered sets. We write $W \prec W'$ if W is order isomorphic to an initial segment of W'. Further, we write $W \preceq W'$ if either $W \prec W'$ or $W \sim W'$.

Theorem 1.7.5 *(Trichotomy theorem for well-ordered sets) For any two well-ordered sets W and W', exactly one of*

$$W \prec W', \ W \sim W', \ \text{and } W' \prec W$$

holds.

Proof. It is easy to see that no two of these can hold simultaneously. For example, if $W \sim W'$ and $W' \prec W$, then W is isomorphic to an initial segment of itself. This is impossible by 1.6.5.

To show that at least one of these holds, take $X = W' \bigcup \{\infty\}$, where ∞ is a point outside W'. Now define a map $f : W \longrightarrow X$ by transfinite induction as follows. Let $w \in W$ and assume that f has been defined on $W(w)$. If $W' \setminus f(W(w)) \neq \emptyset$, then we take $f(w)$ to be the least element of $W' \setminus f(W(w))$; otherwise, $f(w) = \infty$. By 1.7.4, such a function exists.

Let us assume that $\infty \notin f(W)$. Then

(i) the map f is one-to-one and order preserving, and

(ii) the range of f is either whole of W' or an initial segment of W'.

So, in this case at least one of $W \sim W'$ or $W \prec W'$ holds.

If $\infty \in f(W)$, then let w be the first element of W such that $f(w) = \infty$. Then $f|W(w)$ is an order isomorphism from $W(w)$ onto W'. Thus in this case $W' \prec W$. ■

Corollary 1.7.6 *Let $(W, \leq)$, $(W', \leq')$ be well-ordered sets. Then $W \preceq W'$ if and only if there is a one-to-one order-preserving map from W into W'.*

Proof. Suppose there is a one-to-one order-preserving map g from W into W'. Let X and $f : W \longrightarrow X$ be as in the proof of 1.7.5. Then, by induction on w, we easily show that for every $w \in W$, $f(w) \leq' g(w)$. Therefore, $\infty \notin f(W)$. Hence, $W \preceq W'$. The converse is clear. ■

Theorem 1.7.7 *Let $\mathcal{W} = \{(W_i, \leq_i) : i \in I\}$ be a family of pairwise non-isomorphic well-ordered sets. Then there is a $W \in \mathcal{W}$ such that $W \prec W'$ for every $W' \in \mathcal{W}$ different from W.*

Proof. Suppose no such W exists. Then there is a descending sequence

$$\cdots \prec W_n \prec \cdots \prec W_1 \prec W_0$$

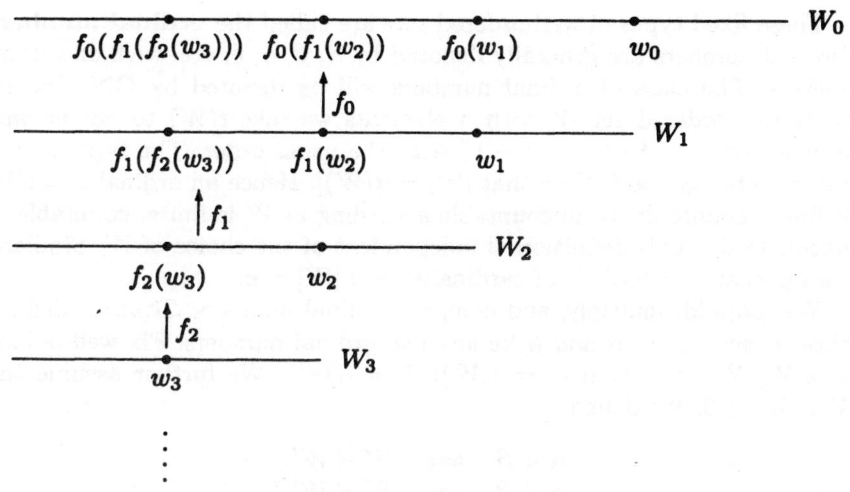

Figure 1.2

in $\mathcal{W}$. For $n \in \mathbf{N}$, choose a $w'_n \in W_n$ such that $W_{n+1} \sim W_n(w'_n)$. Fix an order isomorphism $f_n : W_{n+1} \longrightarrow W_n(w'_n)$. Let $w_0 = w'_0$, and for $n > 0$,

$$w_n = f_0(f_1(\cdots f_{n-1}(w'_n))).$$

(See Figure 1.2.) Then (w_n) is a descending sequence in W_0. This is a contradiction. The result follows. ■

1.8 Ordinal Numbers

Let W, W', and W'' be well-ordered sets. We have

$W \sim W$,

$W \sim W' \Longrightarrow W' \sim W$, and

$(W \sim W' \;\&\; W' \sim W'') \Longrightarrow W \sim W''$.

So, to each well-ordered set W we can associate a well-ordered set $t(W)$, called the **type** of W, such that

$$W \sim t(W),$$

and if W' is any well-ordered set, then

$$W \sim W' \iff t(W) \text{ and } t(W') \text{ are the same.}$$

These fixed types of well-ordered sets are called the **ordinal numbers**. Ordinal numbers are generally denoted by α, β, γ, δ, etc. with or without suffixes. The class of ordinal numbers will be denoted by **ON**. For any finite well-ordered set W with n elements we take $t(W)$ to be the well-ordered set $n = \{0, 1, \dots, n-1\}$ with the usual order. The type of ω_0 is taken to be ω_0 itself. Note that $|W| = |t(W)|$. Hence an ordinal $\alpha = t(W)$ is finite, countable, or uncountable according as W is finite, countable, or uncountable. This definition is independent of the choice of W. Similarly, we say that $\alpha = t(W)$ is of cardinality κ if $|W| = \kappa$.

We can add, multiply, and compare ordinal numbers. Towards defining these concepts, let α and β be any two ordinal numbers. Fix well-ordered sets W, W' such that $\alpha = t(W)$, $\beta = t(W')$. We further assume that $W \bigcap W' = \emptyset$. We define

$$\begin{aligned} \alpha < \beta &\iff W \prec W', \\ \alpha \leq \beta &\iff W \preceq W', \\ \alpha + \beta &= t(W + W'), \\ \alpha \cdot \beta &= t(W \times W'). \end{aligned}$$

Note that these definitions are independent of the choices of W and W'.

An ordinal α is called a **successor ordinal** if $\alpha = \beta + 1$ for some β; otherwise it is called a **limit ordinal**.

Remark 1.8.1 Note that α is a limit ordinal if and only if any well-ordered set W such that $\alpha = t(W)$ has no last element.

Using the results proved in the last section we easily see the following. For ordinals α, β, and γ,

$\alpha \leq \alpha$,

$(\alpha \leq \beta \;\&\; \beta \leq \gamma) \implies \alpha \leq \gamma$,

$(\alpha \leq \beta \;\&\; \beta \leq \alpha) \implies \alpha = \beta$, and exactly one of

$\alpha < \beta, \alpha = \beta$, and $\beta < \alpha$ holds.

Thus $\leq$ is a linear order on any set of ordinal numbers. In fact, by 1.7.7, any set of ordinal numbers is a well-ordered set. Observe that an ordinal is less than ω_0 if and only if it is finite; i.e., ω_0 is the first infinite ordinal. If A is a set of ordinals, then $\sum_{\alpha \in A} \alpha$ is an ordinal greater than or equal to each $\alpha \in A$. The least such ordinal is denoted by $\sup(A)$.

Exercise 1.8.2 Let α be an infinite ordinal and $n > 0$ finite. Show that

$$n + \alpha = \alpha < \alpha + n$$

and

$$n \cdot \alpha = \alpha < \alpha \cdot n.$$

Thus ordinal addition and ordinal multiplication are not commutative.

Theorem 1.8.3 *Every ordinal α can be uniquely written as*

$$\alpha = \beta + n$$

where β is a limit ordinal and n finite.

Proof. Let α be an ordinal number. We first show that there exists a limit ordinal β and an $n \in \omega$ such that $\alpha = \beta + n$. Choose a well-ordered set W such that $t(W) = \alpha$. If W has no last element, then we take $\beta = \alpha$ and $n = 0$. Suppose W has a last element, say w_0. If w_0 has no immediate predecessor, then take $\beta = t(W(w_0))$ and $n = 1$. Now suppose that w_0 does have an immediate predecessor, say w_1. If w_1 has no immediate predecessor, then we take $\beta = t(W(w_1))$ and $n = 2$. Since W has no descending sequence, this process ends after finitely many steps. Thus we get $w_0, w_1, \ldots, w_{k-1}$ such that $w_i = w_{i-1}^-$ for all $i > 0$, and w_{k-1} has no immediate predecessor. We take $\beta = t(W(w_{k-1}))$ and $n = k$.

We now show that α has a unique representation of the type mentioned above. Let W, W' be well-ordered sets with no last element, and A_n, B_m finite well-ordered sets of cardinality n and m respectively such that

$$A_n \bigcap W = B_m \bigcap W' = \emptyset.$$

Let $f : W + A_n \longrightarrow W' + B_m$ be an order isomorphism. It is easy to check that $f(W) = W'$ and $f(A_n) = B_m$. Uniqueness now follows. ∎

Let $\alpha = \beta + n$ with β a limit ordinal and n finite. We call α **even (odd)** if n is even (odd).

Theorem 1.8.4 *Let α be an ordinal. Then*

$$\alpha \sim \{\beta \in \mathbf{ON} : \beta < \alpha\}.$$

Proof. Let $(W, \leq')$ be a well-ordered set such that $t(W) = \alpha$. Fix $\beta < \alpha$. Choose $u \in W$ such that $\beta = t(W(u))$. Note that if $w, v \in W$, then

$$v <' w \Longleftrightarrow W(v) \text{ is an initial segment of } W(w).$$

Therefore, by 1.6.5, there is a unique $u \in W$ such that $\beta = t(W(u))$. Put $u = f(\beta)$. Clearly, the map $f : \{\beta \in \mathbf{ON} : \beta < \alpha\} \longrightarrow W$ is an order isomorphism. ∎

In view of the above theorem, an ordinal α is often identified with $\{\beta : \beta < \alpha\}$ with the ordering of the ordinal numbers. Thus far we have not given an example of an uncountable well-ordered set. We give one now.

Theorem 1.8.5 *The set Ω of all countable ordinals is uncountable.*

Proof. Suppose Ω is countable. Fix an enumeration $\alpha_0, \alpha_1, \ldots$ of Ω. Then

$$\alpha = \sum_n \alpha_n + 1$$

is a countable ordinal strictly larger than each α_n. This is a contradiction. So, Ω is uncountable. ■

This proof shows that if A is a countable set of countable ordinals, then there is a countable ordinal α such that $\beta < \alpha$ for all $\beta \in A$.

The set Ω of all countable ordinals with the ordering of ordinals is an uncountable well-ordered set; this well-ordered set is denoted by ω_1. The type $t(\omega_1)$ is taken to be ω_1 itself. Note that any ordinal less than ω_1 is countable; i.e., ω_1 is the first uncountable ordinal.

Proposition 1.8.6 *Let α be a countable limit ordinal. Then there exist $\alpha_0 < \alpha_1 < \cdots$ such that* $\sup\{\alpha_n : n \in \mathbb{N}\} = \alpha$.

Proof. Since α is countable, $\{\beta \in \mathbf{ON} : \beta < \alpha\}$ is countable. Fix an enumeration $\{\beta_n : n \in \mathbb{N}\}$ of all ordinals less than α. We now define a sequence of ordinals (α_n) by induction on n. Choose α_0 such that $\beta_0 < \alpha_0 < \alpha$. Since α is a limit ordinal, such an ordinal exists. Suppose α_n has been defined. Choose α_{n+1} greater than α_n such that $\beta_{n+1} < \alpha_{n+1} < \alpha$. Clearly,

$$\alpha = \sup\{\beta_n : n \in \mathbb{N}\} \le \sup\{\alpha_n : n \in \mathbb{N}\} \le \alpha.$$

So, $\sup\{\alpha_n : n \in \mathbb{N}\} = \alpha$. ■

1.9 Alephs

In Section 1.5, cardinal numbers were defined as symbols satisfying certain conditions. In this section, assuming **AC**, we give a more specific definition. We also briefly discuss the famous continuum hypothesis.

We put $|\omega_1| = \aleph_1$. (The symbol $\aleph$ is aleph, the first letter of the Hebrew alphabet.)

Exercise 1.9.1 Show that the set Ω' of all ordinals of cardinality less than or equal to $\aleph_1$ is of cardinality greater than $\aleph_1$.

(Hint: For every infinite cardinal κ, $\kappa \cdot \kappa = \kappa$.)

The well-ordered set $(\Omega', \le)$ will be denoted by ω_2. Put $|\omega_2| = \aleph_2$. Further, we take $t(\omega_2)$ to be ω_2 itself. Suppose ω_β, $\aleph_\beta$ have been defined for all $\beta < \alpha$ (α an ordinal). We define

$$\omega_\alpha = \{\gamma \in \mathbf{ON} : |\gamma| \le \aleph_\beta \text{ for some } \beta < \alpha\}.$$

We denote its cardinality by $\aleph_\alpha$. As before we take $t(\omega_\alpha)$ to be ω_α itself. The $\aleph_\alpha$'s are called simply **alephs**.

Exercise 1.9.2 Let α be any ordinal. Show that there is no cardinal κ such that $\aleph_\alpha < \kappa < \aleph_{\alpha+1}$.

An ordinal α is called an **initial ordinal** if $|\beta| < |\alpha|$ for every $\beta < \alpha$. For initial ordinals α, β, note that

$$\alpha < \beta \Longleftrightarrow |\alpha| < |\beta|.$$

Exercise 1.9.3 Show that any infinite initial ordinal is of the form ω_α.

We are now ready to define cardinal numbers. Let X be an infinite set. By **WOP** (which we are assuming), X can be well-ordered. So, $|X| = \aleph_\alpha = |\omega_\alpha|$ for some α. **We identify cardinals with initial ordinals** and put $|X| = \omega_\alpha$.

We can prove all the results on the arithmetic of cardinal numbers obtained in Section 1.5 using ordinal numbers. For instance, the trichotomy theorem for cardinal numbers (1.4.2) follows immediately from the trichotomy theorem for ordinals (applied on initial ordinals). We did not take this path for the simple reason that we do not need any background to understand Zorn's lemma. Interested readers can see [62] for a development of cardinal arithmetic using ordinal numbers.

Exercise 1.9.4 Show that for every cardinal κ there is a cardinal $\kappa^+ > \kappa$ (called the **successor** of κ) such that for no cardinal λ, $\kappa < \lambda < \kappa^+$.

Since every cardinal is an aleph, the question arises; What is $\mathfrak{c}$? That is, for what α is $\mathfrak{c} = \aleph_\alpha$? Cantor conjectured the following.

The Continuum Hypothesis (CH) $\mathfrak{c} = \aleph_1$. ■

CH says that there is no uncountable subset of $\mathbb{R}$ of cardinality less than $\mathfrak{c}$. The following is another famous hypothesis of Cantor on cardinal numbers.

The Generalised Continuum Hypothesis (GCH) *For every ordinal* α, $2^{\aleph_\alpha} = \aleph_{\alpha+1}$. ■

Since $\mathfrak{c} = 2^{\aleph_0}$, **GCH** clearly implies **CH**. Under **GCH** we can describe all the cardinals. Define $\alpha \longrightarrow \beth_\alpha$, $\alpha \in \mathbf{ON}$, by transfinite induction, as follows.

$$\beth_\alpha = \begin{cases} \aleph_0 & \text{if } \alpha = 0, \\ 2^{\beth_\beta} & \text{if } \alpha = \beta + 1 \text{ for some } \beta, \\ \sup_{\beta<\alpha} \beth_\beta & \text{if } \alpha \text{ is a limit ordinal.} \end{cases}$$

Assume **GCH**. By tranfinite induction on α, we can show that for every ordinal α, $\aleph_\alpha = \beth_\alpha$. In particular, it follows that for every infinite cardinal κ, $\kappa^+ = 2^\kappa$.

Are **CH** and/or **GCH** true? These problems, raised by Cantor right at the inception of set theory, turned out to be the central problems of set theory. In 1938 Kurt Gödel obtained deep results on "models of set theory" and produced a "model" of **ZFC** satisfying **GCH**. This was the first time metamathematics entered in a nontrivial way to answer a problem in mathematics. Gödel's result does not say that **CH** or **GCH** can be "proved" in **ZFC**. In 1963 Paul Cohen developed a very powerful technique, known as **forcing**, to build "models of set theory" and constructed "models" of **ZFC** satisfying $\neg$**CH**. The reader is referred to [59] for a very good exposition on the work of Gödel and Cohen.

1.10 Trees

Let A be a nonempty set. If $s \in A^{<\mathbb{N}}$ (the set of all finite sequences of elements of A including the empty sequence e), then $|s|$ will denote the length of s. Let $s = (a_0, a_1, \ldots, a_{n-1}) \in A^{<\mathbb{N}}$. For simplicity sometimes we shall write $a_0 a_1 \cdots a_{n-1}$ instead of $(a_0, a_1, \ldots, a_{n-1})$. We define

$$a^n = \underbrace{aa\cdots a}_{n \text{ times}}, \ a \in A, \ n \geq 0.$$

Note that $a^0 = e$. If $s = (a_0, a_1, \ldots, a_{n-1}) \in A^{<\mathbb{N}}$ and $m < n$, we write

$$s|m = (a_0, a_1, \ldots, a_{m-1}).$$

If $t = s|m$, we say that t is an **initial segment** of s, or s is an **extension** of t, and write $t \prec s$ or $s \succ t$. We write $t \preceq s$ if either $t \prec s$ or $t = s$. We say that s and t are **compatible** if one is an extension of the other; otherwise they are called **incompatible**, written $s \perp t$. Note that $s \perp t$ if and only if there is $i < |s|, |t|$ such that $s(i) \neq t(i)$. The **concatenation** $(a_0, a_1, \ldots, a_{n-1}, b_0, b_1, \ldots, b_{m-1})$ of two finite sequences $s = (a_0, a_1, \ldots, a_{n-1})$ and $t = (b_0, b_1, \ldots, b_{m-1})$ will be denoted by $s\hat{\ }t$. For simplicity of notation we shall write $s\hat{\ }a$ for $s\hat{\ }(a)$. For $s \in A^{<\mathbb{N}}$ and $\alpha \in A^{\mathbb{N}}$, $s\hat{\ }\alpha$ is similarly defined. Let $\alpha = (a_0, a_1, \ldots) \in A^{\mathbb{N}}$. For $k \in \mathbb{N}$, we put $\alpha|k = (\alpha_0, \alpha_1, \ldots \alpha_{k-1})$. If $s \in A^{<\mathbb{N}}$, we shall write $s \prec \alpha$ in case α extends s; i.e., $s = \alpha|k$ for some k.

A **tree** T on A is a nonempty subset of $A^{<\mathbb{N}}$ such that if $s \in T$ and $t \prec s$, then $t \in T$. (See Figure 1.3.)

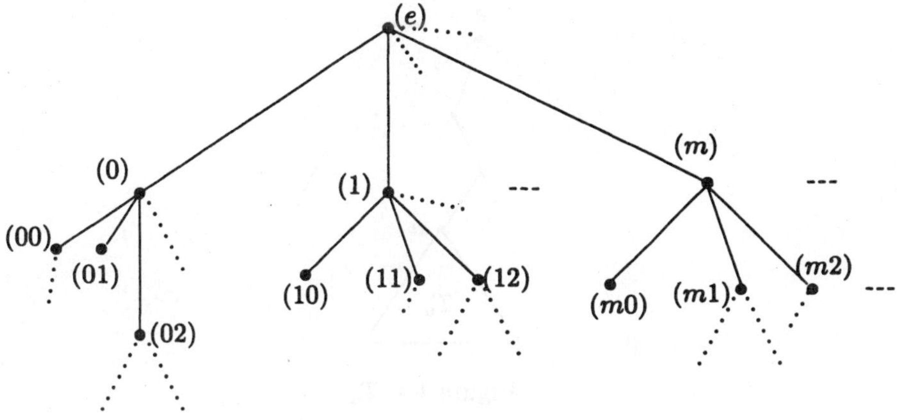

Figure 1.3. A tree on $\mathbb{N}$

Thus the empty sequence e belongs to all trees. Elements of T are often called **nodes** of T. A node u is called **terminal** if for no $a \in A$, $u\hat{}a \in T$. A tree T is called **finitely splitting** if for every node s of T, $\{a \in A : s\hat{}a \in T\}$ is finite. If T is a tree on A, its **body** is the set

$$[T] = \{\alpha \in A^{\mathbb{N}} : \forall k(\alpha|k \in T)\}.$$

Thus, members of $[T]$ are the **infinite branches** of T. A tree T is called **well-founded** if its body is empty; i.e., it has no infinite branch. If $[T] \neq \emptyset$, we call T **ill-founded**.

Exercise 1.10.1 Show that T is well-founded if and only if there is no sequence (s_n) in T such that $\cdots \succ s_n \succ \cdots \succ s_1 \succ s_0$.

Example 1.10.2 The sets $\{e\}$, $\mathbb{N}^{<\mathbb{N}}$, $\{s|i : i < |s|\}$ $(s \in \mathbb{N}^{<\mathbb{N}})$, $\{\alpha|i : i \in \mathbb{N}\}$ $(\alpha \in \mathbb{N}^{\mathbb{N}})$ form trees on $\mathbb{N}$.

Example 1.10.3 The tree

$$T = \{e\} \bigcup \{i0^j : j \leq i, i \in \mathbb{N}\}$$

is infinite and well-founded.

Example 1.10.4 Let T be a tree and u a node of T. The set

$$T_u = \{v \in A^{<\mathbb{N}} : u\hat{}v \in T\}$$

forms a tree. (See Figure 1.4.)

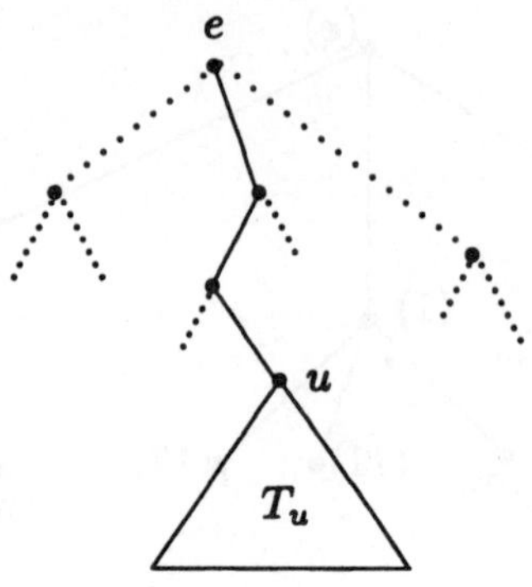

Figure 1.4. T_u

Note that for terminal u, $T_u = \{e\}$.

Example 1.10.5 Let T be a well-founded tree on $\mathbb{N}$ and n a positive integer. Then

$$T_{[n]} = \{0^i{}^\frown s : s \in T, i \leq n\}$$

is a well-founded tree.

Example 1.10.6 Let $T_0, T_1, T_2, \ldots$ be well-founded trees on $\mathbb{N}$. Then

$$\bigvee_n T_n = \{e\} \bigcup \{i^\frown s : s \in T_i, i \in \mathbb{N}\}$$

is a well-founded tree. (See Figure 1.5.)

Proposition 1.10.7 *(König's infinity lemma, [57]) Let T be a finitely splitting, infinite tree on A. Then T is ill-founded.*

Proof. Let T be a finitely splitting, infinite tree on A. Let (a_0) be a node of T with infinitely many extensions in T. Since T is finitely splitting (and $e \in T$), $\{a \in A : (a) \in T\}$ is finite. Further, T is infinite. So, (a_0) exists. By the same argument we get $a_1 \in A$ such that $s_1 = (a_0, a_1)$ has infinitely many extensions in T. Proceeding similarly we get an $\alpha = (a_0, a_1, \ldots)$ such that for all k, $\alpha|k$ has infinitely many extensions in T. In particular, $\alpha \in [T]$, and the result is proved. ■

Proposition 1.10.8 *Let T be a tree on a finite set A. Then*

$$[T] \neq \emptyset \Longleftrightarrow (\forall k \in \mathbb{N})(\exists u \in T)(|u| = k).$$

Proof. "If part": Since A is finite, T is finitely splitting. By our hypothesis, T is infinite. Therefore, by 1.10.7, $[T] \neq \emptyset$. The "only if" part is trivially seen. ■

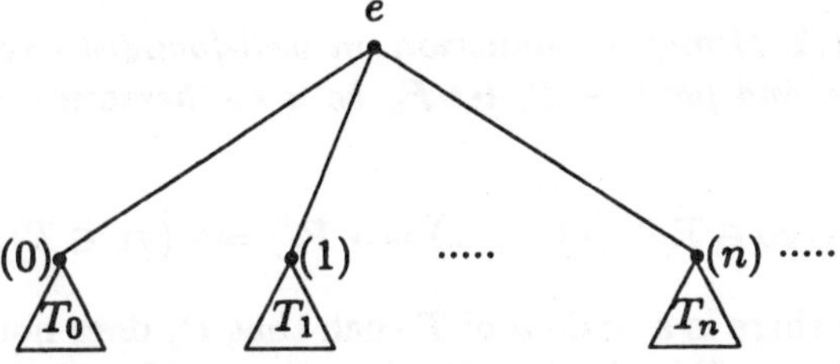

Figure 1.5. $\bigvee_n T_n$

Let T be a tree on a well-ordered set $(A, \leq)$. We define an ordering $<_{KB}$ on T as follows. Fix nodes $s = (a_0, a_1, \ldots, a_{n-1})$ and $t = (b_0, b_1, \ldots, b_{m-1})$ of T. Put $s <_{KB} t$ if either $t \prec s$ (that is, s extends t) or there is an i ¡ min(m, n) such that $a_j = b_j$ for every $j < i$ and $a_i < b_i$. Finally, we define $s \leq_{KB} t$ if either $s <_{KB} t$ or $s = t$. The ordering $\leq_{KB}$ is called the **Kleene – Brouwer ordering** on T.

Exercise 1.10.9 Show that $\leq_{KB}$ is a linear order on T.

Proposition 1.10.10 *A tree T on a well-ordered set A is well-founded if and only if $\leq_{KB}$ is a well-order on T.*

Proof. Let T be ill-founded. Take any α in $[T]$. Then $(\alpha|k)$ is a descending sequence in $(T, \leq_{KB})$. This proves the "if" part of the result.

Conversely, suppose $(T, \leq_{KB})$ is not well-ordered. Since T is linearly ordered by $\leq_{KB}$, there is a descending sequence (s_k) in T. Let

$$s_k = (a_0^k, \ldots, a_{n_k-1}^k),\ k \in N.$$

Since (s_k) is descending,

$$a_0^0 \geq a_0^1 \geq a_0^2 \geq \ldots.$$

Since A is well-ordered, $\{a_0^k\}$ is eventually constant. Let K be such that for all $k \geq K$, $a_0^k = a_0$, say. Note that $(a_0) \in T$. Since $(s_k)_{k \geq K}$ is descending, by the same argument,

$$a_1^K \geq a_1^{K+1} \geq a_1^{K+2} \geq \cdots$$

is eventually constant, say equal to a_1. Again note that $(a_0, a_1) \in T$. Proceeding similarly, we get $\alpha = (a_0, a_1, \ldots)$ such that $\alpha|k \in T$ for all k; i.e., $[T] \neq \emptyset$. ■

1.11 Induction on Trees

The methods of transfinite induction can be extended to induction on well-founded trees, which we describe now.

Proposition 1.11.1 *(Proof by induction on well-founded trees) Let T be a well-founded tree and for $u \in T$, let P_u be a mathematical proposition. Then*

$$(\forall u \in T)(((\forall v \in T_u \setminus \{e\})P_{u^\frown v}) \Longrightarrow P_u) \Longrightarrow (\forall u \in T)P_u.$$

Proof. Suppose there is a node u of T such that P_u does not hold. Take an extension w of u in T such that P_w does not hold and if $v \succ w$ and $v \in T$ then P_v holds. Since T is well-founded, such a w exists. Thus for every extension v of w in T, P_v holds. So, by the hypothesis, P_w holds. This is a contradiction. Therefore, P_u holds for all $u \in T$. ■

Proposition 1.11.2 *(Definition by induction on well-founded trees) Let T be a well-founded tree on a set A, X a set, and $\mathcal{F}$ the set of all maps with domain $T_u \setminus \{e\}$ and range contained in X, where u varies over T. Given any map $G : \mathcal{F} \longrightarrow X$, there is a unique map $f : T \longrightarrow X$ such that for all $v \in T$,*

$$f(v) = G(f_v), \qquad (*)$$

where $f_v : T_v \setminus \{e\} \longrightarrow X$ is the map defined by

$$f_v(u) = f(v^\frown u), \ u \in T_v \setminus \{e\}.$$

Proof. (Existence) For $u \in T$, let P_u be the proposition "*there is a map $f_u : T_u \setminus \{e\} \longrightarrow X$ satisfying ($\star$) for all $v \in T_u \setminus \{e\}$.*" Fix a $u \in T$. Suppose P_w is satisfied for all $w \in T$ such that $w \succ u$. For each $a \in A$ such that $u^\frown a \in T$, let $f_{u^\frown a} : T_{u^\frown a} \setminus \{e\} \longrightarrow X$ be a map satisfying ($\star$) for $v \in T_{u^\frown a} \setminus \{e\}$. Define $f_u : T_u \longrightarrow X$ by

$$f_u(a^\frown w) = \begin{cases} f_{u^\frown a}(w) & \text{if } w \in T_{u^\frown a} \setminus \{e\}, \\ G(f_{u^\frown a}) & \text{if } w = e, \end{cases}$$

where $a \in A$ and $u^\frown a \in T$. So, by 1.11.1, there is an $f_e : T_e \setminus \{e\} \longrightarrow X$ satisfying ($\star$). Put

$$f(w) = \begin{cases} f_e(w) & \text{if } w \in T \ \& \ w \neq e, \\ G(f_e) & \text{if } w = e. \end{cases}$$

(Uniqueness) Let $f, g : T \longrightarrow X$ satisfy ($\star$). By induction on u we easily see that for every $u \in T$, $f(u) = g(u)$. ■

Let T be a well-founded tree. By induction on T, we define a unique map $\rho_T : T \longrightarrow \mathbf{ON}$ by

$$\rho_T(u) = \sup\{\rho_T(v) + 1 : u \prec v, v \in T\}, \ u \in T.$$

(We take $\sup(\emptyset) = 0$.) Note that $\rho_T(u) = 0$ if u is terminal in T. The map ρ_T is called the **rank function** of T. Finally, we define $\rho(T) = \rho_T(e)$ and call it the **rank** of T.

Exercise 1.11.3 Show that

$$\rho_T(u) = \sup\{\rho_T(u\hat{\ }a) + 1 : u\hat{\ }a \in T\}.$$

Example 1.11.4 $\rho(T) = |s|$ if $T = \{s|i : i < |s|\}$, $s \in \mathbb{N}^{<\mathbb{N}}$.

Example 1.11.5 Let

$$T = \{e\} \bigcup \{io^j : j \leq i, i \in \mathbb{N}\}.$$

Then $\rho(T) = \omega_0$.

Example 1.11.6 $\rho(T_{[n]}) = \rho(T) + n$ for all positive integers n and all trees T.

Example 1.11.7 $\rho(\bigvee_n T_n) = \sup\{\rho(T_n) + 1 : n \in \mathbb{N}\}$.

Exercise 1.11.8 Show that for every ordinal $\alpha < \omega_1$, there is a well-founded tree T on $\mathbb{N}$ of rank α.

Exercise 1.11.9 Show that every well-founded tree on $\{0, 1\}$ is of finite rank.

We will sometimes have to deal with trees on sets A that are products of the form $A = B \times C$ or $A = B \times C \times D$. Let $A = B \times C$ and T a tree on A. It will be convenient to identify a node $((b_0, c_0), (b_1, c_1), \ldots, (b_{n-1}, c_{n-1}))$ of T by (u, v), where $u = (b_0, b_1, \ldots, b_{n-1})$ and $v = (c_0, c_1, \ldots, c_{n-1})$. Let (u, v), (u', v') be nodes of T. We write $(u, v) \prec (u', v')$ if $u \prec u'$ and $v \prec v'$. The body of T is identified with

$$[T] = \{(\alpha, \beta) \in B^{\mathbb{N}} \times C^{\mathbb{N}} : \forall k((\alpha|k, \beta|k) \in T)\}.$$

The meaning of $T_{(u,v)}$ is self-explanatory. If T is a tree on $B \times C$ and $\alpha \in B^{\mathbb{N}}$, then the **section** of T at α is defined by

$$T[\alpha] = \{v \in C^{<\mathbb{N}} : (\alpha|\ |v|, v) \in T\}.$$

Note that

$$\alpha \in \pi_1([T]) \iff T[\alpha] \text{ is ill-founded},$$

where $\pi_1 : B^{\mathbb{N}} \times C^{\mathbb{N}} \longrightarrow B^{\mathbb{N}}$ is the projection map. In fact,

$$\alpha \in \pi_1([T]) \iff \exists\beta\forall k((\alpha|k, \beta|k) \in T).$$

1.12 The Souslin Operation

The Souslin operation is an operation on sets that is of fundamental importance to descriptive set theory. It was introduced by Souslin[111]. However, the Souslin operation for $A = \{0, 1\}$ was introduced by Alexandrov in [2] to show that **CH** holds for Borel sets; i.e., every uncountable Borel subset of reals is of cardinality $\mathfrak{c}$.

Let X be a set and $\mathcal{F}$ a family of subsets of X. We put

$$\mathcal{F}_\sigma = \{\bigcup_{n\in\mathbf{N}} A_n : A_n \in \mathcal{F}\}$$

and

$$\mathcal{F}_\delta = \{\bigcap_{n\in\mathbf{N}} A_n : A_n \in \mathcal{F}\}.$$

So, $\mathcal{F}_\sigma$ ($\mathcal{F}_\delta$) is the family of countable unions (resp. countable intersections) of sets in $\mathcal{F}$. The family of finite unions (finite intersections) of sets in $\mathcal{F}$ will be denoted by $\mathcal{F}_s$ (resp. $\mathcal{F}_d$). Finally,

$$\neg\mathcal{F} = \{A \subseteq X : X \setminus A \in \mathcal{F}\}.$$

It is easily seen that

$$\mathcal{F}_s \subseteq \mathcal{F}_\sigma,\ \mathcal{F}_d \subseteq \mathcal{F}_\delta,\ \mathcal{F}_\sigma = \neg(\neg\mathcal{F})_\delta, \text{ and } \mathcal{F}_\delta = \neg(\neg\mathcal{F})_\sigma.$$

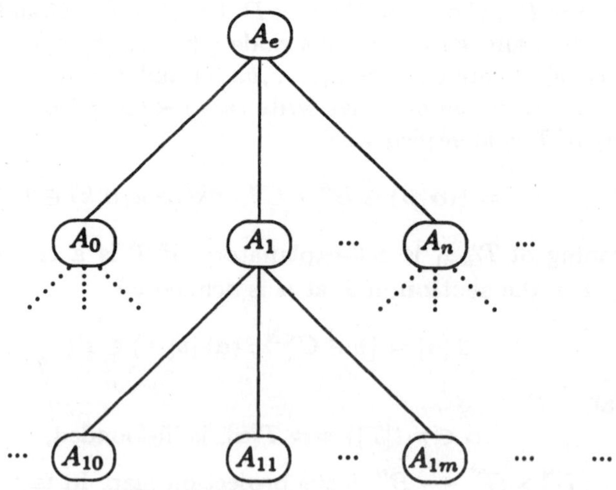

Figure 1.6. A system of sets with $A = \mathbf{N}$

It is essential to become familiar with the above notation, as we shall be using it repeatedly while studying set operations on various pointclasses.

Let A be a nonempty set. A family $\{A_s : s \in A^{<\mathbf{N}}\}$ of subsets of a set X will be called a **system of sets**. For brevity, we shall write $\{A_s\}$ for

$\{A_s : s \in A^{<\mathbb{N}}\}$ when there is no scope for confusion. A system $\{A_s\}$ is called **regular** if $A_s \subseteq A_t$ whenever $s \succ t$. (See Figure 1.6.)

We define

$$\mathcal{A}_A(\{A_s\}) = \bigcup_{\alpha \in A^{\mathbb{N}}} \bigcap_n A_{\alpha|n}.$$

In all the interesting cases A is finite or A equals $\mathbb{N}$. When $A = \mathbb{N}$ we write $\mathcal{A}$ instead of $\mathcal{A}_{\mathbb{N}}$ and call it the **Souslin operation.** If $A = \{0, 1\}$, we write $\mathcal{A}_2$ for $\mathcal{A}_A$. Let $\mathcal{F}$ be a family of subsets of X. Put

$$\mathcal{A}_A(\mathcal{F}) = \{\mathcal{A}_A(\{A_s\}) : A_s \in \mathcal{F}; s \in A^{<\mathbb{N}}\};$$

i.e., $\mathcal{A}_A(\mathcal{F})$ is the family of sets obtained by applying the operation $\mathcal{A}_A$ on a system of sets in $\mathcal{F}$. Note that if $(\mathcal{F})_d = \mathcal{F}$, i.e., if $\mathcal{F}$ is closed under finite intersections, then $\mathcal{A}_A(\mathcal{F})$ consists of sets obtained by performing the operation $\mathcal{A}_A$ on a regular system of sets in $\mathcal{F}$.

It should be noted that the Souslin operation involves uncountable unions. It is closely related to the projection operation, as shown in the following proposition.

For $s \in \mathbb{N}^{<\mathbb{N}}$, let

$$\Sigma(s) = \{\alpha \in \mathbb{N}^{\mathbb{N}} : s \prec \alpha\}.$$

Proposition 1.12.1 *Let $\{A_s : s \in \mathbb{N}^{<\mathbb{N}}\}$ be a system of subsets of a set X. Put*

$$B = \bigcap_k \bigcup_{|s|=k} [A_s \times \Sigma(s)].$$

Then $\mathcal{A}(\{A_s\}) = \pi_X(B)$, where $\pi_X : X \times \mathbb{N}^{\mathbb{N}} \longrightarrow X$ is the projection map.

The proof of the above proposition is routine and is left as an exercise. Our next result shows that the Souslin operation subsumes countable union and countable intersection.

Proposition 1.12.2 *For every family $\mathcal{F}$ of subsets of X,*

$$\mathcal{F}, \mathcal{F}_\sigma, \mathcal{F}_\delta \subseteq \mathcal{A}(\mathcal{F}).$$

Proof. (i) $\mathcal{F} \subseteq \mathcal{A}(\mathcal{F})$. Let $A \in \mathcal{F}$. Take

$$A_s = A, \ s \in \mathbb{N}^{<\mathbb{N}}.$$

Clearly, $A = \mathcal{A}(\{A_s\}) \in \mathcal{A}(\mathcal{F})$.

(ii) $\mathcal{F}_\sigma \subseteq \mathcal{A}(\mathcal{F})$. Let (A_n) be a sequence in $\mathcal{F}$. For $s = (s_0, s_1, \ldots, s_{m-1}) \in \mathbb{N}^{<\mathbb{N}}$, define $B_s = A_{s_0}$. Then $\mathcal{A}(\{B_s\}) = \bigcup A_n$.

(iii) $\mathcal{F}_\delta \subseteq \mathcal{A}(\mathcal{F})$. Let (A_n) be a sequence in $\mathcal{F}$. Take

$$C_s = A_{|s|}, \ s \in \mathbb{N}^{<\mathbb{N}}.$$

Clearly, $\mathcal{A}(\{C_s\}) = \bigcap A_n$. ∎

The next two results give sufficient conditions under which the operation $\mathcal{A}_A$ can be obtained by iterating countable unions and countable intersections. The first one is elementary, but the second one is nontrivial.

Lemma 1.12.3 *Let $\{A_s : s \in A^{<\mathbf{N}}\}$ be a system of sets such that $A_s \bigcap A_t = \emptyset$ whenever $s \perp t$. Then*

$$\mathcal{A}_A(\{A_s\}) = \bigcap_n \bigcup_{|s|=n} A_s.$$

Proof. Let $x \in \mathcal{A}_A(\{A_s\})$. By the definition of $\mathcal{A}_A$, there is an $\alpha \in A^{\mathbf{N}}$ such that $x \in A_{\alpha|n}$ for all n. So $x \in \bigcap_n \bigcup_{|s|=n} A_s$. Conversely, let $x \in \bigcap_n \bigcup_{|s|=n} A_s$. For each n, choose $s_n \in A^{<\mathbf{N}}$ of length n such that $x \in A_{s_n}$. Since $A_s \bigcap A_t = \emptyset$ whenever $s \perp t$, the s_n's are compatible. Therefore, there is an $\alpha \in A^{\mathbf{N}}$ such that $\alpha|n = s_n$ for all n. Thus $x \in \mathcal{A}_A(\{A_s\})$. ■

Proposition 1.12.4 *If A is a finite set and $\{A_s : s \in A^{<\mathbf{N}}\}$ regular, then*

$$\mathcal{A}_A(\{A_s\}) = \bigcap_n \bigcup_{|s|=n} A_s.$$

Proof. We have seen in the proof of 1.12.3 that

$$\mathcal{A}_A(\{A_s\}) \subseteq \bigcap_n \bigcup_{|s|=n} A_s$$

is always true. To prove the other inclusion, take any $x \in \bigcap_n \bigcup_{|s|=n} A_s$. Consider

$$T = \{s \in A^{<\mathbf{N}} : x \in A_s\}.$$

Since $\{A_s\}$ is regular, T is a tree. Since A is finite, the tree T is finitely splitting. By our hypothesis, it is infinite. Therefore, by König's infinity lemma (1.10.7), $[T] \neq \emptyset$. Let $\alpha \in [T]$. Then $x \in A_{\alpha|n}$ for all n. Hence, $x \in \mathcal{A}_A(\{A_s\})$. ■

Corollary 1.12.5 *Let $(\mathcal{F})_s = (\mathcal{F})_d = \mathcal{F}$; i.e., $\mathcal{F}$ is closed under finite intersections and finite unions. Then $\mathcal{A}_2(\mathcal{F}) = \mathcal{F}_\delta$. In particular, $\mathcal{A}_2$ does not subsume the operation of taking countable unions, whereas the Souslin operation does (1.12.2).*

1.13 Idempotence of the Souslin Operation

Another trivial corollary of 1.12.4 is the following: $\mathcal{A}_2$ is idempotent; i.e., if $\mathcal{F}$ is closed under finite intersections and finite unions, then

$$\mathcal{A}_2(\mathcal{A}_2(\mathcal{F})) = \mathcal{A}_2(\mathcal{F}).$$

This is also true for the Souslin operation, though proving it is harder.

Theorem 1.13.1 *Let $\mathcal{F}$ be any family of subsets of X. Then*

$$\mathcal{A}(\mathcal{A}(\mathcal{F})) = \mathcal{A}(\mathcal{F}).$$

Proof. By 1.12.2,

$$\mathcal{A}(\mathcal{A}(\mathcal{F})) \supseteq \mathcal{A}(\mathcal{F}).$$

Therefore, we need to show the other inclusion only. Take a system of sets $\{A_s : s \in \mathbb{N}^{<\mathbb{N}}\}$ in $\mathcal{A}(\mathcal{F})$. Let

$$A = \bigcup_{\alpha \in \mathbb{N}^{\mathbb{N}}} \bigcap_n A_{\alpha|n}.$$

For each $s \in \mathbb{N}^{<\mathbb{N}}$, take a system of sets $\{B_{s,t} : t \in \mathbb{N}^{<\mathbb{N}}\}$ in $\mathcal{F}$ such that

$$A_s = \bigcup_{\gamma \in \mathbb{N}^{\mathbb{N}}} \bigcap_m B_{s,\gamma|m}.$$

We need to define a system of sets $\{C_s : s \in \mathbb{N}^{<\mathbb{N}}\}$ such that for every $x \in X$,

$$x \in A \Longleftrightarrow \exists \beta \in \mathbb{N}^{\mathbb{N}} \forall k (x \in C_{\beta|k}).$$

Let $x \in X$. Note that

$$\begin{aligned} x \in A &\Longleftrightarrow \exists \alpha \in \mathbb{N}^{\mathbb{N}} \forall m (x \in A_{\alpha|m}) \\ &\Longleftrightarrow \exists \alpha \in \mathbb{N}^{\mathbb{N}} \forall m \exists \gamma \in \mathbb{N}^{\mathbb{N}} \forall n (x \in B_{\alpha|m,\gamma|n}) \\ &\Longleftrightarrow \exists \alpha \in \mathbb{N}^{\mathbb{N}} \exists (\gamma_p) \in (\mathbb{N}^{\mathbb{N}})^{\mathbb{N}} \forall m \forall n (x \in B_{\alpha|m,\gamma_m|n}). \end{aligned}$$

We claim that there exist bijections $u : \mathbb{N} \times \mathbb{N} \longrightarrow \mathbb{N}$, $v : \mathbb{N}^{\mathbb{N}} \times (\mathbb{N}^{\mathbb{N}})^{\mathbb{N}} \longrightarrow \mathbb{N}^{\mathbb{N}}$, and maps $\varphi, \psi : \mathbb{N}^{<\mathbb{N}} \longrightarrow \mathbb{N}^{<\mathbb{N}}$ such that for any $(\alpha, (\gamma_p)) \in \mathbb{N}^{\mathbb{N}} \times (\mathbb{N}^{\mathbb{N}})^{\mathbb{N}}$, if $v(\alpha, (\gamma_p)) = \beta$ and $s = \beta|u(m,n)$ for some m, n, then $\varphi(s) = \alpha|m$ and $\psi(s) = \gamma_m|n$.

We first assume that such functions exist and complete the proof. Define

$$C_s = B_{\varphi(s),\psi(s)}, \quad s \in \mathbb{N}^{<\mathbb{N}}.$$

We claim that

$$A = \mathcal{A}(\{C_s\}).$$

This is shown in two steps.

$A \subseteq \mathcal{A}(\{C_s\})$: To see this, take $x \in A$. By the above series of equivalences, there exist $\alpha \in \mathbb{N}^{\mathbb{N}}$ and $(\gamma_p) \in (\mathbb{N}^{\mathbb{N}})^{\mathbb{N}}$ such that for all m and for all n, $x \in B_{\alpha|m,\gamma_m|n}$. Let $\beta = v(\alpha, (\gamma_p))$. Take any k. Let m, n be such that $k = u(m,n)$. So, $\varphi(\beta|k) = \alpha|m$ and $\psi(\beta|k) = \gamma_m|n$. Then $x \in B_{\alpha|m,\gamma_m|n} = C_{\beta|k}$. Thus, $x \in \mathcal{A}(\{C_s\})$.

$A \supseteq \mathcal{A}(\{C_s\})$: To show this, take any $x \in \mathcal{A}(\{C_s\})$. Let $\beta \in \mathbb{N}^{\mathbb{N}}$ be such that $x \in C_{\beta|k}$ for all k. Choose $(\alpha, (\gamma_p))$ such that $v(\alpha, (\gamma_p)) = \beta$. Fix m, n and put $k = u(m,n)$. Then $C_{\beta|k} = B_{\alpha|m,\gamma_m|n}$ by definition. So, $x \in A$ by the above series of equivalences.

It remains to show that the functions u, v, φ, and ψ with the properties stated earlier exist.

The definition of u:

Define $u : \mathbf{N} \times \mathbf{N} \longrightarrow \mathbf{N}$ by

$$u(m,n) = 2^m(2n+1) - 1, \quad m, n \in \mathbf{N}.$$

Then u is a bijection such that for all m, n, and p, $m \leq u(m,n)$ and $u(m,n) < u(m,p))$ if $n < p$.

For $k \in \mathbf{N}$, we define $l(k), r(k)$ to be the natural numbers i, j respectively such that $k = u(i,j)$.

The definition of v:

Let $v : \mathbf{N}^{\mathbf{N}} \times (\mathbf{N}^{\mathbf{N}})^{\mathbf{N}} \longrightarrow \mathbf{N}^{\mathbf{N}}$ be defined by $v(\alpha, (\gamma_n)) = \beta$ where

$$\beta(k) = u(\alpha(k), \gamma_{l(k)}(r(k))), \quad k \in \mathbf{N}.$$

We claim that v is one-to-one. To see this, take $(\alpha, (\gamma_n)) \neq (\alpha', (\gamma'_n))$. If $\alpha \neq \alpha'$, then there is a k such that $\alpha(k) \neq \alpha'(k)$. Since u is one-to-one, it follows that

$$u(\alpha(k), \gamma_{l(k)}(r(k))) \neq u(\alpha'(k), \gamma'_{l(k)}(r(k))).$$

So in this case, $v(\alpha, (\gamma_n))(k) \neq v(\alpha', (\gamma'_n))(k)$. Now assume that for some i, $\gamma_i \neq \gamma'_i$. Choose j such that $\gamma_i(j) \neq \gamma'_i(j)$. Let $k = u(i,j)$. So $l(k) = i$ and $r(k) = j$. Then, as u is one-to-one,

$$v(\alpha, (\gamma_n))(k) = u(\alpha(k), \gamma_i(j)) \neq u(\alpha'(k), \gamma'_i(j)) = v(\alpha', (\gamma'_n))(k).$$

Thus v is one-to-one in this case too.

We now show that v is onto. Towards this, let $\beta \in \mathbf{N}^{\mathbf{N}}$. Define $\alpha \in \mathbf{N}^{\mathbf{N}}$ by

$$\alpha(k) = l(\beta(k)), \quad k \in \mathbf{N}.$$

For any n, define γ_n by

$$\gamma_n(m) = r(\beta(u(n,m))), \quad m \in \mathbf{N}.$$

Fix $k \in \mathbf{N}$. We have

$$\begin{aligned} v(\alpha, (\gamma_n))(k) &= u(\alpha(k), \gamma_{l(k)}(r(k))) \\ &= u(l(\beta(k)), r(\beta(u(l(k), r(k))))) \\ &= u(l(\beta(k)), r(\beta(k))) \\ &= \beta(k). \end{aligned}$$

This shows that $v(\alpha, (\gamma_n)) = \beta$.

Definition of φ:

Fix $s = (s_0, s_1, \ldots, s_{k-1})$. Let $m = l(k) = l(|s|)$. Put

$$\varphi(s) = (l(s_0), l(s_1), \ldots, l(s_{m-1})).$$

Since $i \leq u(i, j)$ for all i, j, this definition makes sense.

Definition of ψ:

Let s and m be as above and $n = r(k) = r(|s|)$. Put $p_i = s_{u(m,i)}$, $i < n$. Since $i < n \Longrightarrow u(m, i) < u(m, n) = k$, p_i is defined. Define

$$\psi(s) = (r(p_0), r(p_1), \ldots, r(p_{n-1})).$$

Let $(\alpha, (\gamma_p)) \in \mathbb{N}^{\mathbb{N}} \times (\mathbb{N}^{\mathbb{N}})^{\mathbb{N}}$, $v(\alpha, (\gamma_p)) = \beta$, $k = u(m, n)$, and $s = \beta|k$. Our proof will be complete if we show that $\varphi(s) = \alpha|m$ and $\psi(s) = \gamma_m|n$. Note the following.

$$\begin{aligned} \varphi(s) &= (l(s_0), l(s_1), \ldots, l(s_{m-1})) \\ &= (l(\beta(0)), l(\beta(1)), \ldots, l(\beta(m-1))) \\ &= \alpha|m \end{aligned}$$

and

$$\begin{aligned} \psi(s) &= (r(p_0), r(p_1), \ldots, r(p_{n-1})) \\ &= (r(\beta(u(m,0))), r(\beta(u(m,1))), \ldots, r(\beta(u(m,n-1)))) \\ &= \gamma_m|n. \end{aligned}$$

■

By 1.12.2 and 1.13.1 we get the following result.

Corollary 1.13.2 *For any family $\mathcal{F}$ of sets, $\mathcal{A}(\mathcal{F})$ is closed under counitable intersections and countable unions.*

The reader is encouraged to give a proof of the above corollary without using 1.13.1. In Chapter 4, we shall see that we may not be able to get $\mathcal{A}(\{A_s\})$ by iterating the operations of countable unions and countable intersections on A_s's. We shall also prove that $\mathcal{A}(\mathcal{F})$ need not be closed under complementation.

2

Topological Preliminaries

As mentioned in the introduction, we shall present the theory of Borel sets in the general context of Polish spaces. In this chapter we give an account of Polish spaces. The space $\mathbb{N}^\mathbb{N}$ of sequences of natural numbers, equipped with the product of discrete topologies on $\mathbb{N}$, is of particular importance to us. Our theory takes a particularly simple form on this space, and it is possible to generalize the results on Borel subsets of $\mathbb{N}^\mathbb{N}$ to general Polish spaces. The relevant results that we shall use to obtain these generalizations are presented in the last section of this chapter.

2.1 Metric Spaces

A **metric** on a set X is a map $d : X \times X \longrightarrow [0, \infty)$ such that for x, y, z in X,

$$d(x, y) = 0 \Longleftrightarrow x = y,$$

$$d(x, y) = d(y, x), \text{ and}$$

$$d(x, z) \leq d(x, y) + d(y, z) \text{ (the triangle inequality).}$$

A metric space is a pair (X, d) where d is a metric on X. When the underlying metric is understood, we shall simply call X a metric space.

Example 2.1.1 Let $X = \mathbb{R}^n$, n a positive integer. For $x = (x_1, \cdots, x_n)$ and $y = (y_1, \cdots, y_n)$ in $\mathbb{R}^n$, let

$$d_1(x, y) = |x - y| = \sqrt{\sum_{i=1}^{n} (x_i - y_i)^2}$$

and

$$d_2(x, y) = \max\{|x_i - y_i| : 1 \leq i \leq n\}.$$

Then d_1 and d_2 are metrics on $\mathbb{R}^n$. The metric d_1 will be referred to as the **usual metric** on $\mathbb{R}^n$.

Example 2.1.2 Let $X = \mathbb{R}^{\mathbb{N}}$, $x = (x_0, x_1, \ldots)$ and $y = (y_0, y_1, \ldots)$. Define

$$d(x, y) = \sum_n \frac{1}{2^{n+1}} \min\{|x_n - y_n|, 1\}.$$

Then d is a metric on $\mathbb{R}^{\mathbb{N}}$.

Example 2.1.3 If X is any set and

$$d(x, y) = \begin{cases} 0 & \text{if } x = y, \\ 1 & \text{otherwise,} \end{cases}$$

then d defines a metric on X, called the **discrete metric**.

Example 2.1.4 Let (X_0, d_0), (X_1, d_1), (X_2, d_2), ... be metric spaces and $X = \prod_n X_n$. Fix $x = (x_0, x_1, \ldots)$ and $y = (y_0, y_1, \ldots)$ in X. Define

$$d(x, y) = \sum_n \frac{1}{2^{n+1}} \min\{d_n(x_n, y_n), 1\}.$$

Then d is a metric on X, which we shall call the **product metric**.

Note that if (X, d) is a metric space and $Y \subseteq X$, then d resticted to Y (in fact to $Y \times Y$) is itself a metric. Thus we can think of a subset of a metric space as a metric space itself and call it a **subspace** of X. Let (X, d) be a metric space, $x \in X$, and $r > 0$. We put

$$B(x, r) = \{y \in X : d(x, y) < r\}$$

and call it the **open ball** with **center** x and **radius** r. The set

$$\{y \in X : d(x, y) \leq r\}$$

will be called the **closed ball** with center x and radius r. Let $\mathcal{T}$ be the set of all subsets U of X such that U is the union of a family (empty or otherwise) of open balls in X. Thus, $U \in \mathcal{T}$ if and only if for every x in U, there exists an $r > 0$ such that $B(x, r) \subseteq U$. Clearly,

(i) $\emptyset, X \in \mathcal{T}$,

(ii) $\mathcal{T}$ is closed under arbitrary unions, i.e., for all $\{U_i : i \in I\} \subseteq \mathcal{T}$, $\bigcup_i U_i \in \mathcal{T}$, and

(iii) $\mathcal{T}$ is closed under finite intersections.

To see (iii), take two open balls $B(x,r)$ and $B(y,s)$ in X. Let $z \in B(x,r) \bigcap B(y,s)$. Take any t such that $0 < t < \min\{r-d(x,z), s-d(y,z)\}$. By the triangle inequality we see that

$$z \in B(z,t) \subseteq B(x,r) \bigcap B(y,s).$$

It follows that the intersection of any two open balls is in $\mathcal{T}$. It is quite easy to see now that $\mathcal{T}$ is closed under finite intersections.

Any family $\mathcal{T}$ of subsets of a set X satisfying (i), (ii), and (iii) is called a **topology** on X; the set X itself will be called a **topological space**. Sets in $\mathcal{T}$ are called **open.** The family $\mathcal{T}$ described above is called **the topology induced by** or **the topology compatible with** d. Most of the results on metric spaces that we need depend only on the topologies induced by their metrics. A topological space whose topology is induced by a metric is called a **metrizable space**. Note that the topology induced by the discrete metric on a set X (2.1.3) consists of all subsets of X. We call this topology the **discrete topology** on X.

Exercise 2.1.5 Show that both the metrics d_1 and d_2 on $\mathbb{R}^n$ defined in 2.1.1 induce the same topology. This topology is called the **usual topology.**

Another such example is obtained as follows. Let d be a metric on X and

$$\rho(x,y) = \min\{d(x,y), 1\}, \quad x, y \in X.$$

Then both d and ρ induce the same topology on X. These examples show that a topology may be induced by more than one metric. Two metrics d and ρ on a set are called **equivalent** if they induce the same topology.

Exercise 2.1.6 Show that two metrics d and ρ on a set X are equivalent if and only if for every sequence (x_n) in X and every $x \in X$,

$$d(x_n, x) \to 0 \iff \rho(x_n, x) \to 0.$$

Exercise 2.1.7 (i) Show that the intersection of any family of topologies on a set X is a topology.

(ii) Let $\mathcal{G} \subseteq \mathcal{P}(X)$. Show that there is a topology $\mathcal{T}$ on X containing $\mathcal{G}$ such that if $\mathcal{T}'$ is any topology containing $\mathcal{G}$, then $\mathcal{T} \subseteq \mathcal{T}'$.

If $\mathcal{G}$ and $\mathcal{T}$ are as in (ii), then we say that $\mathcal{G}$ **generates** $\mathcal{T}$ or that $\mathcal{G}$ is a **subbase** for $\mathcal{T}$. A **base** for a topology $\mathcal{T}$ on X is a family $\mathcal{B}$ of sets in $\mathcal{T}$ such that every $U \in \mathcal{T}$ is a union of elements in $\mathcal{B}$. It is easy to check that if $\mathcal{G}$ is a subbase for a topology $\mathcal{T}$, then $\mathcal{G}_d$, the family of finite intersections of elements of $\mathcal{G}$, is a base for $\mathcal{T}$. The set of all open balls of a metric space (X, d) is a base for the topology on X induced by d. For any X, $\{\{x\} : x \in X\}$ is a base for the discrete topology on X. A topological space X is called **second countable** if it has a countable base.

Exercise 2.1.8 Let $(X, \mathcal{T})$ have a countable subbase. Show that it is second countable.

A set $D \subseteq X$ is called **dense** in X if $U \bigcap D \neq \emptyset$ for every nonempty open set U, or equivalently, D intersects every nonempty open set in some fixed base $\mathcal{B}$. The set of rationals $\mathbb{Q}$ is dense in $\mathbb{R}$, and $\mathbb{Q}^n$ is dense in $\mathbb{R}^n$. A topological space X is called **separable** if it has a countable dense set. Let X be second countable and $\{U_n : n \in \mathbb{N}\}$ a countable base with all U_n's nonempty. Choose $x_n \in U_n$. Clearly, $\{x_n : n \in \mathbb{N}\}$ is dense. On the other hand, let (X, d) be a separable metric space and $\{x_n : n \in \mathbb{N}\}$ a countable dense set in X. Then

$$\mathcal{B} = \{B(x_n, r) : r \in \mathbb{Q}, r > 0 \;\&\; n \in \mathbb{N}\}$$

is a countable base for X. We have proved the following proposition.

Proposition 2.1.9 *A metrizable space is separable if and only if it is second countable.*

A subspace of a second countable space is clearly second countable. It follows that a subspace of a separable metric space is separable.

A subset F of a topological space X is called **closed** if $X \setminus F$ is open. For any $A \subseteq X$, $\text{cl}(A)$ will denote the intersection of all closed sets containing A. Thus $\text{cl}(A)$ is the smallest closed set containing A and is called the **closure** of A. Note that $D \subseteq X$ is dense if and only if $\text{cl}(D) = X$. The largest open set contained in A, denoted by $\text{int}(A)$, will be called the **interior** of A. A set A such that $x \in \text{int}(A)$ is called a **neighborhood** of x.

Exercise 2.1.10 For any $A \subseteq X$, X a topological space, show that

$$X \setminus \text{cl}(A) = \text{int}(X \setminus A).$$

Let (X, d) be a metric space, (x_n) a sequence in X, and $x \in X$. We say that (x_n) converges to x, written $x_n \to x$ or $\lim x_n = x$, if $d(x_n, x) \to 0$ as $n \to \infty$. Such an x is called the **limit** of (x_n). Note that a sequence can have at most one limit. Let $x \in X$. We call x an **accumulation point** of $A \subseteq X$ if every neighborhood of x contains a point of A other than x. Note that x is an accumulation point of A if and only if there is a sequence (x_n)

of distinct elements in A converging to x. The set of all accumulation points of A is called the **derived set**, or simply the **derivative**, of A. It will be denoted by A'. The elements of $A \setminus A'$ are called the **isolated points** of A. So, x is an isolated point of A if and only if there is an open set U such that $A \bigcap U = \{x\}$. A set $A \subseteq X$ is called **dense-in-itself** if it is nonempty and has no isolated point.

Exercise 2.1.11 Let $A \subseteq X$, X metrizable. Show the following.

(i) The set A is closed if and only if the limit of any sequence in A belongs to A.

(ii) The set A is open if and only if for any sequence (x_n) converging to a point in A, there exists an integer $M \geq 0$ such that $x_n \in A$ for all $n \geq M$.

(iii) $\text{cl}(A) = A \bigcup A'$.

Proposition 2.1.12 *Let X be a separable metric space and α an ordinal. Then every nondecreasing family $\{U_\beta : \beta < \alpha\}$ of nonempty open sets is countable.*

Proof. Fix a countable base $\{V_n\}$ for X. Let $\beta < \alpha$ be such that $U_{\beta+1} \setminus U_\beta \neq \emptyset$. Let $n(\beta)$ be the first integer m such that

$$V_m \bigcap U_\beta^c \neq \emptyset \ \& \ V_m \subseteq U_{\beta+1}.$$

Clearly, $\beta \longrightarrow n(\beta)$ is one-to-one and the result is proved. ■

Exercise 2.1.13 Let X be a separable metric space and α an ordinal number. Show that every monotone family $\{E_\beta : \beta < \alpha\}$ of nonempty sets that are all open or all closed is countable.

Let X and Y be topological spaces, $f : X \longrightarrow Y$ a map, and $x \in X$. We say that f is **continuous at** x if for every open V containing $f(x)$, there is an open set U containing x such that $f(U) \subseteq V$. The map f is called **continuous** if it is continuous at every $x \in X$. So, $f : X \longrightarrow Y$ is continuous if and only if $f^{-1}(V)$ is open (closed) in X for every open (closed) set V in Y.

Exercise 2.1.14 Let (X, d) and (Y, ρ) be metric spaces and $f : X \longrightarrow Y$ any map. Show that the following conditions are equivalent.

(i) The function $f : X \longrightarrow Y$ is continuous.

(ii) Whenever a sequence (x_n) in X converges to a point x, $f(x_n) \rightarrow f(x)$.

(iii) For every $\epsilon > 0$, there is a $\delta > 0$ such that $\rho(f(x), f(y)) < \epsilon$ whenever $d(x, y) < \delta$.

A function $f : X \longrightarrow Y$ is called a **homeomorphism** if it is a bijection and both f and f^{-1} are continuous. A homeomorphism f from X onto a subspace of Y will be called an **embedding**. It is easy to see that the composition of any two continuous functions (homeomorphisms) is continuous (a homeomorphism).

A function $f : X \longrightarrow Y$ is called **uniformly continuous** on X if for any $\epsilon > 0$, there exists a $\delta > 0$ satisfying

$$d(x, y) < \delta \Longrightarrow \rho(f(x), f(y)) < \epsilon$$

for any $x, y \in X$. Clearly, any uniformly continuous function is continuous. The converse is not true. For example, $f(x) = \frac{1}{x}$ is continuous but not uniformly continuous on $(0, 1]$.

A function $f : (X, d) \longrightarrow (Y, \rho))$ is called an **isometry** if $\rho(f(x), f(y)) = d(x, y)$ for all x, y in X. An isometry is clearly an embedding.

Exercise 2.1.15 Let (X, d) be a metric space and $\emptyset \neq A \subseteq X$. Define

$$d(x, A) = \inf\{d(x, y) : y \in A\}.$$

Show that for every A, $x \longrightarrow d(x, A)$ is uniformly continuous.

Exercise 2.1.16 Let F be a closed subset of (X, d). Show that

$$F = \bigcap_{n>0} \{x \in X : d(x, F) < \frac{1}{n}\}.$$

A subset of a metrizable space is called a G_δ set if it is a countable intersection of open sets. It follows from 2.1.16 that a closed subset of a metrizable space is a G_δ set. The class of G_δ sets is closed under countable intersections and finite unions. The complement of a G_δ set is called an F_σ set. Clearly, a subset of a metrizable space is an F_σ set if and only if it is a countable union of closed sets. Every open subset of a metric space is an F_σ.

Let $f_n, f : (X, d) \longrightarrow (Y, \rho)$. We say that (f_n) **converges pointwise** (or simply **converges**) to f if for all x, $f_n(x) \rightarrow f(x)$ as $n \rightarrow \infty$. We say f_n **converges uniformly** to f if for any $\epsilon > 0$, there exists $N \in \mathbf{N}$ such that whenever $n \geq N$, $\rho(f_n(x), f(x)) < \epsilon$ for all $x \in X$.

Exercise 2.1.17 Let $f_n : (X, d) \longrightarrow (Y, \rho)$ be a sequence of continuous functions converging uniformly to a function $f : X \longrightarrow Y$. Show that f is continuous. Show that f need not be continuous if f_n converges to f pointwise but not uniformly.

Proposition 2.1.18 *(Urysohn's lemma) Suppose A_0, A_1 are two nonempty, disjoint closed subsets of a metrizable space X. Then there is a continuous function $u : X \longrightarrow [0, 1]$ such that*

$$u(x) = \begin{cases} 0 & \text{if } x \in A_0, \\ 1 & \text{if } x \in A_1. \end{cases}$$

Proof. Let d be a compatible metric on X. Take

$$u(x) = \frac{d(x, A_0)}{d(x, A_0) + d(x, A_1)}.$$

■

A topological space is called **normal** if for every pair of disjoint closed sets A_0, A_1 there exist disjoint open sets U_0, U_1 containing A_0, A_1 respectively. The above proposition shows that every metrizable space is normal.

Proposition 2.1.19 *For every nonempty closed subset A of a metrizable space X there is a continuous function $f : X \longrightarrow [0,1]$ such that $A = f^{-1}(0)$.*

Proof. Write $A = \bigcap_{n=0}^{\infty} U_n$, where the U_n's are open (2.1.16). By 2.1.18, for each $n \in \mathbb{N}$, there is a continuous $f_n : X \longrightarrow [0,1]$ such that

$$f_n(x) = \begin{cases} 0 & \text{if } x \in A, \\ 1 & \text{if } x \in X \setminus U_n. \end{cases}$$

Take $f = \sum_0^{\infty} \frac{1}{2^{n+1}} f_n$. ■

Theorem 2.1.20 *(Tietze extension theorem) Let (X, d) be a metric space, $A \subseteq X$ closed, and $f : A \longrightarrow [1,2]$ continuous. Then there is a continuous extension $F : X \longrightarrow [1,2]$ of f.*

Proof. Define $h : X \longrightarrow [0, \infty)$ by

$$h(x) = \inf\{f(z)d(x,z) : z \in A\}, \quad x \in X.$$

Put

$$F(x) = \begin{cases} h(x)/d(x,A) & \text{if } x \in X \setminus A, \\ f(x) & \text{otherwise.} \end{cases}$$

Since f is continuous on A, F is continuous at each point x of $\text{int}(A)$. It remains to show that F is continuous at each point x of $X \setminus \text{int}(A)$.

First consider the case $x \in X \setminus A$. As $X \setminus A$ is open, it is sufficient to show that $F|(X \setminus A)$ is continuous at x. Since the map $y \longrightarrow d(y, A)$ is continuous, we only need to show that h is continuous at x. Fix $\epsilon > 0$. We have to show that there is a $\delta > 0$ such that whenever $x' \in X \setminus A$ and $d(x, x') < \delta$, $|h(x) - h(x')| < \epsilon$. Take $\delta = \epsilon/2$. Take any $x' \in X \setminus A$ with $d(x, x') < \delta$. For any $z \in A$,

$$d(x,z) \leq d(x,x') + d(x',z) < d(x',z) + \delta.$$

As $f(z) \leq 2$,

$$h(x) = \inf\{f(z)d(x,z) : z \in A\} \leq \inf\{f(z)d(x',z) : z \in A\} + 2\delta = h(x') + \epsilon.$$

Thus, h is continuous at x.

Now consider the case when $x \in A \setminus \operatorname{int}(A)$. Fix any $\epsilon > 0$. As f is continuous on A, there is an $r > 0$ such that whenever $y \in A$ and $d(x,y) < r$, $|f(x) - f(y)| < \epsilon$. Take $\delta = r/4$. If $y \in A$ and $d(x,y) < \delta$, then clearly $|F(x) - F(y)| < \epsilon$. So, assume that $y \in X \setminus A$ and $d(x,y) < \frac{r}{4}$. Our proof will be complete if we show that

$$|f(x) - h(y)/d(y,A)| < \epsilon.$$

We note the following.

(i) $d(y,A) = \inf\{d(y,z) : z \in A \,\&\, d(x,z) < r\}$.

(ii) $h(y) = \inf\{f(z)d(y,z) : z \in A \,\&\, d(x,z) < r\}$.

The assertion (i) is easy to prove. The assertion (ii) follows from the following two observations.

(a) As $f(x) < 2$ and $d(x,y) < \frac{r}{4}$, $f(x)d(x,y) < r/2$. So, the term on the right-hand side of ii) is less than $\frac{r}{2}$.

(b) Suppose $d(x,z) \geq r$. Then

$$d(y,z) \geq d(x,z) - d(x,y) \geq r - \frac{r}{4} = 3r/4.$$

As $f(z) \geq 1$, $f(z)d(y,z) \geq 3r/4$.

Now take any $z \in A$ with $d(x,z) < r$. As

$$f(x) - \epsilon < f(z) < f(x) + \epsilon,$$

it follows that

$$(f(x) - \epsilon)d(y,z) \leq f(z)d(y,z) \leq (f(x) + \epsilon)d(y,z).$$

Taking the infimum over z in A with $d(x,z) < r$, by (i) and (ii) we have

$$|f(x) - h(y)/d(y,A)| \leq \epsilon.$$

■

Exercise 2.1.21 Let X and A be as in the last theorem and $J \subseteq \mathbb{R}$ an interval. Show that every continuous $f : A \longrightarrow J$ admits a continuous extension $F : X \longrightarrow J$.

Exercise 2.1.22 Let X be metrizable, $A \subseteq X$ closed and $K \subseteq \mathbb{R}^n$ a closed, bounded, and convex set. Show that every continuous function f from A to K admits a continuous extension to X.

A real-valued map f defined on a metric space X is called **upper-semicontinuous (lower-semicontinuous)** if for every real number a, the set $\{x \in X : f(x) \geq a\}$ ($\{x \in X : f(x) \leq a\}$) is closed.

Exercise 2.1.23 Let X be a metric space and $f : X \longrightarrow \mathbb{R}$ any map. Show that the following statements are equivalent.

(i) f is upper-semicontinuous.

(ii) For every real number a, $\{x \in X : f(x) < a\}$ is open.

(iii) Whenever a sequence (x_n) in X converges to a point x, $\limsup f(x_n) \leq f(x)$.

Exercise 2.1.24 Let X be a metric space and $f_i : X \longrightarrow \mathbb{R}$, $i \in I$, continuous maps. Show that the map $f : X \longrightarrow \mathbb{R}$ defined by

$$f(x) = \inf\{f_i(x) : i \in I\}, \quad x \in X,$$

is upper-semicontinuous.

Next we show that the converse of this result is also true.

Proposition 2.1.25 *Suppose X is a metric space and $f : X \longrightarrow \mathbb{R}$ an upper-semicontinuousmap such that there is a continuous map $g : X \longrightarrow \mathbb{R}$ such that $f \leq g$; i.e., $f(x) \leq g(x)$ for all x. Then there is a sequence of continuous maps $f_n : X \longrightarrow \mathbb{R}$ such that $f(x) = \inf f_n(x)$ for all x.*

Proof. Let r be any rational number. Set

$$U_r = \{x \in X : f(x) < r < g(x)\}.$$

Since f is upper-semicontinuous and g continuous, U_r is open. Let (F_n^r) be a sequence of closed sets such that $U_r = \bigcup_n F_n^r$. By the Tietze extension theorem, there is a continuous map $f_n^r : X \longrightarrow [r, \infty)$ satisfying

$$f_n^r(x) = \begin{cases} r & \text{if } x \in F_n^r, \\ g(x) & \text{if } x \in X \setminus U_r. \end{cases}$$

We claim that

$$f(x) = \inf\{f_n^r(x) : r \in \mathbb{Q} \text{ and } n \in \mathbb{N}\}$$

for all x. Clearly, $f_n^r(x) \geq f(x)$ for every $x \in X$. Fix any $x_0 \in X$ and $\epsilon > 0$. To complete the proof, we show that for some r and for some n,

$$f_n^r(x_0) < f(x_0) + \epsilon.$$

Take any rational number r such that

$$f(x_0) < r < f(x_0) + \epsilon.$$

Two cases arise: $g(x_0) \leq r$ or $g(x_0) > r$. If $g(x_0) \leq r$, then $x_0 \in X \setminus U_r$. Hence,

$$f_n^r(x_0) = g(x_0) < f(x_0) + \epsilon$$

for all n. If $g(x_0) > r$, then $x_0 \in U_r$. Take any n such that $x_0 \in F_n^r$. Then $f_n^r(x_0) < f(x_0) + \epsilon$, and our result is proved. ∎

We proved the above result under the additional condition that f is dominated by a continuous function. So the question arises; *Is every real-valued upper-semicontinuous function defined on a metric space dominated by a continuous function?* The answer is yes. (See [99].) The proofs of this in some important special cases are given later in this chapter.

Let $\{X_i : i \in I\}$ be a family of topological spaces, $X = \prod_{i \in I} X_i$, and $\pi_i : X \longrightarrow X_i$, $i \in I$, the projection maps. The smallest topology on X making each π_i continuous is called the **product topology**. So,

$$\{\pi_i^{-1}(U) : U \text{ open in } X_i, i \in I\}$$

is a subbase for the product topology.

Exercise 2.1.26 Let $(X_0, d_0), (X_1, d_1), \ldots$ be metric spaces, $X = \prod_n X_n$, and d the product metric on X (2.1.4).

(i) Show that d induces the product topology on X.

(ii) Let $\alpha, \alpha_0, \alpha_1, \alpha_2, \ldots \in X$. Show that

$$(\alpha_n \to \alpha) \iff (\forall k)(\alpha_n(k) \to \alpha(k)).$$

(iii) Let Y be a topological space. Show that $f : Y \longrightarrow X$ is continuous if and only if $\pi_i \circ f$ is continuous for all i, where $\pi_i : X \longrightarrow X_i$ is the projection map.

Proposition 2.1.27 *The product of countably many second countable (equivalently separable) metric spaces is second countable.*

Proof. Let $X_0, X_1, \ldots$ be second countable. Let $X = \prod_i X_i$. We show that X has a countable subbase. The result then follows from 2.1.8. Let $\{U_{in} : n \in \mathbb{N}\}$ be a base for X_i. Then, by the definition of the product topology, $\{\pi_i^{-1}(U_{in}) : i, n \in \mathbb{N}\}$ is a subbase for X. Since $\{\pi_i^{-1}(U_{in}) : i, n \in \mathbb{N}\}$ is countable, the result follows from 2.1.8. ∎

A sequence (x_n) in a metric space (X, d) is called a **Cauchy sequence** if for every $\epsilon > 0$ there is an $N \in \mathbb{N}$ such that $d(x_n, x_m) < \epsilon$ for all $m, n \geq N$. It is easy to see that every convergent sequence is Cauchy and that if a Cauchy sequence (x_n) has a convergent subsequence, then (x_n) itself is convergent. A Cauchy sequence need not be convergent. To see this, let $X = \mathbb{Q}$ with the usual metric and (x_n) a sequence of rationals converging to an irrational number, say $\sqrt{2}$. Then (x_n) is a Cauchy sequence in $\mathbb{Q}$

that does not converge to a point in $\mathbb{Q}$. A metric d on a set X is called **complete** if every Cauchy sequence in (X, d) is convergent. A metric space (X, d) is called **complete** if d is complete on X. It is easy to see that $\mathbb{R}^n$ with the usual metric is complete. We have seen that $\mathbb{Q}$ with the usual metric is not complete. Thus a subspace of a complete metric space need not be complete. However, a closed subspace of a complete metric space is easily seen to be complete. For $A \subseteq X$ we define

$$\text{diameter}(A) = \sup\{d(x, y) : x, y \in A\}.$$

Exercise 2.1.28 Let (X, d) be a metric space. Show that for any $A \subseteq X$,

$$\text{diameter}(A) = \text{diameter}(\text{cl}(A)).$$

Proposition 2.1.29 *(Cantor intersection theorem) A metric space (X, d) is complete if and only if for every decreasing sequence $F_0 \supseteq F_1 \supseteq F_2 \subseteq \cdots$ of nonempty closed subsets of X with diameter$(F_n) \to 0$, the intersection $\bigcap_n F_n$ is a singleton.*

Proof. Assume that (X, d) is complete. Let (F_n) be a decreasing sequence of nonempty closed sets with diameter converging to 0. Choose $x_n \in F_n$. Since diameter$(F_n) \to 0$, (x_n) is Cauchy and so convergent. It is easily seen that $\lim x_n \in \bigcap_n F_n$. Let $x \neq y$. Then $d(x, y) > 0$. Since diameter$(F_n) \to 0$, there is an integer n such that both x and y cannot belong to F_n. It follows that both x and y cannot belong to $\bigcap F_n$.

To show the converse, let (x_n) be a Cauchy sequence. Put

$$F_n = \text{cl}(\{x_m : m \geq n\}).$$

As (x_n) is Cauchy, diam$(F_n) \to 0$. Take $x \in \bigcap_n F_n$. Then $\lim x_n = x$. ■

Exercise 2.1.30 Let d be a metric on $\mathbb{N}$ defined by

$$d(m, n) = \frac{|m - n|}{(m + 1)(n + 1)}.$$

Show the following.

(i) The metric d induces the discrete topology.

(ii) The metric d is not complete on $\mathbb{N}$.

The above exercise shows that a metric equivalent to a complete one need not be complete.

Proposition 2.1.31 *Let (X_0, d_0), (X_1, d_1), (X_2, d_2), ... be complete metric spaces, $X = \prod_n X_n$, and d the product metric on X. Then (X, d) is complete.*

Proof. Let $\alpha_0, \alpha_1, \alpha_2, \ldots$ be a Cauchy sequence in X. Then for each k, $\alpha_0(k), \alpha_1(k), \alpha_2(k), \ldots$ is a Cauchy sequence in X_k. As X_k is complete, we get an $\alpha(k) \in X_k$ such that $\alpha_n(k) \to \alpha(k)$. By 2.1.26, the sequence (α_n) converges to α. ■

Let (X, d) be a metric space and $[X]$ the set of all Cauchy sequences in X. We define a binary relation $\equiv$ on $[X]$ as follows.

$$(x_n) \equiv (y_n) \iff d(x_n, y_n) \to 0 \text{ as } n \to \infty.$$

It is easily checked that $\equiv$ is an equivalence relation. Let $\hat{X}$ denote the set of all equivalence classes. For any Cauchy sequence (x_n), $[x_n]$ will denote the equivalence class containing (x_n). We define a metric $\hat{d}$ on $\hat{X}$ by

$$\hat{d}([x_n], [y_n]) = \lim d(x_n, y_n).$$

Define $f : X \longrightarrow [X]$ by $f(x) = [x_n]$, where $x_n = x$ for all n. We can easily check the following.

(i) $\hat{d}$ is well-defined.

(ii) $\hat{d}$ is a complete metric on $\hat{X}$.

(iii) The function $f : X \longrightarrow \hat{X}$ is an isometry.

(iv) The set $f(X)$ is dense in $\hat{X}$.

(v) If X is separable, so is $\hat{X}$.

Thus we see that every (separable) metric space can be isometrically embedded in a (separable) complete metric space. The metric space $(\hat{X}, \hat{d})$ is called the **completion** of (X, d).

There is another very useful embedding of a separable metric space into a complete separable metric space. The closed unit interval $[0, 1]$ with the usual metric is clearly complete and separable. Therefore, by 2.1.27 and 2.1.29, $\mathbb{H} = [0, 1]^{\mathbb{N}}$ is complete and separable. The topological space $\mathbb{H}$ is generally known as the **Hilbert cube**.

Theorem 2.1.32 *Any second countable metrizable space X can be embedded in the Hilbert cube $\mathbb{H}$.*

Proof. Let (U_n) be a countable base for X. For each pair of integers n, m with $\text{cl}(U_n) \subseteq U_m$, choose a continuous $f_{nm} : X \to [0, 1]$ such that

$$f_{nm}(x) = \begin{cases} 0 & \text{if } x \in \text{cl}(U_n), \\ 1 & \text{if } x \in X \setminus U_m. \end{cases}$$

By 2.1.18, such a function exists. Enumerate $\{f_{nm} : m, n \in \mathbb{N}\}$ as a sequence (f_k). Define f on X by

$$f(x) = (f_0(x), f_1(x), \ldots), \quad x \in X.$$

We can easily check that f embeds X in the Hilbert cube. ■

Exercise 2.1.33 Let $(X_0, d_0), (X_1, d_1), (X_2, d_2), \ldots$ be metric spaces with the X_i's pairwise disjoint and $d_i < 1$ for all i. Let $X = \bigcup_n X_n$. Define d by

$$d(x, y) = \begin{cases} d_i(x, y) & \text{if } x, y \in X_i, \\ 1 & \text{otherwise.} \end{cases}$$

(i) Show that d is a metric on X such that $U \subseteq X$ is open with respect to the induced topology if and only if $U \bigcap X_i$ is open in X_i for all i.

(ii) Further, if each of (X_i, d_i) is complete (separable), then show that (X, d) is complete (separable).

If $X, X_0, X_1, X_2, \ldots$ are as above, then we call X the **topological sum** of the X_i's and write $X = \bigoplus_n X_n$.

Proposition 2.1.34 *Every nonempty open set U in $\mathbb{R}$ is a countable union of pairwise disjoint nonempty open intervals.*

Proof. Let $x \in U$ and let I_x be the union of all open intervals containing x and contained in U. Clearly, for any x, y, either $I_x = I_y$ or $I_x \bigcap I_y = \emptyset$. Since $\mathbb{R}$ is separable, $\{I_x : x \in U\}$ is countable. Further, $U = \bigcup_{x \in U} I_x$. ■

The importance of the next result will become clear in the next chapter.

Proposition 2.1.35 *(Sierpiński) The open unit interval* $(0, 1)$ *cannot be expressed as a countable disjoint union of nonempty closed subsets of* $\mathbb{R}$.

Proof. Let $A_0, A_1, A_2, \ldots$ be a sequence of pairwise disjoint nonempty closed sets in $\mathbb{R}$, each contained in $(0, 1)$. We show that $\bigcup A_i \neq (0, 1)$. Suppose $\bigcup A_i = (0, 1)$. Then for $k \in \mathbb{N}$, we define integers $m_k, n_k \in \mathbb{N}$ and real numbers $a_k, b_k \in (0, 1)$ satisfying the following conditions.

(i) $n_0 < m_0 < \cdots < n_k < m_k$,

(ii) $a_0 < a_1 < \cdots < a_k < b_k < \cdots b_1 < b_0$,

(iii) $a_k \in A_{n_k}$, $b_k \in A_{m_k}$,

(iv) for every $i \leq m_k$, $A_i \bigcap (a_k, b_k) = \emptyset$.

Assume that we have done this. Take $a = \sup a_k$. Then $a_n < a < b_n$ for all n. Hence, $a \notin \bigcup A_i$, which is a contradiction.

We define m_k, n_k, a_k, and b_k by induction. Take $n_0 = 0$ and $a_0 = \sup A_{n_0}$. Let m_0 be the first integer m such that $A_m \bigcap (a_0, 1) \neq \emptyset$. Put $b_0 = \inf[A_{m_0} \bigcap (a_0, 1)]$. Since A_{n_0} and A_{m_0} are disjoint and closed, $a_0 < b_0$. Note that $A_i \bigcap (a_0, b_0) = \emptyset$ for all $i \leq m_0$. Let $k \in \mathbb{N}$ and suppose for every $i \leq k$, m_i, n_i, a_i, and b_i satisfying (i)-(iv) have been defined. Take n_{k+1} to be the first integer n such that $A_n \bigcap (a_k, b_k) \neq \emptyset$. Put $a_{k+1} = \sup[A_{n_{k+1}} \bigcap (a_k, b_k)]$. Clearly, $a_{k+1} < b_k$. Now, let m_{k+1} be the first integer m such that $A_m \bigcap (a_{k+1}, b_k) \neq \emptyset$. Note that $m_{k+1} > n_{k+1}$. Take $b_{k+1} = \inf[A_{m_{k+1}} \bigcap (a_{k+1}, b_k)]$. ■

2.2 Polish Spaces

A topological space is called **completely metrizable** if its topology is induced by a complete metric. A **Polish space** is a separable, completely metrizable topological space.

Some elementary observations.

(i) Any countable discrete space is Polish. In particular, $\mathbb{N}$ and $2 = \{0, 1\}$, with discrete topologies, are Polish.

(ii) The real line $\mathbb{R}$, $\mathbb{R}^n$, $I = [0, 1]$, I^n, etc., with the usual topologies are Polish.

(iii) Any closed subspace of a Polish space is Polish.

(iv) The topological sum of a sequence of Polish spaces is Polish.

(v) The product of countably many Polish spaces is Polish. In particular, $\mathbb{N}^{\mathbb{N}}$, the Hilbert cube $\mathbb{H} = [0, 1]^{\mathbb{N}}$, and the Cantor space $\mathcal{C} = 2^{\mathbb{N}}$ are Polish.

The spaces $\mathbb{N}^{\mathbb{N}}$ and $\mathcal{C}$ are of particular importance to us. A complete metric on $\mathbb{N}^{\mathbb{N}}$ compatible with its topology is given below.

$$\rho(\alpha, \beta) = \begin{cases} \frac{1}{\min\{n:\alpha(n)\neq\beta(n)\}+1} & \text{if } \alpha \neq \beta, \\ 0 & \text{otherwise.} \end{cases}$$

For $s \in \mathbb{N}^{<\mathbb{N}}$, let

$$\Sigma(s) = \{\alpha \in \mathbb{N}^{\mathbb{N}} : s \prec \alpha\}.$$

The family of sets $\{\Sigma(s) : s \in \mathbb{N}^{<\mathbb{N}}\}$ is a base for $\mathbb{N}^{\mathbb{N}}$. Note that the sets $\Sigma(s)$ are both closed and open in $\mathbb{N}^{\mathbb{N}}$. Such sets are called **clopen**. A topological space is called **zero-dimensional** if it has a base consisting of clopen sets. Thus $\mathbb{N}^{\mathbb{N}}$ is a zero-dimensional Polish space. Note that the product of a family of zero-dimensional spaces is zero-dimensional.

A compatible metric and a base for $\mathcal{C}$ can be similarly defined. More generally, let A be a discrete space and $X = A^{\mathbb{N}}$ be equipped with the product topology. Then X is a zero-dimensional completely metrizable space; it is Polish if and only if A is countable. Let $s \in A^{<\mathbb{N}}$. When there is no scope for confusion, we shall also denote the set $\{\alpha \in A^{\mathbb{N}} : s \prec \alpha\}$ by $\Sigma(s)$.

In the next few results we characterize spaces that are Polish: They are the topological spaces that are homeomorphic to G_δ subsets of the Hilbert cube $\mathbb{H}$.

Theorem 2.2.1 *(Alexandrov) Every G_δ subset G of a completely metrizable (Polish) space X is completely metrizable (Polish).*

Proof. Fix a complete metric d on X compatible with its topology. We first prove the result when G is open. Consider the function $f : G \longrightarrow X \times \mathbb{R}$ defined by

$$f(x) = (x, \frac{1}{d(x, X \setminus G)}), \quad x \in G.$$

Note the following.

(i) The function f is one-to-one.

(ii) By 2.1.15 and 2.1.26, f is continuous.

(iii) Since f^{-1} is $\pi_1|f(G)$, it is continuous.

(iv) The set $f(G)$ is closed in $X \times \mathbb{R}$.

To see (iv), let (x_n) be a sequence in G and

$$f(x_n) = (x_n, 1/d(x_n, X \setminus G)) \to (x, y).$$

Then, $x_n \to x$. Hence,

$$d(x_n, X \setminus G) \to d(x, X \setminus G).$$

Since $1/d(x_n, X \setminus G) \to y$, $y = 1/d(x, X \setminus G)$. Hence, $d(x, X \setminus G) \neq 0$. This implies that $x \in G$ and $(x, y) = f(x) \in f(G)$.

So, G is homeomorphic to $f(G)$. As $f(G)$ is closed in the completely metrizable space $X \times \mathbb{R}$, it is completely metrizable. Since f is a homeomorphism, G is completely metrizable.

Now consider the case when G is a G_δ set. Let $G = \bigcap_n G_n$, where the G_n's are open. Define $f : G \longrightarrow X \times \mathbb{R}^{\mathbb{N}}$ by

$$f(x) = (x, \frac{1}{d(x, X \setminus G_0)}, \frac{1}{d(x, X \setminus G_1)}, \ldots), \quad x \in G.$$

Arguing as above, we see that f embeds G onto a closed subspace of $X \times \mathbb{R}^{\mathbb{N}}$, which completes the proof. ■

From the above theorem we see that the spaces J (J an interval) and $\mathbb{R} \setminus \mathbb{Q}$, the set of irrational numbers, with the usual topologies, are completely metrizable, though the usual metrics may not be complete on them.

Exercise 2.2.2 Give complete metrics on $(0, 1)$ and on the set of all irrationals inducing the usual topology.

The converse of 2.2.1 is also true; i.e., every completely metrizable subspace of a completely metrizable space X is a G_δ set in X. To prove this, we need a result on extensions of continuous functions that is interesting in its own right.

Proposition 2.2.3 *Let $f : A \longrightarrow Z$ be a continuous map from a subset A of a metrizable space W to a completely metrizable space Z. Then f can be extended continuously to a G_δ set containing A.*

Proof. Take a bounded complete metric ρ on Z compatible with its topology. For any $x \in \mathrm{cl}(A)$, let

$$O_f(x) = \inf\{\mathrm{diameter}(f(A \cap V)) : V \text{ open}, x \in V\}.$$

We call $O_f(x)$ the **oscillation** of f at x. Put

$$B = \{x \in \mathrm{cl}(A) : O_f(x) = 0\}.$$

The set B is G_δ in W. To see this, take any $t > 0$ and note that for any $x \in \mathrm{cl}(A)$,

$$O_f(x) < t \Longleftrightarrow (\exists \text{ open } V \ni x)(\mathrm{diameter}(f(A \cap V)) < t).$$

Therefore, the set

$$\begin{aligned} &\{x \in \mathrm{cl}(A) : O_f(x) < t\} \\ = &\bigcup\{V \bigcap \mathrm{cl}(A) : V \text{ open and diameter}(f(A \cap V)) < t\}, \end{aligned}$$

and hence it is open in $\mathrm{cl}(A)$. Since

$$B = \bigcap_n \{x \in \mathrm{cl}(A) : O_f(x) < \frac{1}{n+1}\}$$

and $\mathrm{cl}(A)$ is a G_δ set in W, B is a G_δ set. Since f is continuous on A, the oscillation of f at every $x \in A$ is 0. Therefore, $A \subseteq B$.

We now define a continuous map $g : B \longrightarrow Z$ that extends f. Let $x \in B$. Take a sequence (x_n) in A converging to x. Since $O_f(x) = 0$, $(f(x_n))$ is a Cauchy sequence in (Z, ρ). As (Z, ρ) is complete, $(f(x_n))$ is convergent. Put $g(x) = \lim_n f(x_n)$. The following statements are easy to prove.

(i) The map g is well-defined.

(ii) It is continuous.

(iii) It extends f.

■

Remark 2.2.4 Let W, Z be as above and $f : W \longrightarrow Z$ an arbitrary map. The above proof shows that the set $\{x \in W : f \text{ is continuous at } x\}$ is a G_δ set in W.

Exercise 2.2.5 Show that for every G_δ subset A of reals there is a map $f : \mathbb{R} \longrightarrow \mathbb{R}$ whose set of continuity points is precisely A.

Theorem 2.2.6 *(Lavrentiev) Let X, Y be completely metrizable spaces, $A \subseteq X$, $B \subseteq Y$, and $f : A \longrightarrow B$ a homeomorphism onto B. Then f can be extended to a homeomorphism between two G_δ sets containing A and B.*

Proof. Let $g = f^{-1}$. By 2.2.3, choose a G_δ set $A' \supseteq A$ and a contiunuous extension $f' : A' \longrightarrow Y$ of f. Similarly, choose a G_δ set $B' \supseteq B$ and a continuous extension $g' : B' \longrightarrow X$ of g. Let

$$H = \{(x, y) \in A' \times Y : y = f'(x)\} = \text{graph}(f')$$

and

$$K = \{(x, y) \in X \times B' : x = g'(y)\} = \text{graph}(g').$$

Let $A^* = \pi_1(H \bigcap K)$ and $B^* = \pi_2(H \bigcap K)$, where π_1 and π_2 are the two projection functions. Note that

$$A^* = \{x \in A' : (x, f'(x)) \in K)\}$$

and

$$B^* = \{y \in B' : (g'(y), y)) \in H)\}.$$

Since K is closed in $X \times B'$ and B' is a G_δ, K is a G_δ set. As f' is continuous on the G_δ set A', A^* is a G_δ set. Similarly, we can show that B^* is a G_δ set. It is easy to check that $f^* = f'|A^*$ is a homeomorphism from A^* onto B^* that extends f. ■

Theorem 2.2.7 *Let X be a completely metrizable space and Y a completely metrizable subspace. Then Y is a G_δ set in X.*

Proof. The result follows from 2.2.6 by taking $A = B = Y$ and $f : A \longrightarrow B$ the identity map. ■

Remark 2.2.8 In the last section we saw that every second countable metrizable space can be embedded in the Hilbert cube. Thus, *a topological space X is Polish if and only if it is homeomorphic to a G_δ subset of the Hilbert cube.*

We close this section by giving some useful results on zero-dimensional spaces.

Lemma 2.2.9 *Every second countable, zero-dimensional metrizable space X can be embedded in C.*

Proof. Fix a countable base $\{U_n : n \in \mathbb{N}\}$ for X such that each U_n is clopen. Define $f : X \longrightarrow C$ by

$$f(x) = (\chi_{U_0}(x), \chi_{U_1}(x), \chi_{U_2}(x), \ldots), \quad x \in X.$$

Since the characteristic function of a clopen set is continuous and since a map into a product space is continuous if its composition with the projection to each of its coordinate spaces is continuous, f is continuous. Since $\{U_n : n \in \mathbb{N}\}$ is a base for X, f is one-to-one. Further,

$$f(U_n) = f(X) \bigcap \{\alpha \in C : \alpha(n) = 1\}.$$

Therefore, $f^{-1} : f(X) \longrightarrow X$ is also continuous. Thus, f is an embedding of X in C. ∎

Exercise 2.2.10 **(i)** Show that every second countable metrizable space of cardinality less than $\mathfrak{c}$ is zero-dimensional.

(ii) Show that every countable metrizable space is zero-dimensional.

(iii) Show that every countable metrizable space can be embedded into $\mathbb{Q}$.

(iv) Let X be a countable, nonempty metrizable space with no isolated points. Show that X is homeomorphic to $\mathbb{Q}$.

From 2.2.7 we obtain the following.

Proposition 2.2.11 *Every zero-dimensional Polish space is homeomorphic to a G_δ subset of C.*

The Cantor space is clearly embedded in $\mathbb{N}^{\mathbb{N}}$. Hence every zero-dimensional Polish space is homeomorphic to a G_δ subset of $\mathbb{N}^{\mathbb{N}}$.

Exercise 2.2.12 Let $E \subseteq C$ be the set of all sequences $(\epsilon_0, \epsilon_1, \epsilon_2, \ldots)$ with infinitely many 0's and infinitely many 1's. Show that $\mathbb{N}^{\mathbb{N}}$ and E are homeomorphic.

The following result will be used later.

Proposition 2.2.13 *Let A be any set with the discrete topology. Suppose $A^{\mathbb{N}}$ is equipped with the product toplogy and C is any subset of $A^{\mathbb{N}}$. Then C is closed if and only if it is the body of a tree T on A.*

Proof. Let T be a tree on A. We show that $A^{\mathbb{N}} \setminus [T]$ is open. Let $\alpha \notin [T]$. Then there exists a $k \in \mathbb{N}$ such that $\alpha|k \notin T$. So, $\Sigma(\alpha|k) \subseteq A^{\mathbb{N}} \setminus [T]$, whence $A^{\mathbb{N}} \setminus [T]$ is open.

Conversely, let C be closed in $A^{\mathbb{N}}$. Let

$$T = \{\alpha|k : \alpha \in C \text{ and } k \in \mathbb{N}\}.$$

Clearly, $C \subseteq [T]$. Take any $\alpha \notin C$. Since C is closed, choose a $k \in \mathbb{N}$ such that $\Sigma(\alpha|k) \subseteq A^{\mathbb{N}} \setminus C$. Thus $\alpha|k \notin T$. Hence $\alpha \notin [T]$. ∎

Exercise 2.2.14 Let $\mathcal{K}$ be the smallest family of subsets of $\mathbb{N}^{\mathbb{N}}$ satisfying the following conditions.

(a) $\mathcal{K}$ contains $\emptyset$ and $\mathbf{N}^{\mathbf{N}}$.

(b) A set $A \subseteq \mathbf{N}^{\mathbf{N}}$ belongs to $\mathcal{K}$ whenever all its sections A_i, $i \in \mathbf{N}$, belong to $\mathcal{K}$.

For each ordinal $\alpha < \omega_1$, define a family $\mathcal{A}_\alpha$ of subsets of $\mathbf{N}^{\mathbf{N}}$, by induction, as follows.

$$\mathcal{A}_0 = \{\emptyset, \mathbf{N}^{\mathbf{N}}\}.$$

Suppose α is any countable ordinal and for every $\beta < \alpha$, $\mathcal{A}_\beta$ has been defined. Put

$$\mathcal{A}_\alpha = \{A \subseteq \mathbf{N}^{\mathbf{N}} : \text{ for all } i \in \mathbf{N}, A_i \in \bigcup_{\beta<\alpha} \mathcal{A}_\beta\}.$$

Show that

(i) $\mathcal{K} = \bigcup_{\alpha<\omega_1} \mathcal{A}_\alpha$;

(ii) for every $\alpha < \omega_1$, $\mathcal{A}_\alpha \neq \mathcal{A}_{\alpha+1}$;

(iii) $\mathcal{K}$ equals the set of all clopen subsets of $\mathbf{N}^{\mathbf{N}}$.

Remark 2.2.15 The hierarchy $\{\mathcal{A}_\alpha : \alpha < \omega_1\}$ is called the **Kalmar hierarchy**.

2.3 Compact Metric Spaces

Let $(X, \mathcal{T})$ be a topological space and $A \subseteq X$. A family $\mathcal{U}$ of sets whose union contains A is called a **cover** of A. A subfamily of $\mathcal{U}$ that is a cover of A is called a **subcover**. The set A is called **compact** if every open cover of A admits a finite subcover.

Exercise 2.3.1 Let $\mathcal{B}$ be a base for a topology on X. Show that X is compact if and only if every cover $\mathcal{U} \subseteq \mathcal{B}$ admits a finite subcover.

Examples of compact sets are:

(i) any finite subset of a topological space;

(ii) any closed interval $[a, b] \subseteq \mathbb{R}$ with the usual topology;

(iii) any closed cube $\prod_{i=1}^{n}[a_i, b_i] \subseteq \mathbb{R}^n$ with the usual topology.

If X is a compact space, then every closed subset is also compact. Further, a compact subset of a metric space is closed. To see this, let (X, d) be a metric space and $A \subseteq X$ compact. Let $x \in X \setminus A$. Our assertion will be proved if we show that there is an $r > 0$ such that $B(x, r) \bigcap A = \emptyset$. For $a \in A$, set $d(x, a)/2 = r_a$. Then $\{B(a, r_a) : a \in A\}$ covers A. Let $B(a_1, r_1), B(a_2, r_2), \ldots, B(a_n, r_n)$ be a subcover of A. Take $r = \min\{r_1, r_2, \ldots, r_n\}$. This r answers our purpose.

Exercise 2.3.2 Let X be any subset of $\mathbb{R}^n$. Show that X is compact if and only if it is closed and bounded.

The following is an important example of a compact set. It was first considered by Cantorin his study of the sets of uniqueness of trigonometric series [26].

Example 2.3.3 Define a sequence (C_n) of subsets of $[0,1]$ inductively as follows. Take

$$C_0 = [0,1].$$

Suppose C_n has been defined and is a union of 2^n pairwise disjoint closed intervals $\{I_j : 1 \leq j \leq 2^n\}$ of length $1/3^n$ each. Obtain C_{n+1} by removing the open middle third of each I_j. For instance,

$$C_1 = [0, \tfrac{1}{3}] \bigcup [\tfrac{2}{3}, 1],$$

$$C_2 = [0, \tfrac{1}{9}] \bigcup [\tfrac{2}{9}, \tfrac{1}{3}] \bigcup [\tfrac{2}{3}, \tfrac{7}{9}] \bigcup [\tfrac{8}{9}, 1].$$

Finally, put $\mathbf{C} = \bigcap_n C_n$. The set $\mathbf{C}$ is known as the **Cantor ternary set.** As $\mathbf{C}$ is closed and bounded, it is compact. Define a map $f : \{0,1\}^{\mathbb{N}} \longrightarrow \mathbf{C}$ by

$$f((\epsilon_n)) = \sum_{n=0}^{\infty} \frac{2}{3^{n+1}} \epsilon_n, \quad (\epsilon_n) \in \{0,1\}^{\mathbb{N}}.$$

It is easy to check that f is a homeomorphism.

A family $\mathcal{F}$ of nonempty sets is said to have the **finite intersection property** if the intersection of every finite subfamily of $\mathcal{F}$ is nonempty.

Exercise 2.3.4 Show that a topological space X is compact if and only if every family of closed sets with the finite intersection property has nonempty intersection.

Exercise 2.3.5 Show that the topological sum of finitely many compact spaces is compact.

Proposition 2.3.6 *A continuous image of a compact space is compact.*

Proof. Let X be compact and $f : X \longrightarrow Y$ continuous. Suppose $\mathcal{U}$ is an open cover for $f(X)$. Then $\{f^{-1}(U) : U \in \mathcal{U}\}$ is a cover of X. As X is compact, there is a finite subcover of X, say $f^{-1}(U_1), f^{-1}(U_2), \ldots, f^{-1}(U_n)$. Hence $U_1, U_2, \ldots, U_n$ cover $f(X)$. ∎

Corollary 2.3.7 *Every continuous $f : X \longrightarrow \mathbb{R}$, X compact, is bounded and attains its bounds.*

Exercise 2.3.8 Let X be compact, Y metrizable, and $f : X \longrightarrow Y$ a continuous bijection. Show that f is a homeomorphism.

Exercise 2.3.9 Let X be any nonempty set, $\mathcal{T} \subseteq \mathcal{T}'$ two topologies on X such that $(X, \mathcal{T}')$ is compact, and $(X, \mathcal{T})$ metrizable. Show that $\mathcal{T} = \mathcal{T}'$.

Proposition 2.3.10 *If (X, d) is a compact metric space, then every sequence in (X, d) has a convergent subsequence.*

Proof. Suppose (X, d) is compact but that there is a sequence (x_n) in X with no convergent subsequence. Then $\{x_n : n \in \mathbb{N}\}$ is a closed and infinite discrete subspace of X. This contradicts the fact that X is compact. ■

Proposition 2.3.11 *Every compact metric space (X, ρ) is complete.*

Proof. By 2.3.10, every Cauchy sequence in (X, ρ) has a convergent subsequence. So, every Cauchy sequence in (X, ρ) is convergent. ■

Let (X, d) be a metric space and $\epsilon > 0$. An **ϵ-net** in X is a finite subset A of X such that $X = \bigcup_{a \in A} B(a, \epsilon)$; i.e., for every $x \in X$ there is an $a \in A$ such that $d(x, a) < \epsilon$. We call (X, d) **totally bounded** if it has an ϵ-net for every $\epsilon > 0$. The following result is quite easy to prove.

Proposition 2.3.12 *Every compact metric space is totally bounded.*

Exercise 2.3.13 Let (X, d) be a metric space, $A \subseteq X$ totally bounded, and $A \subseteq B \subseteq \text{cl}(A)$. Show that B is totally bounded.

Proposition 2.3.14 *Every compact metrizable space X is separable and hence second countable.*

Proof. Let d be a compatible metric on X. For any $n > 0$, choose a $\frac{1}{n}$-net A_n in X. Then $\bigcup_n A_n$ is a countable, dense set in X. ■

Corollary 2.3.15 *Every zero-dimensional compact metrizable space X is homeomorphic to a closed subset of $\mathcal{C}$.*

Proof. By 2.3.14, X is second countable. Therefore, by 2.2.9, there is an embedding f of X into $\mathcal{C}$. By 2.3.6, the range of f is compact and therefore closed. ■

From 2.3.11 and 2.3.14 it follows that *every compact metrizable space is Polish.* The next few results show that the converse of 2.3.10 is true. A topological space is called **sequentially compact** if every sequence in it has a convergent subsequence.

Proposition 2.3.16 *Let (X, d) be sequentially compact and $\mathcal{U}$ an open cover of X. Then there is a $\delta > 0$ such that every $A \subseteq X$ of diameter less than δ is contained in some $U \in \mathcal{U}$.*

(A δ satisfying the above condition is called a **Lebesgue number** of $\mathcal{U}$.)

Proof. Suppose such a δ does not exist. For every $n > 0$, choose $A_n \subseteq X$ such that diameter$(A_n) < \frac{1}{n}$ and A_n is not a contained in any $U \in \mathcal{U}$.

Choose $x_n \in A_n$. Since X is sequentially convergent, (x_n) has a convergent subsequence, converging to x, say. Choose $U \in \mathcal{U}$ containing x. Fix $r > 0$ such that $B(x,r) \subseteq U$. Note that $x_n \in B(x,r/2)$ for infinitely many n. Choose n_0 such that $1/n_0 < r/2$ and $x_{n_0} \in B(x,r/2)$. As diameter$(A_{n_0}) < 1/n_0 < r/2$,

$$A_{n_0} \subseteq B(x,r) \subseteq U.$$

This contradiction proves the result. ■

Proposition 2.3.17 *Suppose (X,d) and (Y,ρ) are metric spaces with X sequentially compact. Then every continuous $f : X \longrightarrow Y$ is uniformly continuous.*

Proof. Fix $\epsilon > 0$. Let

$$\mathcal{U} = \{f^{-1}(B) : B \text{ an open ball of radius } < \epsilon/2\}.$$

Let δ be a Lebesgue number of $\mathcal{U}$. Plainly, $\rho(f(x), f(y)) < \epsilon$ whenever $d(x,y) < \delta$. ■

Proposition 2.3.18 *Every sequentially compact metric space (X,d) is totally bounded.*

Proof. Let X be not totally bounded. Choose $\epsilon > 0$ such that no finite family of open balls of radius ϵ cover X. Then, by induction on n, we can define a sequence (x_n) in X such that for all $n > 0$, $x_n \notin \bigcup_{i<n} B(x_i,\epsilon)$. Thus for any $m \neq n$, $d(x_m, x_n) \geq \epsilon$. Such a sequence (x_n) has no convergent subsequence. ■

Proposition 2.3.19 *Every sequentially compact metric space is compact.*

Proof. Let (X,d) be sequentially compact and $\mathcal{U}$ an open cover for X. Let $\delta > 0$ be a Lebesgue number of $\mathcal{U}$ and $\{x_1, x_2, \ldots, x_n\}$ a $\delta/3$-net in X. For each $k \leq n$, choose $U_k \in \mathcal{U}$ containing $B(x_k, \delta/3)$. Plainly, $\{U_1, U_2, \ldots, U_n\}$ is a finite subcover of $\mathcal{U}$. ■

Exercise 2.3.20 Let X be any metrizable space. Show that X is compact if and only if every real-valued continuous function f on X is bounded.

Exercise 2.3.21 Let (X,d) be a compact metric space and $f : X \longrightarrow X$ an isometry. Show that f is onto X.

Theorem 2.3.22 *A metric space is compact if and only if it is complete and totally bounded.*

Proof. We have already proved the "only if" part of the result. Let (X,d) be complete and totally bounded. We have to show that X is compact. Take a sequence (x_n) in X. We first show that (x_n) has a Cauchy subsequence. Since X is complete, the "if" part of the result will follow from 2.3.19.

As X is totally bounded, (x_n) has a subsequence $(x_{n_k^0})$ all of whose points lie in some open sphere of radius less than 1. By the same argument, $(x_{n_k^0})$ has a subsequence $(x_{n_k^1})$ all of whose points lie in an open sphere of radius less than 1/2. Proceeding in this manner, for each i we get a sequence $(x_{n_k^i})$ such that

(i) for every i, all of $x_{n_0^i}, x_{n_1^i}, x_{n_2^i}, \ldots$ lie in an open ball of radius less than $1/2^i$, and

(ii) $(x_{n_k^{i+1}})$ is a subsequence of $(x_{n_k^i})$.

Finally, put $y_i = x_{n_i^i}$, $i \in \mathbb{N}$. It is easy to check that (y_i) is a Cauchy subsequence of (x_n). ■

Theorem 2.3.23 *The product of a sequence of compact metric spaces is compact.*

Proof. Let $(X_0, d_0), (X_1, d_1), (X_2, d_2), \ldots$ be a sequence of compact metric spaces, $X = \prod_n X_n$, and d the product metric on X. Fix a sequence (x_n) in X. We show that (x_n) has a convergent subsequence.

Since X_0 is compact, there is a convergent subsequence $(x_{n_k^0}(0))$ of $(x_n(0))$. Similarly, as X_1 is compact, there is a convergent subsequence $(x_{n_k^1}(1))$ of $(x_{n_k^0}(1))$. Proceeding similarly we obtain a double sequence $(x_{n_k^i})$ such that

(i) $(x_{n_k^i}(i))_{k\in\mathbb{N}}$ is convergent for each i, and

(ii) $(x_{n_k^{i+1}})_{k\in\mathbb{N}}$ is a subsequence of $(x_{n_k^i})_{k\in\mathbb{N}}$.

Define $y_i = x_{n_i^i}$, $i \in \mathbb{N}$. As $y_i(k)$ is convergent for each k, (y_i) is a convergent subsequence of (x_n). ■

Exercise 2.3.24 Let X, Y be metrizable spaces with Y compact and $C \subseteq X \times Y$ closed. Show that $\pi_1(C)$ is closed in X.

Exercise 2.3.25 Show that for every real-valued upper-semicontinuous map f defined on a compact metric space X there is an $x_0 \in X$ such that $f(x) \leq f(x_0)$ for every $x \in X$.

Exercise 2.3.26 Show that for every upper-semicontinuous function $f : \mathbb{R} \longrightarrow \mathbb{R}$ there is a continuous $g : \mathbb{R} \longrightarrow \mathbb{R}$ such that $f \leq g$.

Exercise 2.3.27 Let X be a compact metric space and (g_n) a sequence of real-valued, upper-semicontinuous maps decreasing to g pointwise. Show that $g_n \rightarrow g$ uniformly on X.

We prove the next result for future application.

Lemma 2.3.28 *Let X be a compact metric space. Suppose $f, f_n : X \longrightarrow \mathbb{R}$ are upper-semicontinuous and f_n decreases pointwise to f. If $x_n \to x$ in X, then*

$$\limsup_n f_n(x_n) \leq f(x).$$

Proof. Let $\epsilon > 0$. By 2.3.25 and 2.1.25, there is a continuous $h : \mathbb{R} \longrightarrow \mathbb{R}$ such that $f \leq h$ and $h(x) \leq f(x) + \epsilon$. Set

$$h_n = \max(f_n, h), \quad n \in \mathbb{N}.$$

Then h_n is upper-semicontinuous, and (h_n) decreases to h. By 2.3.27, $h_n \to h$ uniformly on X. Hence,

$$\limsup_n f_n(x_n) \leq \lim_n h_n(x_n) = h(x) \leq f(x) + \epsilon.$$

Since $\epsilon > 0$ was arbitrary, our result is proved. ■

A topological space X is called **locally compact** if every point of X has a compact neighborhood. The finite dimensional Euclidean spaces $\mathbb{R}^n$ are locally compact, and so are all compact spaces.

Exercise 2.3.29 Show that the set of rational numbers $\mathbb{Q}$ and the set of irrationals $\mathbb{R} \setminus \mathbb{Q}$, with the usual topologies, are not locally compact.

The following facts are easy to verify.

(i) Every closed subspace of a locally compact space is locally compact.

(ii) The product of finitely many locally compact spaces is locally compact. The product of an infinite family of locally compact spaces is locally compact if and only if all but finitely many of the spaces are compact.

(iii) Every open subspace of a locally compact metrizable space is locally compact.

Theorem 2.3.30 *Every locally compact metrizable space X is completely metrizable.*

Proof. We need a lemma.

Lemma 2.3.31 *Let Y be a locally compact dense subspace of a metrizable space X. Then Y is open in X.*

Assuming the lemma, the proof is completed as follows. Let d be a metric on X inducing its topology and $\hat{X}$ the completion of (X, d). Then X is a locally compact dense subspace of $\hat{X}$. By 2.3.31, X is open in $\hat{X}$. By 2.2.1, X is completely metrizable.

The proof of lemma 2.3.31. Fix $x \in Y$ and choose an open set U in Y containing x such that $\text{cl}(U) \bigcap Y$ is compact, and hence closed in X. Since

$U \subseteq \text{cl}(U) \bigcap Y$, we have $\text{cl}(U) \subseteq \text{cl}(U) \bigcap Y \subseteq Y$. Choose an open set V in X such that $U = V \bigcap Y$. Since Y is dense and V open, $\text{cl}(V) = \text{cl}(V \bigcap Y)$. Thus we have

$$x \in V \subseteq \text{cl}(V) = \text{cl}(V \bigcap Y) = \text{cl}(U) \subseteq Y.$$

We have shown that for every $x \in Y$ there is an open set V in X such that $x \in V \subseteq Y$. Therefore, Y is open. ■

Corollary 2.3.32 *Every locally compact, second countable metrizable space is Polish.*

Exercise 2.3.33 Let X be a second countable, locally compact metrizable space. Show that there exists a sequence (K_n) of compact sets such that $X = \bigcup_n K_n$ and $K_n \subseteq \text{int}(K_{n+1})$ for every n.

A subset of a topological space of the form $\bigcup_n K_n$, K_n compact, is called a K_σ set. From the above exercise it follows that every locally compact, second countable metrizable space is a K_σ set.

2.4 More Examples

In this section we give some interesting examples of Polish spaces.

Spaces of Continuous Functions

Let X be a compact metrizable space and Y a Polish space. Let $C(X, Y)$ be the set of continuous functions from X into Y. Fix a compatible complete metric ρ on Y and define

$$\delta(f, g) = \sup_{x \in X} \rho(f(x), g(x)), \quad f, g \in C(X, Y). \tag{$*$}$$

Exercise 2.4.1 Show that $\delta(f, g)$ is a complete metric on $C(X, Y)$. (Hint: Use 2.1.17.)

The topology on $C(X, Y)$ induced by δ is called the **topology of uniform convergence.**

Exercise 2.4.2 Show that if ρ and ρ' are equivalent metrics on Y, then the corresponding metrics on $C(X, Y)$, defined by the formula ($\star$), are also equivalent.

Theorem 2.4.3 *If (X, d) is a compact metrc space and (Y, ρ) Polish, then $C(X, Y)$, equipped with the topology of uniform convergence, is Polish.*

Proof. We only need to check that $C(X,Y)$ is separable. Let l, m, and n be positive integers. As X is compact, there is a $1/m$-net $X_m = \{x_1, x_2, \ldots, x_k\}$ in X. As Y is separable, there is a countable open cover $\mathcal{U}_l = \{U_0, U_1, \ldots\}$ such that diameter$(U_i) < 1/l$ for each i. Fix such an $\mathcal{U}_l$ for each l. Put

$$C_{m,n} = \{f \in C(X,Y) : \forall x, y(d(x,y) < 1/m \Longrightarrow \rho(f(x), f(y)) < 1/n)\}.$$

For each k-tuple $s = (i_1, i_2, \ldots, i_k)$, whenever possible, choose an $f_s \in C_{m,n}$ such that $f_s(x_j) \in U_{i_j}$ for all $1 \le j \le k$. Let $D_{m,n,l}$ be the collection of all these f_s and set $D_{m,n} = \bigcup_{l>0} D_{m,n,l}$.

We claim that for all $f \in C_{m,n}$ and all $\epsilon > 0$ there is a $g \in D_{m,n}$ such that $\rho(f(y), g(y)) < \epsilon$ for every $y \in X_m$. To see this, take $l > 1/\epsilon$ and choose $i_1, i_2, \ldots, i_k$ such that $f(x_j) \in U_{i_j}$ for all $1 \le j \le k$. Thus f_s exists for $s = (i_1, i_2, \ldots, i_k)$. Take $g = f_s$.

Set $D = \bigcup_{m,n} D_{m,n}$. Note that D is countable. We show that D is dense in $C(X,Y)$. Take $f \in C(X,Y)$ and $\epsilon > 0$. Take any $n > 3/\epsilon$. Since f is uniformly continuous, $f \in C_{m,n}$ for some m . We choose $g \in D_{m,n}$ such that $\rho(f(y), g(y)) < \epsilon/3$ for $y \in X_m$. Since X_m is a $1/m$-net, by the triangle inequality we see that $\rho(f(x), g(x)) < \epsilon$ for all $x \in X$. So, D is dense, and our theorem is proved. ■

The Space of Irreducible Matrices

Fix a positive integer n. Let M_n denote the set of all complex $n \times n$ matrices. As usual, we identify M_n with $\mathbb{C}^{n^2}$, equipped with the usual topology. A matrix $A \in M_n$ is **irreducible** if it commutes with no self-adjoint projections other than the identity and 0. Equivalently, A is irreducible if and only if there is no nontrivial vector subspace of $\mathbb{C}^n$ that is invariant under both A and A^*, the adjoint of A. Let $irr(n)$ denote the set of all irreducible matrices. The following result is a well-known characterization of irreducible matrices, whose proof we omit.

Theorem 2.4.4 *(Jacobson density theorem) A matrix $A \in M_n$ is irreducible if and only if the C^*-algebra generated by A is the whole of M_n. (See [4] for the definition of C^*- algebra.)*

Corollary 2.4.5 *Let $P_0(x,y)$, $P_1(x,y)$, $P_2(x,y)$, ... be an enumeration of all polynomials in two variables with coefficients of the form $p + iq$, where p and q are rational numbers. An $n \times n$ matrix A is irreducible if and only if $\{P_0(A, A^*), P_1(A, A^*), P_2(A, A^*), \ldots\}$ is dense in M_n.*

Proposition 2.4.6 *$irr(n)$ is Polish.*

Proof. By 2.2.1 it is sufficient to show that $irr(n)$ is a G_δ set in M_n. Towards showing this, fix any irreducible matrix A_0. For any matrix A, by 2.4.5 we have

$$A \text{ is irreducible} \Longleftrightarrow \forall m \exists k |A_0 - P_k(A, A^*)| < 2^{-m}.$$

So,

$$irr(n) = \bigcap_m G_m,$$

where

$$G_m = \{A \in M_n : |A_0 - P_k(A, A^*)| < 2^{-m}\} \text{ for some } k\}.$$

Clearly, G_m is open. Hence, $irr(n)$ is a G_δ set. ∎

Polish Groups

A **topological group** is a group $(G, \cdot)$ with a topology such that the maps $(x, y) \longrightarrow x \cdot y$ from $G \times G$ to G and $x \longrightarrow x^{-1}$ from G to G are continuous. If moreover, G is a Polish space, we call it a **Polish group.**

Exercise 2.4.7 Let $(G, \cdot)$ be a topological group and $g \in G$. Show that the following maps from G onto G are homeomorphisms.

(a) $L_g(h) = g \cdot h$;

(b) $R_g(h) = h \cdot g$;

(c) $I(h) = h^{-1}$.

Exercise 2.4.8 Show that the closure of a subgroup of a topological group is a topological group.

Some examples of Polish groups

(i) All countable discrete groups are Polish.

(ii) The additive group of real numbers $(\mathbb{R}, +)$ and the multiplicative group $(\mathbb{T}, \cdot)$ of complex numbers of modulus 1, with usual topologies, are Polish.

(iii) The set $\mathbb{R}_\times$ of nonzero real numbers, being open in $\mathbb{R}$, is Polish. Therefore, the multiplicative group $(\mathbb{R}_\times, \cdot)$ is Polish.

(iv) Let $\mathbf{F}$ denote either the field of real numbers $\mathbb{R}$ or the field of complex numbers $\mathbb{C}$. An $n \times n$ matrix over $\mathbf{F}$ can be identified with a point of $\mathbf{F}^{n^2}$. The set $GL(n, \mathbf{F})$ of nonsingular $n \times n$ matrices is open in $\mathbf{F}^{n^2}$ and hence Polish. Also, the set $SO(n, \mathbb{R})$ of $n \times n$ orthonormal matrices is compact and hence Polish. Similarly, most other matrix groups commonly used in analysis can be seen to be Polish.

The groups described so far are locally compact too. Here is an example of a Polish group that is not locally compact.

(v) Let S_∞ be the set of all bijections from $\mathbb{N}$ onto itself with the composition of functions as the group operation. The elements of S_∞ are called the **permutations** of $\mathbb{N}$. S_∞ is Polish. To see this, first note that

$$\alpha \text{ is one-to-one} \iff \forall m \forall n (m \neq n \implies \alpha(m) \neq \alpha(n)).$$

Let $A = \{\alpha \in \mathbb{N}^{\mathbb{N}} : \alpha \text{ is one-to-one}\}$. As $A = \bigcap_{m \neq n}\{\alpha \in \mathbb{N}^{\mathbb{N}} : \alpha(m) \neq \alpha(n)\}$, it is a G_δ set in $\mathbb{N}^{\mathbb{N}}$. Again, note that

$$\alpha \text{ is onto} \Longleftrightarrow \forall m \exists n(\alpha(n) = m).$$

Therefore, the set $\{\alpha : \alpha \text{ is onto}\}$ equals $\bigcap_m \bigcup_n \{\alpha \in \mathbb{N}^{\mathbb{N}} : \alpha(n) = m\}$ and hence is a G_δ set in $\mathbb{N}^{\mathbb{N}}$. Since the intersection of two G_δ sets is again a G_δ set, S_∞ is a G_δ set in $\mathbb{N}^{\mathbb{N}}$ and therefore Polish.

S_∞ is a topological group.

Let α, β be any two permutations of $\mathbb{N}$ and $m, n \in \mathbb{N}$. Then,

$$\alpha \circ \beta(n) = m \Longleftrightarrow \exists k(\beta(n) = k \;\&\; \alpha(k) = m).$$

This shows that for every n, $(\alpha, \beta) \longrightarrow \alpha \circ \beta(n)$ is continuous. It follows that $(\alpha, \beta) \longrightarrow \alpha \circ \beta$ is continuous.

Next we check that $\alpha \longrightarrow \alpha^{-1}$ is continuous. For any m, n,

$$\alpha^{-1}(n) = m \Longleftrightarrow \alpha(m) = n.$$

Thus $\alpha \longrightarrow \alpha^{-1}(n)$ is continuous for each n. So, the map $\alpha \longrightarrow \alpha^{-1}$ is continuous.

The above arguments prove that S_∞ is a Polish group.

Exercise 2.4.9 Show that S_∞ is not locally compact.

Spaces of Compact Sets

Let X be a topological space and $K(X)$ the family of all nonempty compact subsets of X. The topology on $K(X)$ generated by sets of the form

$$\{K \in K(X) : K \subseteq U\}$$

and

$$\{K \in K(X) : K \bigcap U \neq \emptyset\},$$

U open in X, is known as the **Vietoris topology**. Unless otherwise stated, throughout this section $K(X)$ is equipped with the Vietoris topology.

(i) The sets of the form

$$[U_0; U_1, \ldots, U_n] = \{K \in K(X) : K \subseteq U_0 \;\&\; K \bigcap U_i \neq \emptyset, 1 \leq i \leq n\},$$

where $U_0, U_1, \ldots, U_n$ are open sets in X, form a base for $K(X)$.

(ii) The set of all finite, nonempty subsets of X is dense in $K(X)$.

Proof. Let $[U_0; U_1, \ldots, U_n]$ be a nonempty basic open set. Then $U_0 \bigcap U_i \neq \emptyset$ for $1 \leq i \leq n$. Choose $x_i \in U_0 \bigcap U_i$. Clearly,

$$\{x_1, \ldots, x_n\} \in [U_0; U_1, \ldots, U_n].$$

(iii) If X is separable, then so is $K(X)$.

Proof. Let D be a countable dense set in X and F the set of all finite, nonempty subsets of D. In the proof of (ii), choose x_i such that it also belongs to D. Thus F is dense in $K(X)$. As F is countable, the result follows.

Exercise 2.4.10 Show that if X is zero-dimensional, so is $K(X)$.

Exercise 2.4.11 Let X be metrizable.

(a) Show that the sets

(i) $\{(x, K) \in X \times K(X) : x \in K\}$,
(ii) $\{(K, L) \in K(X) \times K(X) : K \subseteq L\}$, and
(iii) $\{(K, L) \in K(X) \times K(X) : K \bigcap L \neq \emptyset\}$

are closed.

(b) Let $\mathcal{K}$ be a compact subset of $K(X)$. Show that $\bigcup \mathcal{K}$ is compact in X and $\mathcal{K} \longrightarrow \bigcup \mathcal{K}$ is continuous.

Exercise 2.4.12 Let X be a metrizable space. Show the the map $(K_1, K_2) \longrightarrow K_1 \bigcup K_2$ is continuous. Also show that the map $(K_1, K_2) \longrightarrow K_1 \bigcap K_2$ need not be continuous.

Let (X, d) be a metric space. For $K, L \in K(X)$, define

$$\delta_H(K, L) = \max(\max_{x \in K} d(x, L), \max_{y \in L} d(y, K)).$$

Note that for any $\epsilon > 0$,

$$\delta_H(K, L) < \epsilon \Longleftrightarrow K \subseteq B(L, \epsilon) \ \& \ L \subseteq B(K, \epsilon). \qquad (*)$$

(Recall that $B(A, \epsilon) = \{x \in X : d(x, A) < \epsilon\}$.)

Exercise 2.4.13 Show that δ_H is a metric on $K(X)$.

We call δ_H the **Hausdorff metric** on $K(X)$.

Proposition 2.4.14 *The metric δ_H induces the Vietoris topology on $K(X)$.*

Proof. We first show that any open set in $(K(X), \delta_H)$ is open in the Vietoris topology. Take any $K_0 \in K(X)$ and $\epsilon > 0$. As K_0 is compact, there is an $\epsilon/2$-net $\{x_1, x_2, \ldots, x_n\}$ in K_0. Take $U_0 = B(K_0, \epsilon)$ and $U_i = B(x_i, \epsilon/2)$, $1 \leq i \leq n$. It is sufficient to show that

$$K_0 \in [U_0; U_1, \ldots, U_n] \subseteq \{K \in K(X) : \delta_H(K_0, K) < \epsilon\}.$$

Clearly, $K_0 \subseteq U_0$ and $x_i \in K_0 \bigcap U_i$, $1 \leq i \leq n$. Thus, $K_0 \in [U_0; U_1, \ldots, U_n]$.

Now take any $K \in [U_0; U_1, \ldots, U_n]$. We have to show that $\delta_H(K_0, K) < \epsilon$. Since $K \subseteq U_0 = B(K_0, \epsilon)$, by ($\star$), it is sufficient to show that $K_0 \subseteq B(K, \epsilon)$. Let $x \in K_0$. Choose x_i such that $d(x, x_i) < \epsilon/2$. Since $K \bigcap U_i \neq \emptyset$, we get $y \in U_i \bigcap K$. Then $d(x, y) \leq d(x, x_i) + d(x_i, y) < \epsilon$. So, $x \in B(K, \epsilon)$. Since $x \in K_0$ was arbitrary, we have shown that $K_0 \subseteq B(K, \epsilon)$.

We now show that every Vietoris open set is open in $(K(X), \delta_H)$. It is sufficient to show that every subbasic open set is open in $(K(X), \delta_H)$. Fix an open set U in X and a compact set K_0 contained in U. Let

$$\epsilon = \min\{d(x, K_0) : x \in X \setminus U\}.$$

Since K_0 is compact, $\epsilon > 0$. Clearly, for every compact $K \subseteq X$, $\delta_H(K, K_0) < \epsilon \Longrightarrow K \subseteq B(K_0, \epsilon) \subseteq U$. This shows that $\{K \in K(X) : K \subseteq U\}$ is open in $(K(X), \delta_H)$.

Next take any compact K_0 and an open set U with $K_0 \bigcap U \neq \emptyset$. Let $x \in K_0 \bigcap U$ and $\epsilon > 0$ be such that $B(x, \epsilon) \subseteq U$. Suppose $\delta_H(K_0, K) < \epsilon$. Since $x \in K_0$, by ($\star$), $d(x, K) < \epsilon$. So, there exists $y \in K$, $y \in B(x, \epsilon) \subseteq U$ or $K \bigcap U \neq \emptyset$, and the result is proved. ■

Observation 1. Let (X, d) be a complete metric space and (K_n) a Cauchy sequence in $(K(X), \delta_H)$. Let $K = \text{cl}(\bigcup_n K_n)$. We claim that K is compact.

By 2.3.22, it is sufficient to show that (K, d) is totally bounded. Further, by 2.3.13, it is enough to show that $L = \bigcup_n K_n$ is totally bounded. Fix $\epsilon > 0$. Let N be such that $\delta_H(K_n, K_m) < \epsilon/2$ for all $m, n \geq N$. Since $\bigcup_{i \leq N} K_i$ is compact, it is totally bounded. Let $\{x_1, x_2, \ldots, x_k\}$ be an $\epsilon/2$-net in $\bigcup_{i \leq N} K_i$. We now show that $\{x_1, x_2, \ldots, x_k\}$ is an ϵ-net in L. Take any $x \in L$. If $x \in \bigcup_{i \leq N} K_i$, then obviously $d(x, x_i) < \epsilon$ for some i. If $x \in K_i$ for some $i > N$, then as $\delta_H(K_i, K_N) < \epsilon/2$, it follows that $d(x, K_N) < \epsilon/2$. Choose $y \in K_N$ with $d(x, y) < \epsilon/2$. Choose j such that $d(y, x_j) < \epsilon/2$. Then $d(x, x_j) < \epsilon$.

Observation 2. Let $A = \{x_1, x_2, \ldots, x_k\}$ be an ϵ-net in (X, d). Let F be the set of all finite nonempty subsets F of A. Let $K \in K(X)$ and $L = \{x_i \in A : d(x, x_i) < \epsilon \text{ for some } x \in K\}$. Plainly, $\delta_H(K, L) < \epsilon$. Thus F is an ϵ-net in $K(X)$.

Proposition 2.4.15 *If* (X, d) *is a complete metric space, so is* $(K(X), \delta_H)$.

Proof. Let (K_n) be a Cauchy sequence in $K(X)$. Let

$$K = \bigcap_n \text{cl}(\bigcup_{i \geq n} K_i).$$

By Observation 1, $\text{cl}(\bigcup_{i \geq n} K_i)$ are compact. Further, they have the finite intersection property. Therefore, K is nonempty and compact. We show that $\delta_H(K_n, K) \to 0$ as $n \to \infty$.

Fix $\epsilon > 0$. Choose N such that for $m, n \geq N, \delta_H(K_m, K_n) < \epsilon/2$. We show that $\delta_H(K_n, K) < \epsilon$ for every $n \geq N$. Fix $n \geq N$.

(i) Let $x \in K$. As $x \in \text{cl}(\bigcup_{i \geq n} K_i)$, there exist $i \geq n$ and $x_i \in K_i$ such that $d(x, x_i) < \epsilon/2$. Since $\delta_H(K_i, K_n) < \epsilon/2$, take $y \in K_n$ such that $d(y, x_i) < \epsilon/2$. By the triangle inequality $d(x, y) < \epsilon$. Thus, $d(x, K_n) < \epsilon$ for every $x \in K$. So, $K \subseteq B(K_n, \epsilon)$.

(ii) Let $x \in K_n$. We prove that $d(x, K) < \epsilon$. This would show that $K_n \subseteq B(K, \epsilon)$. For each $i \geq N$, $\delta_H(K_i, K_n) < \epsilon/2$. Choose $x_i \in K_i$ such that $d(x, x_i) < \epsilon/2$. Since $\text{cl}(\bigcup_{i \geq N} K_i)$ is compact, (x_i) has a convergent subsequence converging to y, say. Clearly, $y \in K$, and $d(x, y) \leq \epsilon/2 < \epsilon$. ■

Corollary 2.4.16 *If X is a Polish space, so is $K(X)$.*

Proposition 2.4.17 *If X is compact metrizable, so is $K(X)$.*

Proof. Let d be a compatible metric on X. By 2.4.15, $(K(X), \delta_H)$ is completely metrizable. By Observation 2, it is also totally bounded. The result follows. ■

Exercise 2.4.18 Let X be a metrizable space. Show that the set

$$K_f(X) = \{L \in K(X) : L \text{ is finite}\}$$

is an F_σ set.

A compact, dense-in-itself set will be called **perfect**.

Exercise 2.4.19 Let X be separable and metrizable. Show that the set

$$K_p(X) = \{L \in K(X) : L \text{ is perfect}\}$$

is a G_δ set. Also, show that if X is dense-in-itself, so is $K(X)$.

Exercise 2.4.20 Let X be a locally compact Polish space and a base for the topology of X. Give $F(X)$ the topology generated by sets of the form

$$\{F \in F(X) : F \bigcap K = \emptyset \& F \bigcap U_1 \neq \emptyset \& F \bigcap U_2 \neq \emptyset \& \cdots \& F \bigcap U_n \neq \emptyset\},$$

where K ranges over the compact subsets of X and $U_1, U_2, \ldots, U_n$ range over open sets in X. (This topology is called the **Fell topology**.) Show that $F(X)$ with the Fell topology is Polish.

2.5 The Baire Category Theorem

Let X be a topological space. A subset A of X is called **nowhere dense** if $\text{cl}(A)$ has empty interior; i.e., $X \setminus \text{cl}(A)$ is dense. Note that A is nowhere dense if and only if $\text{cl}(A)$ is nowhere dense. For every closed sets F, $F \setminus \text{int}(F)$ is nowhere dense.

Exercise 2.5.1 Show that a set A is nowhere dense if and only if every nonempty open set U contains a nonempty open set V such that $A \bigcap V = \emptyset$.

Exercise 2.5.2 Show that the Cantor ternary set $\mathbb{C}$ (2.3.3) is perfect and nowhere dense in $[0, 1]$.

A set $A \subseteq X$ is called **meager or of first category** in X if it is a countable union of nowhere dense sets. Clearly, every meager set is contained in a meager F_σ set. If A is not meager in X, then we say that it is of **second category** in X. A subset A is called **comeager** in X if $X \setminus A$ is meager in X. Note that $A \subseteq X$ is comeager in X if and only if it contains a countable intersection of dense open sets.

Exercise 2.5.3 (i) Show that the set of rationals $\mathbb{Q}$ with the usual topology is meager in itself.

(ii) Show that every K_σ subset of $\mathbb{N}^\mathbb{N}$ is meager.

Proposition 2.5.4 *Let X be a topological space, U open in X, and $A \subseteq U$. Then A is meager in U if and only if it is meager in X.*

Proof. For the "only if" part, it is sufficient to show that every closed nowhere dense set in U is nowhere dense in X. Let A be a closed nowhere dense subset of U. Suppose A is not nowhere dense in X. Then there exists a nonempty open set V contained in cl(A). Hence, $\emptyset \neq V \bigcap U \subseteq A$. This is a contradiction. (Note that in this part of the proof we did not use the fact that U is open.)

To prove the converse, take any $A \subseteq U$ that is meager in X. Let (U_n) be a sequence of dense open sets in X such that $\bigcap_n U_n \subseteq X \setminus A$. So, $\bigcap_n U_n \bigcap A = \emptyset$. Put $V_n = U_n \bigcap U$. As U is open and U_n dense, V_n is open and dense in U. Clearly, $\bigcap_n V_n \bigcap A = \emptyset$. Thus A is meager in U. ∎

Theorem 2.5.5 *(The Baire category theorem) Let X be a completely metrizable space. Then the intersection of countably many dense open sets in X is dense.*

Proof. Fix a compatible complete metric d on X. Take any sequence (U_n) of dense open sets in X. Let V be a nonempty open set in X. We show that $\bigcap_n U_n \bigcap V \neq \emptyset$. Since U_0 is dense, $U_0 \bigcap V$ is nonempty. Choose an open ball B_0 of diameter ¡ 1 such that cl$(B_0) \subseteq U_0 \bigcap V$. Since U_1 is dense, by the same argument we get an open ball B_1 of diameter $< 1/2$ such that cl$(B_1) \subseteq U_1 \bigcap B_0$. Proceeding similarly, we define a sequence (B_n) of open balls in X such that for each n,

(i) diameter$(B_n) < 1/2^n$,

(ii) cl$(B_0) \subseteq U_0 \bigcap V$, and

(iii) cl$(B_{n+1}) \subseteq U_{n+1} \bigcap B_n$.

Since (X, d) is a complete metric space, by 2.1.29, $\bigcap_n B_n = \bigcap_n \text{cl}(B_n)$ is a singleton, say $\{x\}$. Clearly, $x \in \bigcap_n U_n \bigcap V$. ■

Corollary 2.5.6 *Every completely metrizable space is of second category in itself.*

Proof. Let X be a completely metrizable space. Suppose X is of the first category in itself. Choose a sequence (F_n) of closed and nowhere dense sets such that $X = \bigcup_n F_n$. Then the sets $U_n = X \setminus F_n$ are dense and open, and $\bigcap_n U_n = \emptyset$. This contradicts the Baire category theorem. ■

Corollary 2.5.7 *The set of rationals $\mathbb{Q}$ with the usual topology is not completely metrizable. More generally, no countable dense-in-itself space is completely metrizable.*

Corollary 2.5.8 *Let X be a completely metrizable space and A any subset of X. Then A is comeager in X if and only if it contains a dense G_δ set.*

Corollary 2.5.9 *Let $(G, \cdot)$ be a Polish group. Then G is locally compact if and only if it is a K_σ set.*

Proof. Let G be a Polish space that is a K_σ set. Choose a sequence (K_n) of compact subsets of G such that $G = \bigcup_n K_n$. By the Baire category theorem, $\text{int}(K_n) \neq \emptyset$ for some n. Fix $z \in \text{int}(K_n)$. For any $x \in G$, $(x \cdot z^{-1})K_n$ is a compact neighborhood of x where, for $A \subseteq G$ and $g \in G$, $gA = \{g \cdot h : h \in A\}$. So, G is locally compact.

The converse follows from 2.3.33. ■

Corollary 2.5.10 *Let $(G, \cdot)$ be a completely metrizable group and H any subgroup. Then H is completely metrizable if and only if it is closed in G.*

Proof. Let H be completely metrizable. Consider $G' = \text{cl}(H)$. By 2.4.8, G' is a topological group. It is clearly completely metrizable. We show that $G' = H$, which will complete the proof. By 2.2.7, H is a G_δ set in G'. As it is also dense in G', it is comeager in G'. Suppose $H \neq G'$. Take any $x \in G' \setminus H$. Then the coset xH is comeager in G' and disjoint from H. By the Baire category theorem, G' cannot have two disjoint comeager subsets. This contradiction shows that $H = G'$.

The "if" part of the result is trivially seen. ■

Proposition 2.5.11 *Let $C([0,1])$ be equipped with the uniform convergence topology. The set of all nowhere differentiable continuous functions is comeager in $C([0,1])$. In particular, there exist continuous functions on $[0,1]$ which are nowhere differentiable.*

Proof. For any positive integer n and any $h > 0$, set

$$A_{n,h} = \{(f, x) \in C[0,1] \times [0, 1 - 1/n] : |\frac{f(x+h) - f(x)}{h}| \leq n\}.$$

The set $A_{n,h}$ is closed. To see this, let (f_k, x_k) be a sequence in $A_{n,h}$ converging to (f, x). Then $f_k \to f$ uniformly and $x_k \to x$. Hence, $f_k(x_k+h) \to f(x+h)$ and $f_k(x_k) \to f(x)$. It follows that $|\frac{f(x+h)-f(x)}{h}| \leq n$; i.e., $(f, x) \in A_{n,h}$. Now consider the set N_n defined as follows.

$$N_n = \{f \in C[0,1] : (\exists x \in [0, 1 - \frac{1}{n}])(\forall h \in (0, \frac{1}{n}])(|\frac{f(x+h) - f(x)}{h}| \leq n)\}.$$

Clearly,

$$N_n = \pi_{C[0,1]}(\bigcap_{h \in (0,1/n]} A_{n,h}).$$

Hence, by 2.3.24, N_n is closed.

It is fairly easy to see that each continuous f that is differentiable at some $x \in [0, 1)$ belongs to N_n for some n. Therefore our result will be proved if we show that N_n is a nowhere dense set. Since N_n is closed, it is sufficient to show that $\text{int}(N_n) = \emptyset$. Let $f \in C[0, 1]$, and $\epsilon > 0$. By Weierstrass theorem, there is a polynomial $p(x)$ over $\mathbb{R}$ such that

$$|f(x) - p(x)| < \epsilon/3$$

for every $x \in [0, 1]$. The derivative $p'(x)$ of $p(x)$ is, of course, bounded on $[0, 1]$. Set

$$M = \sup\{|p'(x)| : 0 \leq x \leq 1\}.$$

Let $l(x)$ be a piecewise linear, nonnegative function such that the absolute value of the slope of each segment of $l(x)$ is precisely $M + n + 1$ and $|l(x)| \leq \epsilon/3$ for all $x \in [0, 1]$. Put $g(x) = l(x) + p(x)$, $0 \leq x \leq 1$. Clearly, $|f(x) - g(x)| < \epsilon$ for every $x \in [0, 1]$.

We now show that $g \notin N_n$. Suppose not. Then there is a $x \in [0, 1 - 1/n]$ such that for every $h \in (0, 1/n]$, $|\frac{g(x+h)-g(x)}{h}| \leq n$. We shall get a contradiction now. Choose a positive $h < 1/n$ such that the map l is affine between x and $x + h$. Now

$$\begin{array}{rcl} |\frac{g(x+h)-g(x)}{h}| & \geq & |\frac{l(x+h)-l(x)}{h}| - |\frac{p(x+h)-p(x)}{h}| \\ & \geq & (M + n + 1) - |p'(x + \theta h)| \quad 0 < \theta < 1 \\ & > & n, \end{array}$$

and we have arrived at a contradiction. ∎

Exercise 2.5.12 Let X be a completely metrizable space and A a nonempty subset of X that is simultaneously F_σ and G_δ in X. Show that there is an open set U such that $U \bigcap A$ is nonempty and closed in U.

Example 2.5.13 Let X be a Polish space and $K \subseteq X$ compact. For $\alpha < \omega_1$, we define K^α by transfinite induction.

$$K^\alpha = \begin{cases} K & \text{if } \alpha = 0, \\ (K^\beta)' & \text{if } \alpha = \beta + 1, \\ \bigcap_{\beta < \alpha} K^\beta & \text{if } \alpha \text{ is limit.} \end{cases}$$

(Recall that for any $A \subseteq X$, A' denotes the derived set of A.) The set K^α is called the α th **Cantor – Bendixson derivative** of K. By 2.1.13, there is an $\alpha < \omega_1$ such that $K^\alpha = K^{\alpha+1}$. The first such α will be denoted by $\rho(K)$. Note that $K^{\rho(K)}$ has no isolated points.

Exercise 2.5.14 Let X be a countable Polish space. Show that X has no dense-in-itself subset.

Exercise 2.5.15 Let $\alpha < \omega_1$ be a successor ordinal. Show that there is a countable, compact $K \subseteq \mathbb{R}$ such that $\rho(K) = \alpha$.

Theorem 2.5.16 *(The Banach category theorem) Let X be a topological space, $\mathcal{U} = \{U_i : i \in I\}$, and $U = \bigcup\{U_i : i \in I\}$. Assume that each U_i is open in U.*

(i) *If each U_i is nowhere dense in X, so is U.*

(ii) *If each U_i is meager in X, so is U.*

Proof. Assertion (i) immediately follows from the following lemma.

Lemma 2.5.17 *Let X, U_i $(i \in I)$, and U satisfy the hypothesis of the theorem. Then*

$$cl(int(cl(U))) = cl(\bigcup_i int(cl(U_i))).$$

Proof of the lemma. Since $U_i \subseteq U$, $(i \in I)$,

$$\mathrm{int}(\mathrm{cl}(U_i)) \subseteq \mathrm{int}(\mathrm{cl}(U)).$$

Therefore,

$$\mathrm{cl}(\mathrm{int}(\mathrm{cl}(U))) \supseteq \mathrm{cl}(\bigcup_i \mathrm{int}(\mathrm{cl}(U_i))).$$

The reverse inclusion follows from

$$\mathrm{int}(\mathrm{cl}(U)) \subseteq \mathrm{cl}(\bigcup_i \mathrm{int}(\mathrm{cl}(U_i))),$$

which we show now. We make two observations first.

(i) Take any $i \in I$. Since U_i is open in U,

$$U_i = U \setminus \mathrm{cl}(U \setminus U_i) \subseteq X \setminus \mathrm{cl}(U \setminus U_i).$$

Therefore,

$$U \subseteq \bigcup_i (X \setminus \mathrm{cl}(U \setminus U_i)).$$

(ii) Since $\mathrm{int}(\mathrm{cl}(U)) \setminus \mathrm{cl}(U \setminus U_i) \subseteq \mathrm{cl}(U) \setminus \mathrm{cl}(U \setminus U_i) \subseteq \mathrm{cl}(U_i)$,

$$\mathrm{int}(\mathrm{cl}(U)) \setminus \mathrm{cl}(U \setminus U_i) \subseteq \mathrm{int}(\mathrm{cl}(U_i)).$$

Now,

$$\begin{aligned}
\text{int}(\text{cl}(U)) &= \text{int}(\text{cl}(U))\bigcap \text{cl}(U) \\
&\subseteq \text{cl}(\text{int}(\text{cl}(U))\bigcap U) \\
&\subseteq \text{cl}(\text{int}(\text{cl}(U))\bigcap \textstyle\bigcup_i (X\setminus \text{cl}(U\setminus U_i))) \quad \text{(by (i))} \\
&= \text{cl}(\textstyle\bigcup_i (\text{int}(\text{cl}(U))\setminus \text{cl}(U\setminus U_i)) \\
&\subseteq \text{cl}(\textstyle\bigcup_i \text{int}(\text{cl}(U_i))) \quad \text{(by (ii))}
\end{aligned}$$

The proof of the lemma is complete.

Proof of (ii). Let $\mathcal{V} = \{V_j : j \in J\}$ be a maximal family of pairwise disjoint nonempty open sets such that $U\bigcap V_j$ is meager. Put $V = \bigcup V_j$. We show that

(a) $U\bigcap V$ is meager, and

(b) V^c is nowhere dense.

The result will then follow.

Proof of (a). Write $U\bigcap V_j = \bigcup_{n\in\mathbb{N}} N_{jn}$, N_{jn} nowhere dense. Let $N_n = \bigcup_j N_{jn}$. As $N_{jn} = N_n\bigcap V_j$, it is open in N_n. Therefore, by (i), N_n is nowhere dense.

Proof of (b). Suppose V^c is not nowhere dense. Choose a nonempty open set W contained in V^c. By the maximality of $\mathcal{V}$, $U\bigcap W$ is nonmeager. In particular, $W\bigcap U_i \neq \emptyset$ for some i. Since U_i is open in U, $U_i = U\setminus \text{cl}(U\setminus U_i)$. Set $G = W\setminus \text{cl}(U\setminus U_i)$. Now note the following.

$$U\bigcap G = (U\bigcap W)\setminus \text{cl}(U\setminus U_i) \subseteq U_i.$$

Thus, $U\bigcap G$ is meager. Further, $\emptyset \neq W\bigcap U_i \subseteq G$. Thus G is a nonempty open set disjoint from V whose intersection with U is meager. This contradicts the maximality of $\mathcal{V}$, and (b) is proved. ■

2.6 Transfer Theorems

Let X be a Polish space and d a compatible complete metric with diameter$(X) < 1$. Fix any nonempty set A. A **Souslin scheme** on X is a system $\{F_s : s \in A^{<\mathbb{N}}\}$ of subsets of X such that

(i) $\text{cl}(F_{s^\frown a}) \subseteq F_s$ for all s and a, and

(ii) for every $\alpha \in A^{\mathbb{N}}$, diameter$(F_{\alpha|n}) \to 0$ as $n \to \infty$.

A Souslin scheme $\{F_s : s \in A^{<\mathbb{N}}\}$ is called a **Lusin scheme** if in addition to (i) and (ii) the following condition is also satisfied:

(iii) for every $s, t \in A^{<\mathbb{N}}$,

$$s \perp t \Longrightarrow F_s \bigcap F_t = \emptyset.$$

A Cantor scheme is a Lusin scheme $\{F_s : s \in A^{<\mathbb{N}}\}$ such that $A = \{0, 1\}$ and each F_s is closed and nonempty.

Let $\{F_s : s \in A^{<\mathbb{N}}\}$ be a Souslin scheme. Equip $A^{\mathbb{N}}$ with the product of discrete topologies on A. Below we make a series of simple observations that will be freely used in the sequel.

(i) Set

$$D = \{\alpha \in A^{\mathbb{N}} : \forall n(F_{\alpha|n} \neq \emptyset)\}.$$

Then D is closed. To see this, let $\alpha \in A^{\mathbb{N}} \setminus D$. By the definition of D, $F_{\alpha|n} = \emptyset$ for some n. So, $\Sigma(\alpha|n) \subseteq A^{\mathbb{N}} \setminus D$.

(ii) By 2.1.29, $\bigcap_n F_{\alpha|n} = \bigcap_n \text{cl}(F_{\alpha|n})$ is a singleton for each $\alpha \in D$. Define $f : D \longrightarrow X$ such that

$$\{f(\alpha)\} = \bigcap_n F_{\alpha|n}.$$

We call f the **associated map** of $\{F_s : s \in A^{<\mathbb{N}}\}$. The map f is continuous. To see this, take any $\alpha \in D$ and $\epsilon > 0$. Choose n such that diameter$(F_{\alpha|n}) < \epsilon$. Then

$$f(D \bigcap \Sigma(\alpha|n)) \subseteq B(f(\alpha), \epsilon).$$

Hence, f is continuous.

(iii) Further, assume that

$$F_e = X \;\&\; \forall s(F_s = \bigcup_n F_{s^\frown n}).$$

It is easy to check that the associated map f is onto X.

(iv) If $\{F_s : s \in A^{<\mathbb{N}}\}$ is a Lusin scheme, f is easily seen to be one-to-one. It follows that if $\{F_s : s \in 2^{<\mathbb{N}}\}$ is a Cantor scheme, then $D = \mathcal{C}$, and f is an embedding in X.

Proposition 2.6.1 *Every dense-in-itself Polish space X contains a homeomorph of $\mathcal{C}$.*

Proof. Let $d \leq 1$ be a compatible complete metric on X. We show that there is a Souslin scheme $\{U_s : s \in 2^{<\mathbb{N}}\}$ of nonempty open sets such that

$$s \perp t \Longrightarrow \text{cl}(U_s) \bigcap \text{cl}(U_t) = \emptyset.$$

Assuming that such a system of sets $\{U_s : s \in 2^{<\mathbb{N}}\}$ exists, define $F_s = \text{cl}(U_s)$, $s \in 2^{<\mathbb{N}}$. Then $\{F_s : s \in 2^{<\mathbb{N}}\}$ is a Cantor scheme on X, and so X contains a homeomorph of the Cantor set by (iv).

We define $\{U_s : s \in 2^{<\mathbb{N}}\}$ by induction on the length of s. Take $U_e = X$. Suppose for some $s \in 2^{<\mathbb{N}}$, U_s has been defined and is a nonempty open set. Since X is dense-in-itself, there exist two distinct points x_0, x_1 in U_s. Choose open sets $U_{s^\frown 0}$, $U_{s^\frown 1}$, containing x_0, x_1 respectively, of diameters $\leq 2^{-(|s|+1)}$ whose closures are disjoint and contained in U_s. ■

Proposition 2.6.2 *(Cantor – Bendixson theorem) Every separable metric space X can be written as $X = Y \bigcup Z$ where Z is countable, Y closed with no isolated point, and $Y \bigcap Z = \emptyset$.*

Proof. Let (U_n) be a countable base for X. Take

$$Z = \bigcup\{U_n : U_n \text{ countable}\}$$

and $Y = X \setminus Z$. ■

From 2.6.1 and 2.6.2 we have the following result.

Theorem 2.6.3 *Every uncountable Polish space contains a homeomorph of C, and hence is of cardinality $\mathfrak{c}$.*

Exercise 2.6.4 (i) Show that the cardinality of the set of all open subsets of an infinite separable metric space X is $\mathfrak{c}$.

(ii) Show that the cardinality of the set of all uncountable closed subsets of an uncountable Polish space X is $\mathfrak{c}$.

Remark 2.6.5 Since C contains a homeomorph of $\mathbb{N}^{\mathbb{N}}$ (2.2.12), we see that every uncountantable Polish space X contains a homeomorph of $\mathbb{N}^{\mathbb{N}}$, which, by 2.2.7, is a G_δ set in X.

Exercise 2.6.6 Let X be a second countable metrizable space and Y an uncountable Polish space. Show that $C(X, Y)$, the space of all continuous functions from X to Y, is of cardinality $\mathfrak{c}$.

Here is an interesting generalization of 2.6.3. Let X be a Polish space and E an equivalence relation on X. In particular, $E \subseteq X \times X$. We call the relation E closed (open, F_σ, G_δ, etc.) if E is a closed (open, F_σ, G_δ, etc.) subset of $X \times X$.

Theorem 2.6.7 *Let E be a closed equivalence relation on a Polish space X with uncountably many equivalence classes. Then there is a homeomorph D of the Cantor set in X consisting of pairwise inequivalent elements. In particular, there are exactly $\mathfrak{c}$ equivalence classes.*

Proof. Fix a compatible complete metric $d \leq 1$ on X and a countable base (V_n) for X. Let

$$Z = \bigcup\{V_n : E|V_n \text{ has countably many equivalence classes}\},$$

and $Y = X\setminus Z$. Note that every nonempty open set U in Y has uncountably many inequivalent elements. If necessary, we replace X by Y and assume that every nonempty open set has uncountably many inequivalent elements.

We now define a system $\{U_s : s \in 2^{<\mathbb{N}}\}$ of nonempty open sets such that

(i) diameter(cl(U_s)) $\leq \frac{1}{2^{|s|}}$;

(ii) $\text{cl}(U_{s^\frown\epsilon}) \subseteq U_s$ for $\epsilon = 0$ or 1; and

(iii) $s \perp t \Longrightarrow E\bigcap(F_s \times F_t) = \emptyset$, where, $F_s = \text{cl}(U_s)$.

Suppose such a system has been defined. Take $D = \mathcal{A}_2(\{F_s\})$. Then D is a homeomorph of the Cantor set. Let $\alpha \neq \beta$ be two elements of D. So there exists an n such that $\alpha|n \neq \beta|n$. As $\alpha \in F_{\alpha|n}$ and $\beta \in F_{\beta|n}$, they are inequivalent by (iii).

The definition of $\{U_s : s \in 2^{<\mathbb{N}}\}$. Put $U_e = X$. Take two inequivalent elements x_0 and x_1. Then $(x_0, x_1) \notin E$. Since E is closed, we get open sets $U_0 \ni x_0$ and $U_1 \ni x_1$ of diameters less than 1/2 such that

$$(\text{cl}(U_0) \times \text{cl}(U_1))\bigcap E = \emptyset.$$

Suppose U_s has been defined for all s of length less than or equal to n satisfying conditions (i) to (iii). Fix an s of length n. Choose inequivalent elements y_0 and y_1 in U_s. Using the same arguments, choose open sets $U_{s^\frown 0}$ and $U_{s^\frown 1}$ of diameters less than $1/2^{n+1}$ such that

$$y_\epsilon \in U_{s^\frown\epsilon} \subseteq \text{cl}(U_{s^\frown\epsilon}) \subseteq U_s,$$

$\epsilon = 0$ or 1, and

$$(\text{cl}(U_{s^\frown 0}) \times \text{cl}(U_{s^\frown 1}))\bigcap E = \emptyset.$$

Our construction of the system $\{U_s : s \in 2^{<\mathbb{N}}\}$ is complete. ■

Note that if we take $E = \{(x, x) : x \in X\}$ in the above result, we get 2.6.3.

Exercise 2.6.8 Show that 2.6.7 is true even when E is an F_σ equivalence relation.

Theorem 2.6.9 *Every Polish space X is a one-to-one, continuous image of a closed subset D of $\mathbb{N}^\mathbb{N}$.*

Proof. Fix a complete metric $d \leq 1$ on X compatible with its topology. It is enough to define a Lusin scheme $\{F_s : s \in \mathbb{N}^{<\mathbb{N}}\}$ on X such that

$$F_e = X \;\&\; F_s = \bigcup_i F_{s^\frown i}.$$

We construct such a family $\{F_s : s \in \mathbb{N}^{<\mathbb{N}}\}$ by induction on $|s|$ such that each F_s is an F_σ set. Suppose F_s has been defined. Write $F_s = \bigcup_i C_i$

where $\{C_i\}$ is a sequence of closed sets of diameter less than $2^{-(|s|+1)}$. Put $F_{s^\frown i} = C_i \setminus C_{i-1}$. (We take $C_{-1} = \emptyset$.) Since an open set in a metrizable space is an F_σ set, so is $F_{s^\frown i}$. The proof is complete. ■

Theorem 2.6.10 *Every compact metric space X is a continuous image of a zero-dimensional compact metric space Z.*

Proof. Fix a metric $d \leq 1$ on X compatible with its topology. We define a sequence (n_i) of positive integers and for each k and for each $s \in \{0,1,\ldots,n_0\} \times \cdots \times \{0,1,\ldots,n_k\}$, a nonempty closed set F_s such that

(i) $F_e = X$;

(ii) $F_s = \bigcup_{i \leq n_{|s|}} F_{s^\frown i}$;

(iii) diameter$(F_s) \leq 2^{-|s|}$.

To define such a family we proceed by induction. As X is compact, there is an $n_0 \in \mathbb{N}$ and a finite open cover $\{U_0^e, U_1^e, \ldots, U_{n_0}^e\}$ of X such that the diameter of each U_i^e is less than 1. Take

$$F_i = \text{cl}(U_i^e), \quad 1 \leq i \leq n_0.$$

Let $k \in \mathbb{N}$. Suppose $n_0, n_1, \ldots, n_k$ and sets F_s for $s \in \{0,1,\ldots,n_0\} \times \cdots \times \{0,1,\ldots,n_k\}$ satisfying conditions (i)-(iii) have been defined. Fix $s \in \{0,1,\ldots,n_0\} \times \cdots \times \{0,1,\ldots,n_k\}$. As F_s is compact, we obtain a finite open cover $\{U_i^s : i \leq n_s\}$ of F_s such that diameter$(U_i^s) < 2^{-(k+1)}$. Since there are only finitely many sequences of length k, we can assume that there exist n_{k+1} such that $n_s = n_{k+1}$ for all s. Put $F_{s^\frown i} = \text{cl}(U_i^s) \bigcap F_s$.

To complete the proof, take

$$Z = \{0,1,\ldots,n_0\} \times \{0,1,\ldots,n_1\} \times \cdots$$

with the product of discrete topologies. For $\alpha \in Z$, take $f(\alpha)$ to be the unique element of $\bigcap_n F_{\alpha|n}$. As before, we see that $f : Z \longrightarrow X$ is continuous and onto. ■

A subset A of a topological space X is called a **retract** of X if there is a continuous function $f : X \rightarrow A$ such that $f|A$ is the identity map. In such a case, the map f is called a **retraction**. Let X be metrizable, A a retract of X, and $f : X \longrightarrow A$ a retraction. As $A = \{x \in X : f(x) = x\}$, it is closed. Below we give a useful converse of this.

Proposition 2.6.11 *Let A be a discrete space and $X = A^{\mathbb{N}}$. Then every nonempty closed subset of X is a retract of X.*

Proof. Let C be a nonempty closed set in X. For each $s \in A^{<\mathbb{N}}$ such that $C \bigcap \Sigma(s) \neq \emptyset$, choose and fix $x_s \in C \bigcap \Sigma(s)$. Let $\alpha \in X$. Define $f(\alpha) = \alpha$ for $\alpha \in C$. Suppose $\alpha \notin C$. As C is closed, there is an integer k

such that $\Sigma(\alpha|k)\bigcap C = \emptyset$. Let k be the largest natural number such that $C\bigcap\Sigma(\alpha|k) \neq \emptyset$. Define $f(\alpha) = x_{\alpha|k}$.

We now show that f is continuous at every $\alpha \in X$. Let $\alpha \notin C$ and $f(\alpha) = \beta$. Let k be the least natural number such that $C\bigcap\Sigma(\alpha|k) = \emptyset$. Then $f \equiv \beta$ on $\Sigma(\alpha|k)$. So, f is continuous at α.

Now assume that $\alpha \in C$. Then $f(\alpha) = \alpha$ and $f(\Sigma(\alpha|k)) \subseteq \Sigma(\alpha|k)$ for all k. So, f is continuous at α, and our result is proved. ■

From 2.6.9 and 2.6.11, we immediately get the following.

Theorem 2.6.12 *Every Polish space X is a continuous image of $\mathbf{N}^{\mathbf{N}}$.*

From 2.2.9, 2.6.10, and 2.6.11 we have the following result.

Theorem 2.6.13 *Every compact metric space is a continuous image of C.*

Theorem 2.6.14 *Every zero-dimensional compact, dense-in-itself metric space is homeomorphic to C.*

Proof. It is sufficient to show that there is a Cantor scheme $\{C_s : s \in 2^{<\mathbf{N}}\}$ on X of clopen sets such that $C_e = X$ and $C_s = C_{s\hat{}0}\bigcup C_{s\hat{}1}$ for all s.

Construction of $\{C_s : s \in 2^{<\mathbf{N}}\}$. Since X is perfect and zero-dimensional, we can write $X = X_1\bigcup\cdots\bigcup X_n$, where $n > 1$ and the X_i are pairwise-disjoint nonempty clopen sets of diameter less than 1/2. Put $C_e = X$, $C_0 = \bigcup_{i>1} X_i$, $C_1 = X_1$, $C_{00} = \bigcup_{i>2} X_i$, $C_{01} = X_2$, etc. Thus we have

$$C_s = \begin{cases} \bigcup_{i>j} X_i & \text{if } s = 0^j \;\&\; j < n, \\ X_{j+1} & \text{if } s = 0^j\hat{}1 \;\&\; j < n-1. \end{cases}$$

For the next stage of construction, fix i, $1 \leq i \leq n$. Let $C_s = X_i$. Note that C_s is perfect and zero-dimensional. Write C_s as a finite union of pairwise-disjoint nonempty clopen sets $Y_1, Y_2, \ldots, Y_m$ of diameter less than 1/3. Repeat the above process replacing X by C_s and $X_1, X_2, \ldots, X_n$ by $Y_1, Y_2, \ldots, Y_m$ to get $C_{s\hat{}t}$ for $t = 0^j, 1 \leq j \leq m-1$, or $t = 0^j\hat{}1, 0 \leq j \leq m-2$. Continuing this process, we get the required Cantor scheme. ■

The space $\mathbf{N}^{\mathbf{N}}$ is a zero-dimensional, dense-in-itself Polish space such that every compact subset of $\mathbf{N}^{\mathbf{N}}$ is nowhere dense. The next exercise is to show that this characterizes $\mathbf{N}^{\mathbf{N}}$ topologically.

Exercise 2.6.15 Let X be a zero-dimensional Polish space with no isolated points such that every compact subset is nowhere dense. Show that X is homeomorphic to $\mathbf{N}^{\mathbf{N}}$.

3

Standard Borel Spaces

In this chapter we introduce Borel sets and Borel functions—the main topics of this monograph. However, many of the deep results on Borel sets and Borel functions require the theory of analytic and coanalytic sets, which is developed in the next chapter. So, this chapter, though quite important, should be seen mainly as an introduction to these topics.

3.1 Measurable Sets and Functions

An **algebra** on a set X is a collection $\mathcal{A}$ of subsets of X such that

(i) $X \in \mathcal{A}$;

(ii) whenever A belongs to $\mathcal{A}$ so does $A^c = X \setminus A$; i.e., $\mathcal{A}$ is closed under complementation; and

(iii) $\mathcal{A}$ is closed under finite unions.

Note that $\emptyset \in \mathcal{A}$ if $\mathcal{A}$ is an algebra. An algebra closed under countable unions is called a **σ-algebra**. A **measurable space** is an ordered pair $(X, \mathcal{A})$ where X is a set and $\mathcal{A}$ a σ-algebra on X. We sometimes write X instead of $(X, \mathcal{A})$ if there is no scope for confusion. Sets in $\mathcal{A}$ are called **measurable**. Let $(X, \mathcal{A})$ be a σ-algebra and $A_0, A_1, A_2, \ldots \in \mathcal{A}$. Then, as

$\bigcap_n A_n = (\bigcup_n A_n^c)^c$,

$\limsup_n A_n = \bigcap_n \bigcup_{m \geq n} A_m$, and

$\liminf_n A_n = \bigcup_n \bigcap_{m \geq n} A_m,$

these sets all belong to $\mathcal{A}$.

Example 3.1.1 Let X be any set, $\mathcal{B}_1 = \{\emptyset, X\}$, and $\mathcal{B}_2 = \mathcal{P}(X)$. Then $\mathcal{B}_1$ and $\mathcal{B}_2$ are σ-algebras, called the **indiscrete** and **discrete** σ-algebras respectively. These are the trivial σ-algebras and are not very interesting.

Example 3.1.2 Let X be an infinite set and

$$\mathcal{A} = \{A \subseteq X : \text{ either} A \text{ or} A^c \text{ is finite}\}.$$

Then $\mathcal{A}$ is an algebra that is not a σ-algebra.

Example 3.1.3 Let X be an uncountable set and

$$\mathcal{A} = \{A \subseteq X : \text{ either} A \text{ or} A^c \text{ is countable}\}.$$

Then $\mathcal{A}$ is a σ-algebra, called the **countable-cocountable** σ-algebra.

Example 3.1.4 The family of finite disjoint unions of nondegenerate intervals including the empty set is an algebra on $\mathbb{R}$.

Example 3.1.5 Let $(X, \mathcal{A})$ and $(Y, \mathcal{B})$ be measurable spaces, $Z = X \times Y$, and $\mathcal{D}$ the family of finite disjoint unions of "measurable rectangles," i.e., sets of the form $A \times B$, $A \in \mathcal{A}$ and $B \in \mathcal{B}$. Then $\mathcal{D}$ is an algebra on Z.

It is easy to see that the intersection of a nonempty family of σ-algebras on a set X is a σ-algebra. Let $\mathcal{G}$ be any family of subsets of a set X. Let $\mathcal{S}$ be the family of all σ-algebras containing $\mathcal{G}$. Note that $\mathcal{S}$ contains the discrete σ-algebra and hence is not empty. Let $\sigma(\mathcal{G})$ be the intersection of all members of $\mathcal{S}$. Then $\sigma(\mathcal{G})$ is the smallest σ-algebra on X containing $\mathcal{G}$. We say $\sigma(\mathcal{G})$ is **generated** by $\mathcal{G}$ or $\mathcal{G}$ is a **generator** of $\sigma(\mathcal{G})$. For example, the family $\mathcal{G} = \{\{x\} : x \in X\}$ generates the countable-cocountable σ-algebra on X. A σ-algebra $\mathcal{A}$ is called **countably generated** if it has a countable generator.

Lemma 3.1.6 *Let $(X, \mathcal{A})$ be a measurable space, where $\mathcal{A} = \sigma(\mathcal{G})$. Suppose $x, y \in X$ are such that for every $G \in \mathcal{G}$, $x \in G$ if and only if $y \in G$. Then for all $A \in \mathcal{A}$, $x \in A$ if and only if $y \in A$.*

Proof. Let
$$\mathcal{B} = \{A \subseteq X : x \in A \Longleftrightarrow y \in A\}.$$
It is easy to see that $\mathcal{B}$ is a σ-algebra. By our assumption, it contains $\mathcal{G}$. The result follows. ■

Proposition 3.1.7 *Let $(X, \mathcal{B})$ be a measurable space, $\mathcal{G}$ a generator of $\mathcal{B}$, and $A \in \mathcal{B}$. Then there exists a countable $\mathcal{G}' \subseteq \mathcal{G}$ such that $A \in \sigma(\mathcal{G}')$.*

Proof. Let $\mathcal{A}$ be the collection of all subsets A of X such that $A \in \sigma(\mathcal{G}')$ for some countable $\mathcal{G}' \subseteq \mathcal{G}$.

Clearly, $\mathcal{A}$ is closed under complementation, and $\mathcal{G} \subseteq \mathcal{A}$.

Let $A_0, A_1, A_2, \ldots \in \mathcal{A}$. Choose countable $\mathcal{G}_n \subseteq \mathcal{G}$ such that $A_n \in \sigma(\mathcal{G}_n)$. Set $\mathcal{G}' = \bigcup_n \mathcal{G}_n$. Then $\mathcal{G}'$ is countable, and $\bigcup_n A_n \in \sigma(\mathcal{G}')$. Thus $\mathcal{A}$ is closed under countable unions. The proof is complete. ∎

Let $\mathcal{D} \subseteq \mathcal{P}(X)$ and $Y \subseteq X$. We set

$$\mathcal{D}|Y = \{B \bigcap Y : B \in \mathcal{D}\}.$$

Let $(X, \mathcal{B})$ be a measurable space and $Y \subseteq X$. Then $\mathcal{B}|Y$ is a σ-algebra on Y, called the **trace** of $\mathcal{B}$. It is easy to see that if $\mathcal{G}$ generates $\mathcal{B}$, then $\mathcal{G}|Y$ generates $\mathcal{B}|Y$.

From now on, unless otherwise stated, a subset of a measurable space will be equipped with the trace σ-algebra.

Let X be a metrizable space. The σ-algebra generated by the topology of X is called the **Borel σ- algebra** of X. It will be denoted by $\mathcal{B}_X$. Sets in $\mathcal{B}_X$ are called **Borel** in X.

Exercise 3.1.8 Let X be a second countable metrizable space and $\mathcal{G}$ a subbase for the topology of X. Show that $\mathcal{G}$ generates $\mathcal{B}_X$. Also show that this need not be true if X is not second countable.

From now on, unless otherwise stated, a metrizable space will be equipped with its Borel σ-algebra.

Proposition 3.1.9 *The Borel σ-algebra $\mathcal{B}_X$ of a metrizable space X equals the smallest family $\mathcal{B}$ of subsets of X that contains all open sets and that is closed under countable intersections and countable unions.*

Proof. Since $\mathcal{B}$ is the smallest family of subsets of X containing all open sets, closed under countable intersections and countable unions, and $\mathcal{B}_X$ is one such family, $\mathcal{B} \subseteq \mathcal{B}_X$. The reverse inclusion will be shown if we show that $\mathcal{B}$ is closed under complementation. Towards proving this, consider

$$\mathcal{D} = \{A \in \mathcal{B} : A^c \in \mathcal{B}\}.$$

We need to show that $\mathcal{B} \subseteq \mathcal{D}$. Since every closed set in a metrizable space is a G_δ set, open sets are in $\mathcal{D}$. Now suppose $A_0, A_1, A_2, \ldots$ are in $\mathcal{D}$. Then $A_i, A_i^c \in \mathcal{B}$ for all i. As

$$(\bigcup_i A_i)^c = \bigcap_i A_i^c \text{ and } (\bigcap_i A_i)^c = \bigcup_i A_i^c,$$

$\bigcup_i A_i$ and $\bigcap_i A_i$ belong to $\mathcal{D}$. Thus $\mathcal{D}$ contains open sets and is closed under countable unions and countable intersections. Since $\mathcal{B}$ is the smallest such family, $\mathcal{B} \subseteq \mathcal{D}$. ∎

Since every open set is an F_σ set, the above argument also shows the following.

Proposition 3.1.10 *The Borel σ-algebra $\mathcal{B}_X$ of a metrizable space X equals the smallest family $\mathcal{B}$ of subsets of X that contains all closed sets and that is closed under countable intersections and countable unions.*

A slight modification of the above arguments gives us the following useful result.

Proposition 3.1.11 *The Borel σ-algebra $\mathcal{B}_X$ of a metrizable space X equals the smallest family $\mathcal{B}$ that contains all open subsets of X and that is closed under countable intersections and countable disjoint unions.*

Proof. By the argument contained in the proof of 3.1.9, it is sufficient to prove that $\mathcal{B}$ is closed under complementation. Let

$$\mathcal{D} = \{B \in \mathcal{B} : B^c \in \mathcal{B}\}.$$

Since every closed set in X is a G_δ set, all open sets belong to $\mathcal{D}$. We now show that $\mathcal{D}$ is closed under countable disjoint unions and countable intersections.

Fix $A_0, A_1, A_2, \ldots \in \mathcal{D}$. Then $A_i, A_i^c \in \mathcal{B}$ for all i. We have $\bigcap_i A_i \in \mathcal{B}$. Note that the sets $B_0 = A_0^c$, $B_1 = A_1^c \bigcap A_0$, $B_2 = A_2^c \bigcap A_0 \bigcap A_1, \ldots$ are pairwise disjoint and belong to $\mathcal{B}$. Further, $(\bigcap_i A_i)^c = \bigcup_i B_i \in \mathcal{B}$. Thus $\mathcal{D}$ is closed under countable intersections. Similarly, we show that $\mathcal{D}$ is closed under countable disjoint unions. As before, we conclude that $\mathcal{B} \subseteq \mathcal{D}$; i.e., $\mathcal{B}$ is closed under complementation. ■

It is interesting to note that 3.1.11 remains true even if we replace "open" by "closed" in its statement, though its proof is fairly sophisticated.

Proposition 3.1.12 *(Sierpiński) The Borel σ-algebra $\mathcal{B}_X$ of a metrizable space X equals the smallest family $\mathcal{B}$ that contains all closed subsets of X and that is closed under countable intersections and countable disjoint unions.*

Proof. By 3.1.11, it is sufficient to show that every open set belongs to $\mathcal{B}$. The main difficulty lies here. Recall that in 2.1.35 we showed that $(0,1)$ cannot be expressed as a countable disjoint union of closed subsets of $\mathbb{R}$. We need a lemma.

Notation. For any family $\mathcal{F}$, let $\mathcal{F}_+$ denote the family of countable disjoint unions of sets in $\mathcal{F}$.

Lemma 3.1.13 *Let $\mathcal{F}$ be the set of closed subsets of $\mathbb{R}$. Then* $(0,1] \in \mathcal{F}_{+\delta+}$.

Assuming the lemma, we complete the proof as follows. Given an open set $U \subseteq X$, by 2.1.19, choose a continuous map $f : X \longrightarrow [0,1]$ such that $U = f^{-1}((0,1])$. The lemma immediately implies that $U \in \mathcal{B}$.

Proof of the lemma. Let D be the set of all endpoints of the middle-third intervals removed from $[0,1]$ to construct the Cantor ternary set $\mathbf{C}$

(2.3.3). So, $D = \{\frac{1}{3}, \frac{2}{3}, \frac{1}{9}, \frac{2}{9}, \frac{7}{9}, \frac{8}{9}, \ldots\}$. Let $E = D \bigcup \{0\}$ and $P = \mathbf{C} \setminus E$. Note that

$$(0,1] \setminus P = [\frac{1}{3}, \frac{2}{3}] \cup [\frac{1}{9}, \frac{2}{9}] \cup [\frac{7}{9}, \frac{8}{9}] \cup \cdots;$$

i.e., $(0,1] \setminus P$ is the union of the closures of the middle-third intervals removed to form $\mathbf{C}$. These interval are, of course, disjoint. Therefore, the lemma will be proved if we show that P is in $\mathcal{F}_{+\delta}$. Now,

$$P = \bigcap_{x \in E} (\mathbf{C} \setminus \{x\}). \qquad (\star)$$

Since $\mathbf{C}$ is a zero-dimensional compact metric space, each $\mathbf{C} \setminus \{x\}$ is a countable disjoint union of clopen subsets of $\mathbf{C}$, which, being compact, are closed in $\mathbb{R}$; i.e., $C \setminus \{x\} \in \mathcal{F}_{+}$. Since E is countable, $P \in \mathcal{F}_{+\delta}$ by $(\star)$. ■

A collection $\mathcal{M}$ of subsets of a set X is called a **monotone class** if it is closed under countable nonincreasing intersections and countable nondecreasing unions.

Proposition 3.1.14 *(The monotone class theorem) The smallest monotone class $\mathcal{M}$ containing an algebra $\mathcal{A}$ on a set X equals $\sigma(\mathcal{A})$, the σ-algebra generated by $\mathcal{A}$.*

Proof. Since every σ-algebra is a monotone class, $\mathcal{M} \subseteq \sigma(\mathcal{A})$.

To show the other inclusion, we first show that $\mathcal{M}$ is closed under finite intersections. For $A \subset X$, let

$$\mathcal{M}(A) = \{B \in \mathcal{M} : A \bigcap B \in \mathcal{M}\}.$$

As $\mathcal{M}$ is a monotone class, $\mathcal{M}(A)$ is a monotone class. As $\mathcal{A}$ is an algebra, $\mathcal{A} \subseteq \mathcal{M}(A)$ for every $A \in \mathcal{A}$. Therefore, $\mathcal{M} \subseteq \mathcal{M}(A)$ for $A \in \mathcal{A}$. Thus for every $A \in \mathcal{A}$ and every $B \in \mathcal{M}$, $A \bigcap B \in \mathcal{M}$. Using this and following the above argument, we see that for every $A \in \mathcal{M}$, $\mathcal{M}(A)$ is a monotone class containing $\mathcal{A}$. So, $\mathcal{M} \subseteq \mathcal{M}(A)$. This proves our claim.

As $\mathcal{M}$ is a monotone class closed under finite intersections, it is closed under countable intersections. Our proof will be complete if we show that $\mathcal{M}$ is closed under complementation. Consider

$$\mathcal{D} = \{A \in \mathcal{M} : A^c \in \mathcal{M}\}.$$

It is routine to show that $\mathcal{D}$ is a monotone class. Clearly, $\mathcal{D} \supseteq \mathcal{A}$. So, $\mathcal{M} \subseteq \mathcal{D}$; i.e., $\mathcal{M}$ is closed under complementation. ■

Let $(X, \mathcal{A})$ be a measurable space. A nonempty measurable set A is called an **$\mathcal{A}$-atom** if it has no nonempty measurable proper subset. Note that no two distinct atoms intersect. A measurable space X is called **atomic** if X is the union of its atoms. If X is metrizable, then $(X, \mathcal{B}_X)$ is atomic, the atoms being singletons.

Proposition 3.1.15 *Every countably generated measurable space is atomic.*

Proof. Let $\mathcal{A}$ be a countably generated σ-algebra on X. Fix a countable generator $\mathcal{G} = \{A_n : n \in \mathbb{N}\}$ for $\mathcal{A}$. For any $B \subseteq X$, set $B^0 = B$ and $B^1 = X \setminus B$. For every sequence $\alpha = (\epsilon_0, \epsilon_1, \epsilon_2, \ldots)$ of 0's and 1's, define

$$A(\alpha) = \bigcap_n A_n^{\epsilon_n}.$$

Each $A(\alpha)$ is clearly measurable. Let $x \in X$. Put $\epsilon_n = \chi_{A_n}(x)$ and $\alpha = (\epsilon_0, \epsilon_1, \epsilon_2, \ldots)$. Then $x \in A(\alpha)$. Thus X is the union of $A(\alpha)$'s. Note that x, y belong to the same $A(\alpha)$ if and only if for every n, either both x and y belong to A_n or neither does.

We now show that each $A(\alpha)$ is an atom of $\mathcal{A}$. Suppose this is not the case. Thus, there is an $\alpha = (\epsilon_0, \epsilon_1, \epsilon_2, \ldots)$ such that $A(\alpha)$ contains a nonempty, proper, measurable subset, say B. Choose $x \in B$ and $y \in A(\alpha) \setminus B$. By 3.1.6, there is an n such that A_n contains exactly one of x and y. Since both $x, y \in A(\alpha)$, for every m, $x \in A_m$ if and only if $y \in A_m$. This contradicts 3.1.6. ∎

Exercise 3.1.16 Let X be uncountable. Show that the countable-cocountable σ-algebra on X is atomic but not countably generated.

The next exercise is to show that a sub σ-algebra of a countably generated σ-algebra need not be countably generated.

Exercise 3.1.17 Let X be a metrizable space and A a nonBorel subset. Show that

$$\mathcal{B} = \{C \in \mathcal{B}_X : \text{ either } A \subseteq C \text{ or } A \bigcap C = \emptyset\}$$

is not countably generated.

Exercise 3.1.18 Show that a σ-algebra is either finite or of cardinality at least $\mathfrak{c}$.

Let $(X, \mathcal{A})$ and $(Y, \mathcal{B})$ be measurable spaces. A map $f : (X, \mathcal{A}) \to (Y, \mathcal{B})$ is called **measurable** if $f^{-1}(B) \in \mathcal{A}$ for every $B \in \mathcal{B}$. If $\mathcal{A} = \mathcal{P}(X)$ and Y any measurable space, then every $f : X \longrightarrow Y$ is measurable. Let $\mathcal{G}$ generate $\mathcal{B}$. Then f is measurable if and only if $f^{-1}(B) \in \mathcal{A}$ for every $B \in \mathcal{G}$. To see this, note that the family

$$\{B \subseteq Y : f^{-1}(B) \in \mathcal{A}\}$$

is a σ-algebra containing $\mathcal{G}$. So, it contains $\mathcal{B}$.

A measurable function $f : (X, \mathcal{B}_X) \longrightarrow (Y, \mathcal{B}_Y)$ will be called **Borel measurable**, or simply **Borel**. If X and Y are metrizable spaces, then every continuous function $f : X \longrightarrow Y$ is Borel. Further, if Y is second countable, $\mathcal{G}$ a subbase of Y, and $f^{-1}(V)$ is Borel for all $V \in \mathcal{G}$, then f is Borel.

Let $(X_i, \mathcal{A}_i)$, $i \in I$, be a family of measurable spaces and $X = \prod_i X_i$. The σ-algebra on X generated by

$$\{\pi_i^{-1}(B) : B \in \mathcal{A}_i, i \in I\},$$

where $\pi_i : X \longrightarrow X_i$ are the projection maps, is called the **product σ-algebra**. It is denoted by $\bigotimes_i \mathcal{A}_i$. Note that $\bigotimes_i \mathcal{A}_i$ is the smallest σ-algebra such that each π_i is measurable. Let $(X, \mathcal{A})$ and $(Y, \mathcal{B})$ be two measurable spaces. The product σ-algebra on $X \times Y$ will be denoted simply by $\mathcal{A} \bigotimes \mathcal{B}$.

From now on, unless otherwise stated, the product of measurable spaces will be equipped with the product σ-algebra.

Exercise 3.1.19 Let $(X_i, \mathcal{A}_i)$, $i \in I$, be measurable spaces and $\sigma(\mathcal{G}_i) = \mathcal{A}_i$. Show that

$$\bigotimes_i \mathcal{A}_i = \sigma(\{\pi_i^{-1}(B) : B \in \mathcal{G}_i, i \in I\}).$$

In particular, the product of countably many countably generated measurable spaces is countably generated.

Exercise 3.1.20 Let $(X, \mathcal{A})$ and $(Y, \mathcal{B})$ be measurable spaces and $B \in \mathcal{A} \bigotimes \mathcal{B}$. Show that for every $x \in X$, the section $B_x = \{y \in Y : (x, y) \in B\} \in \mathcal{B}$.

Proposition 3.1.21 *Let $(X, \mathcal{A})$ be a measurable space and Y a second countable metrizable space. If $f : X \longrightarrow Y$ is a measurable function, then* graph(f) *is in $\mathcal{A} \bigotimes \mathcal{B}_Y$.*

Proof. Let (U_n) be a countable base for Y. Note that

$$y \neq f(x) \iff \exists n (f(x) \in U_n \;\&\; y \notin U_n).$$

Therefore,

$$\text{graph}(f) = [\bigcup_n (f^{-1}(U_n) \times U_n^c)]^c,$$

and the result follows. ■

Corollary 3.1.22 *Let $(X, \mathcal{A})$ be a measurable space and Y a discrete measurable space of cardinality at most $\mathfrak{c}$. Then the graph of every measurable function $f : X \longrightarrow Y$ is measurable.*

Proof. Without loss of generality, assume $Y \subseteq \mathbb{R}$. Let $f : (X, \mathcal{A}) \longrightarrow (Y, \mathcal{P}(Y))$ be measurable. In particular, $f : (X, \mathcal{A}) \longrightarrow (Y, \mathcal{B}_Y)$ is also measurable. By 3.1.21, graph(f) $\in \mathcal{A} \bigotimes \mathcal{B}_Y \subseteq \mathcal{A} \bigotimes \mathcal{P}(Y)$. ■

Proposition 3.1.23 *Let $X_i, i = 0, 1, \ldots$, be a sequence of second countable metrizable spaces and $X = \prod_i X_i$. Then*

$$\mathcal{B}_X = \bigotimes_i \mathcal{B}_{X_i}.$$

Proof. Fix a countable base $\mathcal{B}_i$ for X_i, $i \in \mathbb{N}$, and put

$$\mathcal{G} = \{\pi_i^{-1}(B) : B \in \mathcal{B}_i, i \in \mathbb{N}\}.$$

Then $\mathcal{G}$ generates $\bigotimes_i \mathcal{B}_{X_i}$. On the other hand, since $\mathcal{G}$ is a subbase for the topology on X, by 3.1.8, it generates $\mathcal{B}_X$. ■

Here is an interesting question raised by Ulam[121]. Is

$$\mathcal{P}(\mathbb{R}) \bigotimes \mathcal{P}(\mathbb{R}) = \mathcal{P}(\mathbb{R} \times \mathbb{R})?$$

We show that under **CH** the answer to this question is yes. The solution presented here is due to B. V. Rao[94].

Theorem 3.1.24 $\mathcal{P}(\omega_1) \bigotimes \mathcal{P}(\omega_1) = \mathcal{P}(\omega_1 \times \omega_1)$.

Proof. Let $A \subseteq \omega_1 \times \omega_1$. Write $A = B \bigcup C$, where

$$B = A \bigcap \{(\alpha, \beta) \in \omega_1 \times \omega_1 : \alpha \geq \beta\}$$

and

$$C = A \bigcap \{(\alpha, \beta) \in \omega_1 \times \omega_1 : \alpha \leq \beta\}.$$

We shall show that B is in the product σ-algebra. By symmetry it will follow that C is in the product σ-algebra. The result will then follow.

For each $\alpha < \omega_1$, B_α is countable, say $B_\alpha = \{\alpha_0, \alpha_1, \alpha_2, \ldots\}$. By 3.1.22,

$$G_n = \{(\alpha, \alpha_n) : \alpha \in \omega_1\}$$

is in the product of discrete σ-algebras. Now note that $B = \bigcup_n G_n$. ■

Exercise 3.1.25 Show that if $|X| > \mathfrak{c}$, then

$$\mathcal{P}(X) \bigotimes \mathcal{P}(X) \neq \mathcal{P}(X \times X).$$

Corollary 3.1.26 *Under* **CH**,

$$\mathcal{P}(X) \bigotimes \mathcal{P}(X) = \mathcal{P}(X \times X) \Longleftrightarrow |X| \leq \mathfrak{c}.$$

Proposition 3.1.27 *Let (f_n) be a sequence of measurable maps from a measurable space X to a metrizable space Y converging pointwise to f. Then $f : X \to Y$ is measurable.*

Proof. Let d be a compatible metric on Y. Fix any open set U in Y. For each positive integer k, set

$$U_k = \{x \in U : d(x, U^c) > 1/k\}.$$

Since U is open,

$$U = \bigcup_k U_k = \bigcup_k \text{cl}(U_k).$$

Note that for every $x \in X$, we have

$$\begin{aligned} f(x) \in U & \Longrightarrow \exists k \lim_n f_n(x) \in U_k \\ & \Longrightarrow \exists k \exists N \forall n \geq N f_n(x) \in U_k \\ & \Longrightarrow \exists k f(x) \in \text{cl}(U_k) \\ & \Longrightarrow f(x) \in U. \end{aligned}$$

Thus,

$$f^{-1}(U) = \bigcup_k \liminf_n f_n^{-1}(U_k).$$

Since each f_n is measurable, it follows from the above observation that f is measurable. ■

A function $f : X \longrightarrow \mathbb{R}$ is **simple** if its range is finite.

Proposition 3.1.28 *Let X be metrizable. Then every Borel function $f : X \longrightarrow \mathbb{R}$ is the pointwise limit of a sequence of simple Borel functions.*

Proof. Fix $n \geq 1$. For $-n2^n \leq j < n2^n$, let

$$B_j^n = f^{-1}([j/2^n, (j+1)/2^n)).$$

As f is Borel, each B_j^n is Borel. Set

$$f_n = \sum_{j=-n2^n}^{(n-1)2^n} \frac{j}{2^n} \chi_{B_j^n}.$$

Clearly, f_n is a simple Borel function. It is easy to check that $f_n \to f$ pointwise. ■

The following proposition is quite easy to prove.

Proposition 3.1.29 (i) *If $f : (X, \mathcal{A}) \longrightarrow (Y, \mathcal{B})$ and $g : (Y, \mathcal{B}) \longrightarrow (Z, \mathcal{C})$ are measurable, then so is $g \circ f : (X, \mathcal{A}) \longrightarrow (Z, \mathcal{C})$.*

(ii) *A map $f : (X, \mathcal{A}) \longrightarrow (\prod X_i, \bigotimes_i \mathcal{A}_i)$ is measurable if and only if its composition with each projection map is measurable.*

Theorem 3.1.30 *Let $(X, \mathcal{A})$ be a measurable space, Y and Z metrizable spaces with Y second countable. Suppose D is a countable dense set in Y and $f : X \times Y \longrightarrow Z$ a map such that*

(i) *the map $y \longrightarrow f(x,y)$ from Y to Z is continuous for every $x \in X$;*

(ii) *$x \longrightarrow f(x,y)$ is measurable for all $y \in D$.*

Then $f : X \times Y \to Z$ is measurable.

Proof. Fix compatible metrics d and ρ on Y and Z respectively. Take any closed set C in Z. For $(x,y) \in X \times Y$, it is routine to check that

$$f(x,y) \in C \Longleftrightarrow (\forall n \geq 1)(\exists y' \in D)[d(y,y') \leq \frac{1}{n} \ \&\ \rho(f(x,y'),C) \leq \frac{1}{n}].$$

Therefore,

$$f^{-1}(C) = \bigcap_n \bigcup_{y' \in D} [\{x \in X : \rho(f(x,y'),C) \leq \frac{1}{n}\} \times \{y \in Y : d(y,y') \leq \frac{1}{n}\}].$$

By our hypothesis, $f^{-1}(C) \in \mathcal{A} \bigotimes \mathcal{B}_Y$. ∎

Example 3.1.31 We shall see later (3.3.18) that for every uncountable Polish space E, $|\mathcal{B}_E| = \mathfrak{c}$. So there exists a nonBorel set $A \subseteq S^1$. Let $f = \chi_A : \mathbb{R}^2 \longrightarrow \mathbb{R}$. Then f is separately Borel (in fact, of class 2 (see Section 3.6)) in each variable, but f is not Borel measurable.

Let X and Y be metrizable spaces and $\mathcal{B}(X,Y)$ the smallest class of functions from X to Y containing all continuous functions and closed under taking pointwise limits of sequences of functions. Functions belonging to $\mathcal{B}(X,Y)$ are called **Baire functions.**

Proposition 3.1.32 *Let X and Y be metrizable spaces. Then every Baire function $f : X \longrightarrow Y$ is Borel.*

Proof. Since every continuous function is Borel and since the limit of a pointwise convergent sequence of Borel functions is Borel (3.1.27), Baire functions are Borel. ∎

Remark 3.1.33 Every Baire function $f : \mathbb{R} \longrightarrow \mathbb{N}$ is a constant. (Prove it.) So the converse of the above proposition is not true even when X and Y are Polish.

However, we shall show in 3.1.36 that for every metrizable X, every Borel $f : X \longrightarrow \mathbb{R}$ is Baire.

Exercise 3.1.34 (i) Let $f : X \longrightarrow Y$ and $g : Y \longrightarrow Z$ be Baire functions with at least one of them continuous. Then $g \circ f : X \longrightarrow Z$ is Baire.

(ii) If $a, b \in \mathbb{R}$ and $f, g : X \longrightarrow \mathbb{R}$ are Baire, then so is $af + bg$.

Lemma 3.1.35 *Let X be a metrizable space and $B \subseteq X$ Borel. Then $\chi_B : X \longrightarrow \mathbb{R}$ is Baire.*

Proof. Let
$$\mathcal{B} = \{B \subseteq X : \chi_B \text{ is Baire}\}.$$
(a) Let U be open in X. Write $U = \bigcup_n F_n$, where the F_n's are closed and $F_n \subseteq F_{n+1}$. By 2.1.18, there is a continuous function $f_n : X \longrightarrow [0,1]$ identically equal to 1 on F_n and equal to 0 on $X \setminus U$. Then the sequence (f_n) converges pointwise to χ_U. Thus, $U \in \mathcal{B}$.

(b) Let $B_0, B_1, B_2, \ldots$ be pairwise disjoint and belong to $\mathcal{B}$. Set
$$f_n = \sum_{i \leq n} \chi_{B_i}.$$
By our hypothesis and 3.1.34(ii), f_n is Baire. Since (f_n) converges pointwise to the characteristic function of $\bigcup_n B_n$, we see that $\bigcup_n B_n \in \mathcal{B}$.

(c) Let $B_0, B_1, B_2, \ldots$ belong to $\mathcal{B}$. Put
$$f_n = \min_{i \leq n} \chi_{B_i}.$$
By our hypothesis and 3.1.34, f_n is Baire. As (f_n) converges pointwise to the characteristic function of $\bigcap_n B_n$, it follows that $\bigcap_n B_n \in \mathcal{B}$.

The result now follows from 3.1.11. ■

Theorem 3.1.36 *(Lebesgue – Hausdorff theorem) Every real-valued Borel function defined on a metrizable space is Baire.*

Proof. By 3.1.35 the characteristic function of every Borel set is Baire. Hence, by 3.1.34(ii), every simple Borel function is Baire. Now the result follows from 3.1.28. ■

3.2 Borel-Generated Topologies

In this section we prove some results that often help in reducing measurability problems to topological ones.

Lemma 3.2.1 *Let $(X, \mathcal{T})$ be a (zero-dimensional, second countable) metrizable space and (B_n) a sequence of Borel subsets of X. Then there is a (respectively zero-dimensional, second countable) metrizable topology $\mathcal{T}'$ such that $\mathcal{T} \subseteq \mathcal{T}' \subseteq \mathcal{B}_X$ and each $B_n \in \mathcal{T}'$.*

(If $\mathcal{T}$ and $\mathcal{T}'$ are topologies on a set X such that $\mathcal{T} \subseteq \mathcal{T}'$, we say that $\mathcal{T}'$ is finer than $\mathcal{T}$.)

Proof. Define $f : X \longrightarrow X \times \mathcal{C}$ by
$$f(x) = (x, \chi_{B_0}(x), \chi_{B_1}(x), \chi_{B_2}(x), \ldots).$$
This map is clearly one-to-one. Let
$$\mathcal{T}' = \{f^{-1}(U) : U \text{ open in } X \times \mathcal{C}\}.$$

As $(X, \mathcal{T}')$ is homeomorphic to a subset of $X \times C$, it is metrizable. Further, if X is zero-dimensional (separable), so is $(X, \mathcal{T}')$.

Let $U \subseteq X$ be open with respect to the original topology $\mathcal{T}$. Then

$$U = f^{-1}(\{(x, \alpha) \in X \times C : x \in U\})$$

and hence belongs to $\mathcal{T}'$. Thus, $\mathcal{T}'$ is finer that $\mathcal{T}$. By 3.1.29, f is Borel measurable. Therefore, $\mathcal{T}' \subseteq \mathcal{B}_X$. It remains to show that each $B_n \in \mathcal{T}'$. Let

$$V_n = \{(x, \alpha) \in X \times C : \alpha(n) = 1\}.$$

Then V_n is open in $X \times C$. Since $B_n = f^{-1}(V_n)$, $B_n \in \mathcal{T}'$. ∎

Remark 3.2.2 The topology $\mathcal{T}'$ defined above is the topology generated by $\mathcal{T} \bigcup \{B_n : n \in \mathbb{N}\} \bigcup \{B_n^c : n \in \mathbb{N}\}$.

Proposition 3.2.3 *Let $(X, \mathcal{T})$ be a metrizable space, $A \subseteq X$, Y Polish, and $f : A \longrightarrow Y$ any Borel map. Then*

(i) *there is a finer metrizable topology $\mathcal{T}'$ on X generating the same Borel σ-algebra such that $f : A \longrightarrow Y$ is continuous with respect to the new topology $\mathcal{T}'$, and*

(ii) *the map $f : A \longrightarrow Y$ admits a Borel extension $g : X \longrightarrow Y$.*

Proof. Fix a countable base (U_n) for Y. Let $n \in \mathbb{N}$. As $f^{-1}(U_n)$ is Borel in A, there is a Borel set B_n in X such that $f^{-1}(U_n) = A \bigcap B_n$. Take $\mathcal{T}'$ as in 3.2.1. This answers (i).

To prove (ii), take $\mathcal{T}'$ as above. By 2.2.3, there is a G_δ set $C \supseteq A$ and a continuous extension $g' : C \longrightarrow Y$ of f. Here we are assuming that X is equipped with the finer topology $\mathcal{T}'$. As $\mathcal{T}$ and $\mathcal{T}'$ generate the same σ-algebra $\mathcal{B}_X$, C is Borel in X and g' measurable relative to the original topology $\mathcal{T}$. Extend g' to the whole space X by defining it to be a constant on $X \setminus C$. ∎

We see that Theorem 3.2.1, though elementary, is already quite useful. However, with some extra care we get the following much deeper generalization of 3.2.1 with significant applications.

Theorem 3.2.4 *Suppose $(X, \mathcal{T})$ is a Polish space. Then for every Borel set B in X there is a finer Polish topology $\mathcal{T}_B$ on X such that B is clopen with respect to $\mathcal{T}_B$ and $\sigma(\mathcal{T}) = \sigma(\mathcal{T}_B)$.*

We make a few observations first.

Observation 1. Let F be a closed set in a Polish space $(X, \mathcal{T})$. Let $(X, \mathcal{T}')$ be the direct sum $F \bigoplus F^c$ of $(F, \mathcal{T}|F)$ and $(F^c, \mathcal{T}|F^c)$; i.e., $\mathcal{T}'$ is the topology generated by $\mathcal{T} \bigcup \{F\}$. By 2.2.1, $\mathcal{T}'$ is a Polish topology on X. It clearly generates the same Borel σ-algebra and makes F clopen.

Observation 2. Let $(\mathcal{T}_n)$ be a sequence of Polish topologies on X such that for any two distinct elements x, y of X, there exist disjoint sets $U, V \in \bigcap_n \mathcal{T}_n$ such that $x \in U$ and $y \in V$. Then the topology $\mathcal{T}_\infty$ generated by $\bigcup_n \mathcal{T}_n$ is Polish. This can be seen as follows.

Define $f : X \longrightarrow X^{\mathbf{N}}$ by

$$f(x) = (x, x, x, \ldots), \quad x \in X.$$

It is easy to see that f is an embedding of $(X, \mathcal{T}_\infty)$ in $\prod(X, \mathcal{T}_n)$. Further, the range of f is closed in $\prod(X, \mathcal{T}_n)$. To see this, let $(x_0, x_1, x_2, \ldots)$ be not in the range of f. Take m, n such that $x_n \neq x_m$. By our hypothesis, there exist disjoint sets U_n and U_m in $\bigcap \mathcal{T}_i$ such that $x_n \in U_n$ and $x_m \in U_m$. Then

$$(x_i) \in \pi_n^{-1}(U_n) \bigcap \pi_m^{-1}(U_m) \subseteq X^{\mathbf{N}} \setminus \text{ range}(f).$$

Proof of 3.2.4. Let $\mathcal{B}$ be the class of all Borel subsets B of X such that there is a finer Polish topology $\mathcal{T}_B$ generating $\mathcal{B}_X$ and making B clopen.

By Observation 1, $\mathcal{B}$ contains all closed sets, and it is clearly closed under complementation.

To show $\mathcal{B} = \mathcal{B}_X$, we need to prove only that $\mathcal{B}$ is closed under countable unions. Let B_n belong to $\mathcal{B}$ and $B = \bigcup B_n$. Let $\mathcal{T}_n$ be a finer Polish topology on X making B_n clopen and generating the same Borel σ-algebra. Then $B \in \mathcal{T}_\infty$, where $\mathcal{T}_\infty$ is the topology generated by $\bigcup \mathcal{T}_n$. By Observation 2, $\mathcal{T}_\infty$ is Polish. Take $\mathcal{T}_B$ to be the topology generated by $\mathcal{T}_\infty \bigcup \{B^c\}$. By Observation 1, $\mathcal{T}_B$ is Polish. ■

Corollary 3.2.5 *Suppose $(X, \mathcal{T})$ is a Polish space. Then for every sequence (B_n) of Borel sets in X there is a finer Polish topology $\mathcal{T}'$ on X generating the same Borel σ-algebra and making each B_n clopen.*

Corollary 3.2.6 *Suppose $(X, \mathcal{T})$ is a Polish space, Y a separable metric space, and $f : X \longrightarrow Y$ a Borel map. Then there is a finer Polish topology $\mathcal{T}'$ on X generating the same Borel σ-algebra such that $f : (X, \mathcal{T}') \longrightarrow Y$ is continuous.*

We shall see many applications of these results later. At the moment we show the following.

Theorem 3.2.7 *Every uncountable Borel subset of a Polish space contains a homeomorph of the Cantor set. In particular, it is of cardinality $\mathfrak{c}$.*

Proof. Let $(X, \mathcal{T})$ be Polish and B an uncountable Borel subset of X. By 3.2.4, let $\mathcal{T}'$ be a finer Polish topology on X making B closed. By 2.6.3, $(B, \mathcal{T}'|B)$ contains a homeomorph of the Cantor set, say K. By 2.3.9, $\mathcal{T}'|K = \mathcal{T}|K$, and the result follows. ■

In 3.2.7, we saw that every uncountable Borel subset of a Polish space contains a homeomorph of the Cantor set. The following example shows that this is not true for all sets. More precisely, we show that *there is a set A of real numbers such that for any uncountable closed subset C of $\mathbb{R}$, both $A \bigcap C$ and $A^c \bigcap C$ are uncountable.* Such a set will be called a **Bernstein set**.

Example 3.2.8 By 2.6.4, there are exactly $\mathfrak{c}$ uncountable closed subsets of $\mathbb{R}$. Let $\{C_\alpha : \alpha < \mathfrak{c}\}$ be an enumeration of these. We shall get distinct points $x_\alpha, y_\alpha, \alpha < \mathfrak{c}$, such that $x_\alpha, y_\alpha \in C_\alpha$. Then the set $A = \{x_\alpha : \alpha < \mathfrak{c}\}$ is easily seen to be a Bernstein set.

To define the x_α's and y_α's, we proceed by transfinite induction. Choose $x_0, y_0 \in C_0$ with $x_0 \neq y_0$. Let $\alpha < \mathfrak{c}$. Suppose x_β, y_β has been chosen for all $\beta < \alpha$. Let $D = \{x_\beta : \beta < \alpha\} \bigcup \{y_\beta : \beta < \alpha\}$. Note that $|D| = |\alpha| + |\alpha| < \mathfrak{c}$. As $|C_\alpha| = \mathfrak{c}$, we choose distinct points x_α, y_α in $C_\alpha \setminus D$.

3.3 The Borel Isomorphism Theorem

A map f from a measurable space X to a measurable space Y is called **bimeasurable** if it is measurable and $f(A)$ is measurable for every measurable subset A of X. A bimeasurable bijection will be called an **isomorphism**. Thus a bijection $f : X \longrightarrow Y$ is an isomorphism if and only if both f and f^{-1} are measurable. In the special case when X, Y are metrizable spaces equipped with Borel σ-algebras and $f : X \longrightarrow Y$ is an isomorphism, f will be called a **Borel isomorphism** and X and Y **Borel isomorphic**. The Borel σ-algebra of a countable metrizable space is discrete. Hence two countable metrizable spaces are Borel isomorphic if and only if they are of the same cardinality.

Example 3.3.1 The closed unit interval $I = [0, 1]$ and the Cantor set C are Borel isomorphic.

Proof. Let D be the set of all dyadic rationals in I and $E \subset C$ the set of all sequences of 0's and 1's that are eventually constant. Define $f : C \setminus E \longrightarrow I \setminus D$ by

$$f(\epsilon_0, \epsilon_1, \epsilon_2, \ldots) = \sum_{n \in \mathbb{N}} \epsilon_n / 2^{n+1}.$$

It is easy to check that f is a homeomorphism from $C \setminus E$ onto $I \setminus D$. Since both D and E are countably infinite, there is a bijection $g : E \longrightarrow D$. The function $h : I \longrightarrow C$ obtained by piecing f and g together is clearly a Borel isomorphism from I onto C. ■

Proposition 3.3.2 *Suppose $(X, \mathcal{A})$ is a measurable space with $\mathcal{A}$ countably generated. Then there is a subset Z of C and a bimeasurable map $g : X \longrightarrow Z$ such that for any x, y in X, $g(x) = g(y)$ if and only if x and y belong to the same atom of $\mathcal{A}$.*

Proof. Let $\mathcal{G} = \{A_n : n \in \mathbb{N}\}$ be a countable generator of $\mathcal{A}$. Define $g : X \longrightarrow \mathcal{C}$ by

$$g(x) = (\chi_{A_0}(x), \chi_{A_1}(x), \chi_{A_2}(x), \ldots).$$

Take $Z = g(X)$. By 3.1.29, g is measurable. Also note that for any two x, y in X, $g(x) = g(y)$ if and only if x and y belong to the same A_i's. Recall that the atoms of $\mathcal{A}$ are precisely the sets of the form $\bigcap_n A^{\epsilon(n)}$, $(\epsilon(0), \epsilon(1), \ldots) \in \mathcal{C}$. (See the proof of 3.1.15.) It follows that $g(x) = g(y)$ if and only if x and y belong to the same atom of $\mathcal{A}$. As

$$g(A_n) = Z \bigcap \{\alpha \in \mathcal{C} : \alpha(n) = 1\},$$

it is Borel in Z. Now observe that

$$\mathcal{B} = \{B \in \mathcal{A} : g(B) \text{ is Borel in } Z\}$$

is a σ-algebra containing A_n for all n. So, g^{-1} is also measurable. ■

Remark 3.3.3 In the above proposition, further assume that the σ-algebra $\mathcal{A}$ separates points; i.e., for $x \neq y$ there is a measurable set containing exactly one of x and y. In particular, $\mathcal{A}$ is atomic, and its atoms are singletons. Then the g obtained in 3.3.2 is an isomorphism. So, X can be given a topology making it homeomorphic to Z such that $\mathcal{B}_X = \mathcal{A}$.

Proposition 3.3.4 *Let $(X, \mathcal{A})$ be a measurable space, Y a Polish space, $A \subseteq X$, and $f : A \longrightarrow Y$ a measurable map. Then f admits a measurable extension to X.*

Proof. Fix a countable base (U_n) for Y. For every n, choose $B_n \in \mathcal{A}$ such that $f^{-1}(U_n) = B_n \bigcap A$. Without loss of generality, we assume that $\mathcal{A} = \sigma((B_n))$. By 3.3.2, get a metrizable space Z and a bimeasurable map $g : X \longrightarrow Z$ such that for any $x, x' \in X$, $g(x) = g(x')$ if and only if x and x' belong to the same atom of $\sigma((B_n))$. Hence, for $x, x' \in A$, $f(x) = f(x')$ if and only if $g(x) = g(x')$. Set $B = g(A)$ and define $h : B \longrightarrow Y$ by

$$h(z) = f(x),$$

where $x \in A$ is such that $g(x) = z$. It is easy to see that h is well-defined and $h^{-1}(U_n) = g(B_n) \bigcap B$. Hence, h is Borel. By 3.2.3, there is a Borel extension $h' : Z \longrightarrow Y$ of h. The composition $h' \circ g$ is clearly a measurable extension of f to X. ■

Exercise 3.3.5 Let X and Y be Polish spaces, $A \subset X$, $B \subset Y$, and $f : A \longrightarrow B$ a Borel isomorphism. Show that f can be extended to a Borel isomorphism between two Borel sets containing A and B.

(Hint: Use 3.2.5.)

Proposition 3.3.6 *Let X and Y be measurable spaces and $f : X \longrightarrow Y$, $g : Y \longrightarrow X$ one-to-one, bimeasurable maps. Then X and Y are isomorphic.*

Proof. As f and g are bimeasurable, the set E described in the proof of the Schröder – Bernstein theorem (1.2.3) is measurable. So the bijection $h : X \longrightarrow Y$ obtained there is bimeasurable. ■

A **standard Borel space** is a measurable space isomorphic to a Borel subset of a Polish space. In particular, a metrizable space X is standard Borel if $(X, \mathcal{B}_X)$ is standard Borel.

Proposition 3.3.7 *Let X be a second countable metrizable space. Then the following statements are equivalent.*

(i) *X is standard Borel.*

(ii) *X is Borel in its completion $\hat{X}$.*

(iii) *X is homeomorphic to a Borel subset of a Polish space.*

Proof. Clearly, (ii) implies (iii), and (i) follows from (iii). We show that (i) implies (ii).

Let X be standard Borel. Then, there is a Polish space Z, a Borel subset Y of Z, and a Borel isomorphism $f : X \longrightarrow Y$. By 3.3.5, there is a Borel isomorphism $g : X' \longrightarrow Y'$ extending f between Borel subsets X' and Y' of $\hat{X}$ (the completion of X) and Z respectively. Since $X = g^{-1}(Y)$, it is Borel in X' and hence in $\hat{X}$. ■

Remark 3.3.8 In 4.3.8 we shall show that if X is a second countable metrizable space that is standard Borel, Y a metrizable space, and f a Borel map from X onto Y, then Y is separable. Hence a metrizable space that is standard Borel is separable. Therefore, the second countability condition can be dropped from the above proposition.

Let X be a compact metric space. Then $K(X)$, the space of nonempty compact sets with Vietoris topology, being Polish (2.4.15), is standard Borel. It is interesting to note that $\mathcal{B}_{K(X)}$ is generated by sets of the form

$$\{K \in K(X) : K \bigcap U \neq \emptyset\},$$

where U varies over open sets in X. To prove this, let $\mathcal{B}$ be the σ-algebra generated by sets of the form $\{K \in K(X) : K \bigcap U \neq \emptyset\}$, U open in X. It is enough to check that for open U,

$$\{K \in K(X) : K \subseteq U\} \in \mathcal{B}.$$

Let (U_n) be a countable base for the topology of X that is closed under finite unions. Then for any open U and compact K,

$$\begin{aligned} K \subseteq U &\iff K \bigcap U^c = \emptyset \\ &\iff (\exists n)(U^c \subseteq U_n \ \& \ K \bigcap U_n = \emptyset). \end{aligned}$$

Thus,

$$\{K \in K(X) : K \subseteq U\} = \bigcup_{U^c \subset U_n} \{K \in K(X) : K \bigcap U_n = \emptyset\}.$$

Therefore, $\{K \in K(X) : K \subseteq U\}$ belongs to $\mathcal{B}$.

Exercise 3.3.9 Let X be a Polish space. Show that the maps

(a) $K \longrightarrow K'$ from $K(X)$ to $K(X)$,

(b) $(K_1, K_2) \longrightarrow K_1 \bigcap K_2$ from $K(X) \times K(X)$ to $K(X)$, and

(c) $(K_n) \longrightarrow \bigcap K_n$ from $K(X)^{\mathbb{N}}$ to $K(X)$

are Borel.

Effros Borel Space

Let X be a Polish space and $F(X)$ the set of all nonempty closed subsets of X. Equip $F(X)$ with the σ-algebra $\mathcal{E}(X)$ generated by sets of the form

$$\{F \in F(X) : F \bigcap U \neq \emptyset\},$$

where U varies over open sets in X. $(F(X), \mathcal{E}(X))$ is called the **Effros Borel space** of X. We proved above that $\mathcal{E}(X) = \mathcal{B}_{K(X)}$ if X is compact. Therefore, the Effros Borel space of a compact metrizable space is standard Borel. In fact we can prove more.

Theorem 3.3.10 *The Effros Borel space of a Polish space is standard Borel.*

Proof. Let Y be a compact metric space containing X as a dense subspace. By 2.2.7, X is a G_δ set in Y. Write $X = \bigcap U_n$, U_n open in Y. Let (V_n) be a countable base for Y. Now consider

$$\mathcal{Z} = \{\, \mathrm{cl}(F) \in F(Y) : F \in F(X)\},$$

where closure is relative to Y.

Note that $\mathcal{Z} \subseteq K(Y)$ and

$$K \in \mathcal{Z} \Longleftrightarrow K \bigcap X \text{ is dense in } K.$$

The result will be proved if we show the following.

(i) The map $F \longrightarrow \mathrm{cl}(F)$ from $(F(X), \mathcal{E}(X))$ onto $\mathcal{Z}$ is an isomorphism, and

(ii) $\mathcal{Z}$ is a G_δ set in $K(Y)$.

Clearly, $F \longrightarrow \mathrm{cl}(F)$ is one-to-one on $F(X)$. Further, for any $F \in F(X)$ and any U open in Y,

$$\mathrm{cl}(F)\bigcap U \neq \emptyset \Longleftrightarrow F\bigcap(U\bigcap X) \neq \emptyset.$$

Hence, (i) follows.

We now prove (ii). We have

$$K \in \mathcal{Z} \Longleftrightarrow K\bigcap\bigcap_n U_n \text{ is dense in } K.$$

Therefore, by the Baire category theorem,

$$\begin{aligned} K \in \mathcal{Z} &\Longleftrightarrow \forall n(K\bigcap U_n \text{ is dense in} K) \\ &\Longleftrightarrow \forall n\forall m(K\bigcap V_m \neq \emptyset \Longrightarrow K\bigcap V_m\bigcap U_n \neq \emptyset). \end{aligned}$$

Thus,

$$\mathcal{Z} = \bigcap_n\bigcap_m\{K \in F(Y) : K\bigcap V_m = \emptyset \text{ or } K\bigcap V_m\bigcap U_n \neq \emptyset\},$$

and the result follows. ■

Exercise 3.3.11 Let X, Y be Polish spaces. Show the following.

(i) $\{(F_1, F_2) \in F(X) \times F(X) : F_1 \subseteq F_2\}$ is Borel.

(ii) The map $(F_1, F_2) \longrightarrow F_1 \bigcup F_2$ from $F(X) \times F(X)$ to $F(X)$ is Borel.

(iii) The map $(F_1, F_2) \longrightarrow F_1 \times F_2$ from $F(X) \times F(Y)$ to $F(X \times Y)$ is Borel.

(iv) $\{K \in F(X) : K \text{ is compact}\}$ is Borel.

(v) For every continuous map $g : X \longrightarrow Y$, the map $F \longrightarrow \mathrm{cl}(g(F))$ from $F(X)$ to $F(Y)$ is measurable.

In the next chapter we shall show that the map $(F_1, F_2) \longrightarrow F_1 \bigcap F_2$ from $F(X) \times F(X)$ to $F(X)$ need not be Borel.

Exercise 3.3.12 Let X be a Polish space. Show that the Borel space of $F(X)$ equipped with the Fell topology is exactly the same as the Effros Borel space.

We now proceed to prove one of the main results on standard Borel spaces—the Borel isomrphism theorem, due to K. Kuratowski[61]. It classifies standard Borel spaces. More specifically, it says that two standard Borel spaces X and Y are isomorphic if and only if they are of the same cardinality. For countable spaces this is, of course, trivial. The proof presented here is due to B. V. Rao and S. M. Srivastava[96].

Theorem 3.3.13 *(The Borel isomorphism theorem) Any two uncountable standard Borel spaces are Borel isomorphic.*

We first prove a few auxiliary results.

Lemma 3.3.14 *Every standard Borel space B is Borel isomorphic to a Borel subset of $\mathcal{C}$.*

Proof. By 3.3.1, I and $\mathcal{C}$ are Borel isomorphic. Therefore, the Hilbert cube $I^{\mathbb{N}}$ and $\mathcal{C}^{\mathbb{N}}$ are isomorphic. But $\mathcal{C}^{\mathbb{N}}$ is homeomorphic to $\mathcal{C}$. Thus, the Hilbert cube and the Cantor set are Borel isomorphic. By 2.1.32, every standard Borel space is isomorphic to a Borel subset of the Hilbert cube, and hence of $\mathcal{C}$. ■

Proposition 3.3.15 *For every Borel subset B of a Polish space X, there is a Polish space Z and a continuous bijection from Z onto B.*

Proof. Let $\mathcal{B}$ be the set of all $B \subseteq X$ such that there is a continuous bijection from a Polish space Z onto B. We show that $\mathcal{B} = \mathcal{B}_X$. Since every open subset of X is Polish, (2.2.1), open sets belong to $\mathcal{B}$. By 3.1.11, it is sufficient to show that $\mathcal{B}$ is closed under countable intersections and countable disjoint unions. Let $B_0, B_1, B_2, \ldots$ belong to $\mathcal{B}$. Fix Polish spaces $Z_0, Z_1, \ldots$ and continuous bijections $g_i : Z_i \longrightarrow B_i$. Let

$$Z = \{z \in \prod Z_i : g_0(z_0) = g_1(z_1) = \cdots\}.$$

Then Z is closed in $\prod_i Z_i$. Therefore, Z is Polish. Define $g : Z \longrightarrow X$ by

$$g(z) = g_0(z_0).$$

Then g is a one-to-one, continuous map from Z onto $\bigcap B_i$. Thus, $\mathcal{B}$ is closed under countable intersections.

Let us next assume that $B_0, B_1, \ldots$ are pairwise disjoint. Choose g_i, Z_i as before. Take $Z = \bigoplus Z_i$, the direct sum of the Z_i's. Define $g : Z \longrightarrow X$ by

$$g(z) = g_i(z) \text{ if } z \in Z_i.$$

Then Z is a Polish space, and g is a one-to-one, continuous map from Z onto $\bigcup B_i$. So, $\mathcal{B}$ is also closed under countable disjoint unions. ■

Proof of 3.3.13. Let B be an uncountable standard Borel space. Without loss of generality, we assume that B is a Borel subset of some Polish space. By 3.3.14, there is a bimeasurable bijection from B into $\mathcal{C}$. By 3.2.7, B contains a homeomorph of the Cantor set. By 3.3.6, B is Borel isomorphic to $\mathcal{C}$, and the proof is complete. ■

Corollary 3.3.16 *Two standard Borel spaces are Borel isomorphic if and only if they are of the same cardinality.*

Theorem 3.3.17 *Every Borel subset of a Polish space is a continuous image of* $\mathbb{N}^{\mathbb{N}}$ *and a one-to-one, continuous image of a closed subset of* $\mathbb{N}^{\mathbb{N}}$.

Proof. The result follows directly from 3.3.15, 2.6.9, and 2.6.13. ∎

Theorem 3.3.18 *For every infinite Borel subset* X *of a Polish space,* $|\mathcal{B}_X| = \mathfrak{c}$.

Proof. Without loss of generality, we assume that X is uncountable. Since X contains a countable infinite set, $|\mathcal{B}_X| \geq \mathfrak{c}$. By 2.6.6, the cardinality of the set of continuous maps from $\mathbb{N}^{\mathbb{N}}$ to X is $\mathfrak{c}$. Therefore, by 3.3.17, $|\mathcal{B}_X| \leq \mathfrak{c}$. The result follows from the Schröder – Bernstein Theorem. ∎

Exercise 3.3.19 Let X and Y be uncountable Polish spaces. Show that the set of all Borel maps from X to Y is of cardinality $\mathfrak{c}$.

Exercise 3.3.20 Let X be a Polish space, $A \subseteq X$, and $f : A \longrightarrow A$ a Borel isomorphism. Show that f can be extended to a Borel isomorphism $g : X \longrightarrow X$.

Exercise 3.3.21 Let X be an uncountable Polish space. Give an example of a map $f : X \longrightarrow \mathbb{R}$ such that there is no Borel $g : X \longrightarrow \mathbb{R}$ satisfying $g(x) \leq f(x)$ for all x.

Theorem 3.3.22 *(Ramsey – Mackey theorem) Suppose* $(X, \mathcal{B})$ *is a standard Borel space and* $f : X \longrightarrow X$ *a Borel isomorphism. Then there is a Polish topology* $\mathcal{T}$ *on* X *generating* $\mathcal{B}$ *and making* f *a homeomorphism.*

Proof. If X is countable, we equip X with the disrete topology, and the result follows. So, we assume that X is uncountable. By the Borel isomorphism theorem, there is a Polish topology $\mathcal{T}_0$ generating $\mathcal{B}$. Suppose for some $n \in \mathbb{N}$, a Polish topology $\mathcal{T}_n$ generating $\mathcal{B}$ has been defined. Let $\{B_i^n : i \in \mathbb{N}\}$ be a countable base for $(X, \mathcal{T}_n)$. Consider

$$\mathcal{D} = \{f(B_i^n) : i \in \mathbb{N}\} \bigcup \{f^{-1}(B_i^n) : i \in \mathbb{N}\}.$$

By 3.2.5, there is a Polish topology $\mathcal{T}_{n+1}$ finer than $\mathcal{T}_n$ making each element of $\mathcal{D}$ open. Now take $\mathcal{T}$ to be the topology generated by $\bigcup \mathcal{T}_n$. By Observation 2 following 3.2.4, $\mathcal{T}$ is Polish. A routine argument now completes the proof. ∎

3.4 Measures

Let X be a nonempty set and $\mathcal{A}$ an algebra on X. A **measure** on $\mathcal{A}$ is a map $\mu : \mathcal{A} \longrightarrow [0, \infty]$ such that

(i) $\mu(\emptyset) = 0$; and

(ii) μ **is countably additive**; i.e., if $A_0, A_1, A_2, \ldots$ are pairwise disjoint sets in $\mathcal{A}$ such that $\bigcup_n A_n \in \mathcal{A}$, then $\mu(\bigcup_n A_n) = \sum_0^\infty \mu(A_n)$.

When $\mathcal{A}$ is understood from the context, we shall simply say that μ is a measure on X.

The measure μ is called **finite** if $\mu(X) < \infty$; it is σ**-finite** if X can be written as a countable union of sets in $\mathcal{A}$ of finite measure. It is called a **probability measure** if $\mu(X) = 1$. A **measure space** is a triple $(X, \mathcal{A}, \mu)$ where $\mathcal{A}$ is a σ-algebra on X and μ a measure; it is called a **probabilty space** if μ is a probability. **Finite measure spaces** and σ**-finite measure spaces** are analogously defined.

Example 3.4.1 Let X be uncountable and $\mathcal{A}$ the countable-cocountable σ-algebra. For $A \in \mathcal{A}$, let

$$\mu(A) = \begin{cases} 1 & \text{if } A \text{ is uncountable,} \\ 0 & \text{otherwise.} \end{cases}$$

Then μ is a measure on $\mathcal{A}$.

Example 3.4.2 Let $(X, \mathcal{A})$ be a measurable space and $x \in X$. For $A \in \mathcal{A}$, let

$$\delta_x(A) = \begin{cases} 1 & \text{if } x \in A, \\ 0 & \text{otherwise.} \end{cases}$$

Then δ_x is a measure on $\mathcal{A}$, called the **Dirac measure** at x.

Example 3.4.3 Let X be a finite set with n elements ($n > 0$) and $\mathcal{A} = \mathcal{P}(X)$. The **uniform measure** on X is the measure μ on $\mathcal{A}$ such that $\mu(\{x\}) = 1/n$ for every $x \in X$.

Example 3.4.4 Let X be a nonempty set. For $A \subseteq X$, let $\mu(A)$ denote the number of elements in A. ($\mu(A)$ is ∞ if A is infinite.) Then μ is a measure on $\mathcal{P}(X)$, called the **counting measure.**

Let $(X, \mathcal{A}, \mu)$ be a measure space. The following are easy to check.

(i) μ **is monotone**: If A and B are measurable sets with $A \subseteq B$, then $\mu(A) \leq \mu(B)$.

(ii) μ **is countably subadditive**: For any sequence (A_n) of measurable sets,

$$\mu(\bigcup_n A_n) \leq \sum_0^\infty \mu(A_n).$$

(iii) If the A_n's are measurable and nondecreasing, then

$$\mu(\bigcup_n A_n) = \lim \mu(A_n).$$

(iv) If μ is finite and (A_n) a nonincreasing sequence of measurable sets, then

$$\mu(\bigcap_n A_n) = \lim \mu(A_n).$$

Lemma 3.4.5 *Let $(X, \mathcal{B})$ be a measurable space and $\mathcal{A}$ an algebra such that $\sigma(\mathcal{A}) = \mathcal{B}$. Suppose μ_1 and μ_2 are finite measures on $(X, \mathcal{B})$ such that $\mu_1(A) = \mu_2(A)$ for every $A \in \mathcal{A}$. Then $\mu_1(A) = \mu_2(A)$ for every $A \in \mathcal{B}$.*

Proof. Let

$$\mathcal{M} = \{A \in \mathcal{B} : \mu_1(A) = \mu_2(A)\}.$$

By our hypothesis $\mathcal{A} \subseteq \mathcal{M}$. By (iii) and (iv) above, $\mathcal{M}$ is a monotone class. The result follows from 3.1.14. ■

The following is a standard result from measure theory. Its proof can be found in any textbook on the subject.

Theorem 3.4.6 *Let $\mathcal{A}$ be an algebra on X and μ a σ-finite measure on $\mathcal{A}$. Then there is a unique measure ν on $\sigma(\mathcal{A})$ that extends μ.*

Example 3.4.7 Let $\mathcal{A}$ be the algebra on $\mathbb{R}$ consisting of finite disjoint unions of nondegenerate intervals (3.1.4). For any interval I, let $|I|$ denote the length of I. Let $I_0, I_1, \ldots, I_n$ be pairwise disjoint intervals and $A = \bigcup_{k=0}^n I_k$. Set

$$\lambda(A) = \sum_{k=0}^n |I_k|.$$

Then λ is a σ-finite measure on $\mathcal{A}$. By 3.4.6, there is a unique measure on $\sigma(\mathcal{A}) = \mathcal{B}_{\mathbb{R}}$ extending λ. We call this measure the **Lebesgue measure** on $\mathbb{R}$ and denote it by λ itself.

Example 3.4.8 Let $(X, \mathcal{A}, \mu)$ and $(Y, \mathcal{B}, \nu)$ be σ-finite measure spaces. Let $Z = X \times Y$ and let $\mathcal{D}$ be the algebra of finite disjoint unions of measurable rectangles (3.1.5). Let $\mu \times \nu$ be the finitely additive measure on $\mathcal{D}$ satisfying

$$\mu \times \nu(A \times B) = \mu(A) \cdot \nu(B).$$

Then $\mu \times \nu$ is countably additive. (Show this.) By 3.4.6, there is a unique measure extending $\mu \times \nu$ to $\sigma(\mathcal{D}) = \mathcal{A} \bigotimes \mathcal{B}$. We call this extension the **product measure** and denote it by $\mu \times \nu$ itself. Similarly we define the product of finitely many σ-finite measures.

Example 3.4.9 Let $(X_n, \mathcal{A}_n, \mu_n)$, $n \in \mathbb{N}$, be a sequence of probability spaces and $X = \prod_n X_n$. For any nonempty, finite $F \subseteq \mathbb{N}$, let $\pi_F : X \longrightarrow \prod_{n \in F} X_n$ be the canonical projection map. Define

$$\mathcal{A} = \{\pi_F^{-1}(R) : R \in \bigotimes_{n \in F} \mathcal{A}_n, F \text{ finite}\}.$$

Then $\mathcal{A}$ is an algebra that generates the product σ-algebra $\bigotimes_n \mathcal{A}_n$. Define $\prod_n \mu_n$ on $\mathcal{A}$ by

$$\prod_n \mu_n(\pi_F^{-1}(R)) = (\times_{i\in F}\mu_i)(R).$$

Then $\prod_n \mu_n$ defines a probability on $\mathcal{A}$. By 3.4.6, there is a unique probability on $\bigotimes_n \mathcal{A}_n$ that extends $\prod_n \mu_n$. We call it the **product** of the μ_n's and denote it by $\prod_n \mu_n$. If $(X_n, \mathcal{A}_n, \mu_n)$ are the same, say $\mu_n = \mu$ for all n, then we shall denote the product measure simply by $\mu^{\mathbf{N}}$.

Example 3.4.10 Let μ be the uniform probability measure on $\{0,1\}$. We call the product measure $\mu^{\mathbf{N}}$ on C the **Lebesgue measure** on C and denote it by λ.

Let $(X, \mathcal{A}, \mu)$ be a measure space. A subset A of X is called **μ-null** or simply **null** if there is a measurable set B containing A such that $\mu(B) = 0$. The measure space $(X, \mathcal{A}, \mu)$ is called **complete** if every null set is measurable. The counting measure and the uniform measure on a finite set are complete.

An **ideal** on a nonempty set X is a nonempty family $\mathcal{I}$ of subsets of X such that

(i) $X \notin \mathcal{I}$,

(ii) whenever $A \in \mathcal{I}$, $\mathcal{P}(A) \subseteq \mathcal{I}$, and

(iii) I is closed under finite unions.

A **σ-ideal** is an ideal closed under countable unions. Let $\mathcal{I}$ be a nonempty family of subsets of X such that $X \notin \mathcal{I}$, and let

$$\mathcal{J} = \{A \subseteq X : A \subseteq \bigcup_n B_n, B_n \in \mathcal{I}\}.$$

Then $\mathcal{J}$ is the smallest σ-ideal containing $\mathcal{I}$. We call it the σ-ideal generated by $\mathcal{I}$.

Let $(X, \mathcal{A}, \mu)$ be a measure space and $\mathcal{N}_\mu$ the family of all μ-null sets. Then $\mathcal{N}_\mu$ is a σ-ideal. The σ-algebra generated by $\mathcal{A} \bigcup \mathcal{N}_\mu$ is called the **μ-completion** or simply the **completion** of the measure space X. We denote it by $\overline{\mathcal{A}}^\mu$. Sets in $\overline{\mathcal{A}}^\mu$ are called **μ-measurable.**

Exercise 3.4.11 Show that $\overline{\mathcal{A}}^\mu$ consists of all sets of the form $A \Delta N$ where $A \in \mathcal{A}$ and N is null. Further, $\overline{\mu}(A \Delta N) = \mu(A)$ defines a measure on the completion.

Exercise 3.4.12 Show that a set A is μ-measurable if and only if there exist measurable sets B and C such that $B \subseteq A \subseteq C$ and $C \setminus B$ is null.

An **outer measure** μ^* on a set X is a countably subadditive, monotone set function $\mu^* : \mathcal{P}(X) \longrightarrow [0, \infty]$ such that $\mu^*(\emptyset) = 0$.

Example 3.4.13 Let $(X, \mathcal{A}, \mu)$ be a σ-finite measure space. Define $\mu^* : \mathcal{P}(X) \longrightarrow [0, \infty]$ by

$$\mu^*(A) = \inf\{\mu(B) : B \in \mathcal{A} \,\&\, A \subseteq B\}.$$

It is routine to check that μ^* is an outer measure on X. The set function μ^* is called the **outer measure induced by** μ. Clearly, for every set A there is a set $B \in \mathcal{A}$ such that $A \subseteq B$ and $\mu(B) = \mu^*(A)$. Note that if B' is another measurable set containing A then $B \setminus B'$ is null.

Lemma 3.4.14 *Let X be a metrizable space and μ a finite measure on X. Then μ is* **regular**; *i.e., for every Borel set B,*

$$\begin{aligned} \mu(B) &= \sup\{\mu(F) : F \subseteq B, F \text{ closed}\} \\ &= \inf\{\mu(U) : U \supseteq B, U \text{ open}\}. \end{aligned}$$

Proof. Consider the class $\mathcal{D}$ of all sets B satisfying the above conditions. We show that $\mathcal{D} = \mathcal{B}_X$. Let B be closed. Therefore, it is a G_δ set. Write $B = \bigcap_n U_n$, the U_n's open and nonincreasing. Since μ is finite,

$$\mu(B) = \inf \mu(U_n) = \lim \mu(U_n).$$

Thus every closed set has the above property. $\mathcal{D}$ is clearly closed under complementation.

Now let $B_0, B_1, B_2, \ldots$ belong to $\mathcal{D}$, and $B = \bigcup_n B_n$. Fix $\epsilon > 0$. Choose N such that $\mu(B \setminus \bigcup_{i \le N} B_i) < \epsilon/2$. For each $0 \le i \le N$, there is a closed set $F_i \subseteq B_i$ such that $\mu(B_i \setminus F_i) < \epsilon/(2(N+1))$. It is easy to check that $\mu(B \setminus \bigcup_{i \le N} F_i) < \epsilon$.

To show the other equality, choose closed sets $F_i \subseteq B_i^c$ such that $\mu(B_i^c \setminus F_i) < \epsilon/2^{i+1}$. As $B^c \setminus \bigcap F_i \subseteq \bigcup(B_i^c \setminus F_i)$, it follows that $\mu(B^c \setminus \bigcap F_i) < \epsilon$. Take $U = (\bigcap F_i)^c$. Then U is an open set containing B such that $\mu(U \setminus B) < \epsilon$. It follows that $\mathcal{D}$ is closed under countable unions too. The result follows. ■

Sets in $\overline{\mathcal{B}_{\mathbb{R}}}^{\lambda}$, λ being the Lebesgue measure on reals, are called **Lebesgue measurable**. It is easy to see that the Cantor ternary set $\mathbf{C}$ is null with respect to the Lebesgue measure. So, every subset of $\mathbf{C}$, and there are $2^{\mathfrak{c}}$ of them, is Lebesgue measurable. As $|\mathcal{B}_{\mathbb{R}}| = \mathfrak{c} < 2^{\mathfrak{c}}$, there are Lebesgue measurable sets that are not Borel.

Remark 3.4.15 Let A be a Bernstein set (3.2.8). We claim that A is not Lebesgue measurable. Suppose not. We shall get a contradiction. Clearly, both A and $\mathbb{R} \setminus A$ cannot be null. Without any loss of generality, let A be not null. So, A contains an uncountable Borel set, and hence an uncountable closed set (3.2.7). We have arrived at a contradiction.

Exercise 3.4.16 Show the following.

(i) The Lebesgue measure on $\mathbb{R}$ is translation invariant; i.e., for every Lebesgue measurable set E and every real number x,

$$\lambda(E) = \lambda(E + x),$$

where $E + x = \{y + x : y \in E\}$.

(ii) For every Lebesgue measurable set E, the map $x \dashrightarrow \lambda(E \bigcap (E + x))$ is continuous.

(Hint: Use the monotone class theorem (3.1.14).)

Theorem 3.4.17 *If $E \subseteq \mathbb{R}$ is a Lebesgue measurable set of positive Lebesgue measure, then the set*

$$E - E = \{x - y : x, y \in E\}$$

is a neighborhood of 0.

Proof. By 3.4.16 (ii), the function $f(x) = \lambda(E \bigcap (E + x))$, $x \in \mathbb{R}$, is continuous. Since $f(0) = \lambda(E) > 0$, there is a nonempty open interval $(-a, a)$ such that $f(x) > 0$ for every $x \in (-a, a)$. In particular, $E \bigcap (E + x) \neq \emptyset$ for every $x \in (-a, a)$. It follows that $(-a, a) \subseteq E - E$. ∎

Using the above theorem, below we give another proof of the existence of a non-Lebesgue measurable set.

Example 3.4.18 Let G be the additive group $\mathbb{R}$ of real numbers, $\mathbb{Q}$ the subgroup of rationals, and Π the partition of $\mathbb{R}$ consisting of all the cosets of $\mathbb{Q}$. The partition Π is known as the **Vitali partition**. By **AC**, there exists a set S intersecting each coset in exactly one point. We claim that S is not Lebesgue measurable. Suppose not. Two cases arise. Either $\lambda(S) = 0$ or $\lambda(S) > 0$. Assume first that $\lambda(S) = 0$. Then, as $\mathbb{R} = \bigcup_{r \in \mathbb{Q}}(r + S)$, $\lambda(\mathbb{R}) = 0$, which is a contradiction. Now, let $\lambda(S) > 0$. By 3.4.17, $S - S$ contains a nonempty open interval. Hence, there are distinct points x, y in S such that $x - y$ is rational. We have arrived at a contradiction again.

It should be remarked that we have used **AC** to show the existence of non-Lebegue measurable sets. In a significant contribution to the theory, Solovay([110] or [9]) gave a model of **ZF** + $\neg$**AC** where every subset of the reals is Lebesgue measurable.

A **Borel measure** is a measure on some standard Borel space.

Theorem 3.4.19 *Let X be a Polish space, μ a finite Borel measure on X, and $\epsilon > 0$. Then there is a compact set K such that $\mu(X \setminus K) < \epsilon$.*

Proof. Fix a compatible complete metric $d \leq 1$ on X. Take a regular system $\{F_s : s \in \mathbb{N}^{<\mathbb{N}}\}$ of nonempty closed sets such that

(i) $F_e = X$,

(ii) $F_s = \bigcup_n F_{s^\frown n}$, and

(iii) diameter$(F_s) \leq 1/2^{|s|}$.

To see that such a system exists, we proceed by induction on $|s|$. Suppose F_s has been defined. Since X is second countable, there is a sequence (U_n) of open sets of diameter $\leq 2^{-|s|}$ covering F_s, and further, $F_s \bigcap U_n \neq \emptyset$ for all n. Take $F_{s^\frown n} = \text{cl}(F_s \bigcap U_n)$.

By an easy induction, we now define positive integers $n_0, n_1, n_2, \ldots$ such that the following conditions hold: for every $s = (m_0, m_1, \ldots, m_{k-1})$ with $m_i \leq n_i$,

$$\mu(F_s \setminus \bigcup_{j \leq n_k} F_{s^\frown j}) < \frac{\epsilon}{2^{k+1}.n_0.\cdots.n_{k-1}}.$$

Set

$$K = \bigcap_k \bigcup_s F_s,$$

where the union varies over all $s = (m_0, m_1, \ldots, m_{k-1})$ with $m_i \leq n_i$. It is easy to check that K is closed and totally bounded and hence compact. Further, $\mu(X \setminus K) < \epsilon$. ∎

Theorem 3.4.20 *Let $(X, \mathcal{T})$ be a Polish space and μ a finite Borel measure on X. Then for every Borel subset B of X and every $\epsilon > 0$, there is a compact $K \subseteq B$ such that $\mu(B \setminus K) < \epsilon$.*

Proof. By 3.2.4, there is a Polish topology $\mathcal{T}_B$ on X finer than $\mathcal{T}$ generating the same Borel σ-algebra such that B is clopen with respect to $\mathcal{T}_B$. By 3.4.19, there is a compact set K relative to $\mathcal{T}_B$ contained in B such that $\mu(B \setminus K) < \epsilon$. Since K is compact with respect to the original topology too, the result follows. ∎

Let μ be a probability on $I = [0, 1]$. Define

$$F(x) = \mu([0, x]), \ x \in I.$$

The function F is called the **distribution function** of μ. It is a monotonically increasing, right-continuous function such that $F(1) = 1$.

Exercise 3.4.21 Show that a monotonically increasing, right-continuous $F : [0, 1] \longrightarrow [0, 1]$ with $F(1) = 1$ is the distribution function of a probability on $[0, 1]$.

A measure μ on a standard Borel space X is called **continuous** if $\mu(\{x\}) = 0$ for every $x \in X$.

Exercise 3.4.22 Show that a probability on $[0, 1]$ is continuous if and only if its distribution function is continuous.

Theorem 3.4.23 *(The isomorphism theorem for measure spaces) If μ is a continuous probability on a standard Borel space X, then there is a Borel isomorphism $h : X \longrightarrow I$ such that for every Borel subset B of I, $\lambda(B) = \mu(h^{-1}(B))$.*

Proof. By the Borel isomorphism theorem (3.3.13), we can assume that $X = I$. Let $F : I \to I$ be the distribution function of μ. So, F is a continuous, nondecreasing map with $F(0) = 0$ and $F(1) = 1$. Let

$$N = \{y \in I : F^{-1}(\{y\}) \text{ contains more than one point}\}.$$

Since F is monotone, N is countable. If N is empty, take $h = F$. Otherwise, we take an uncountable Borel set $M \subset I \setminus N$ of Lebesgue measure 0, e.g., $\mathcal{C} \setminus N$. So, $\mu(F^{-1}(M)) = 0$. Put $Q = M \bigcup N$ and $P = F^{-1}(Q)$. Both P and Q are uncountable Borel sets with $\mu(P) = \lambda(Q) = 0$. Fix a Borel isomorphism $g : P \longrightarrow Q$. Define

$$h(x) = \begin{cases} g(x) & \text{if } x \in P, \\ F(x) & \text{if } x \in I \setminus P. \end{cases}$$

The map h has the desired properties. ■

Let $(X, \mathcal{A})$ be a measurable space and Y a second countable metrizable space. A **transition probability** on $X \times Y$ is a map $P : X \times \mathcal{B}_Y \longrightarrow [0,1]$ such that

(i) for every $x \in X$, $P(x, .)$ is a probability on Y, and

(ii) for every $B \in \mathcal{B}_Y$, the map $x \longrightarrow P(x, B)$ is measurable.

Proposition 3.4.24 *Let X, Y, and P be as above. Then for every $A \in \mathcal{A} \bigotimes \mathcal{B}_Y$, the map $x \longrightarrow P(x, A_x)$ is measurable.*

In particular, for every $A \in \mathcal{A} \bigotimes \mathcal{B}_Y$ such that $P(x, A_x) > 0$, $\pi_X(A)$ is measurable.

Proof. Let

$$\mathcal{B} = \{A \in \mathcal{A} \bigotimes \mathcal{B}_Y : \text{ the map } x \longrightarrow P(x, A_x) \text{ is measurable}\}.$$

It is obvious that $\mathcal{B}$ contains all the measurable rectangles and is closed under finite disjoint unions. Clearly, $\mathcal{B}$ is a monotone class. As finite disjoint unions of measurable rectangles form an algebra generating $\mathcal{A} \bigotimes \mathcal{B}_Y$, the result follows from 3.1.14. ■

3.5 Category

Let X be a topological space. A subset A of X is said to have the **Baire property** (in short **BP**) if there is an open set U such that the symmetric difference $A \Delta U$ is of first category in X. Clearly, open sets and meager sets have BP.

Proposition 3.5.1 *The collection $\mathcal{D}$ of all subsets of a topological space X having the Baire property forms a σ-algebra.*

Proof. Closure under countable unions: Let $A_0, A_1, A_2, \ldots$ belong to $\mathcal{D}$. Take open sets $U_0, U_1, U_2, \ldots$ such that $A_n \Delta U_n$ is meager for each n. Since

$$(\bigcup_n A_n)\Delta(\bigcup_n U_n) \subseteq \bigcup_n (A_n \Delta U_n)$$

and the union of a sequence of meager sets is meager, $\bigcup_n A_n \in \mathcal{D}$.

Closure under complementation: Let $A \in \mathcal{D}$ and let U be an open set such that $A\Delta U$ is meager. We have

$$\begin{aligned}&(X \setminus A)\Delta \operatorname{int}(X \setminus U)\\ &\subseteq ((X \setminus A)\Delta(X \setminus U))\bigcup((X \setminus U) \setminus \operatorname{Int}(X \setminus U)).\end{aligned}$$

As $(X \setminus A)\Delta(X \setminus U) = A\Delta U$,

$$(X \setminus A)\Delta \operatorname{int}(X \setminus U) \subseteq (A\Delta U)\bigcup((X \setminus U) \setminus \operatorname{Int}(X \setminus U)).$$

Since for any closed set F, $F \setminus \operatorname{int}(F)$ is nowhere dense, $(X \setminus A)\Delta \operatorname{int}(X \setminus U)$ is meager.

The result follows. ■

The σ-algebra $\mathcal{D}$ defined above is called the **Baire σ-algebra** of X.

Corollary 3.5.2 *Every Borel subset of a metrizable space has the Baire property.*

The Cantor ternary set is nowhere dense and so are all its subsets. Therefore, there are subsets of reals with BP that are not Borel. Since every meager set is contained in a meager F_σ set, every nonmeager set with BP contains a nonmeager G_δ set. Hence, a Bernstein set does not have the Baire property. We cannot show the existence of a subset of the reals not having the Baire property without **AC**. In fact, in Solovay'smodel mentioned in the last section, every subset of the reals has the Baire property.

The Lebesgue σ-algebra on $\mathbb{R}$ is the smallest σ-algebra containing all open sets and all null sets. Is every Lebesgue measurable set the symmetric difference of an open set and a null set? The answer is no.

Exercise 3.5.3 (a) Give an example of a dense G_δ subset of $\mathbb{R}$ of Lebesgue measure zero.

(b) For every $0 < r < 1$, construct a closed nowhere dense set $C \subseteq [0, 1]$ such that $\lambda(C) > r$.

A topological space X is called a **Baire space** if no nonempty open subset of X is of first category (in X or equivalently in itself). The following proposition is very simple to prove.

Proposition 3.5.4 *The following statements are equivalent.*

(i) *X is a Baire space.*

(ii) *Every comeager set in X is dense in X.*

(iii) *The intersection of countably many dense open sets in X is dense in X.*

Every open subset of a Baire space is clearly a Baire space. By 2.5.6, we see that every completely metrizable space is a Baire space. The converse need not be true.

Exercise 3.5.5 Give an example of a metrizable Baire space that is not completely metrizable. Also, show that a closed subspace of a Baire space need not be Baire

Here are some elementary but useful observations. Let X be a topological space, A and U subsets of X with U open. We say that A is **meager (nonmeager, comeager) in** U if $A \bigcap U$ is meager (respectively nonmeager, comeager) in U.

Proposition 3.5.6 *Let X be a second countable Baire space and (U_n) a countable base for X. Let U be an open set in X.*

(i) *For every sequence (A_n) of subsets of X, $\bigcap A_n$ is comeager in U if and only if A_n is comeager in U for each n.*

(ii) *Let $A \subseteq X$ be a nonmeager set with BP. Then A is comeager in U_n for some n.*

(iii) *A set A with BP is comeager if and only if A is nonmeager in each U_n.*

Proof. Suppose $\bigcap A_n$ is comeager in U. Then clearly each of A_n is comeager in U. Conversely, if each of A_n is comeager in U, then $U \setminus A_n$ is meager in U for all n. So, $\bigcup_n (U \setminus A_n) = U \setminus \bigcap_n A_n$ is meager in U. Thus we have proved (i).

To prove (ii), take A with BP. Write $A = V \Delta I$, V open, I meager. If A is nonmeager, V must be nonempty. Then A is comeager in every U_n contained in V.

We now prove (iii). Let A be comeager. Then trivially $U_n \setminus A$ is meager for all n. As U_n is open, it follows that $U_n \setminus A$ is meager in U_n. Since X is a Baire space, this implies that A is nonmeager in U_n. Conversely, let A be not comeager; i.e., A^c is not meager. So, there is U_n such that A^c is comeager in U_n; i.e., A is meager in U_n. ∎

Proposition 3.5.7 *A topological group is Baire if and only if it is of second category in itself.*

Proof. The "only if" part of the result is trivial. For the converse, let G be a topological group that is not Baire. Take a nonempty, meager, open set U. Then each $g \cdot U$ is open and meager, and $G = \bigcup g \cdot U$. By the Banach category theorem (2.5.16), G is meager. ■

Let X, Y be metrizable spaces. A function $f : X \longrightarrow Y$ is called **Baire measurable** if for every open subset U of Y, $f^{-1}(U)$ has BP.

(**Caution:** Baire measurable functions are not the same as Baire functions.) Clearly, every Borel function is Baire measurable.

Proposition 3.5.8 *Let Y be a second countable topological space and $f : X \longrightarrow Y$ Baire measurable. Then there is a comeager set A in X such that $f|A$ is continuous.*

Proof. Take a countable base (V_n) for Y. Since f is Baire measurable, for each n there is a meager set I_n in X such that $f^{-1}(V_n)\Delta I_n$ is open. Let $I = \bigcup_n I_n$. Plainly, $f|(X \setminus I)$ is continuous. ■

Proposition 3.5.9 *Let G be a completely metrizable group and H a second countable group. Then every Baire measurable homomorphism $\varphi : G \longrightarrow H$ is continuous. In particular, every Borel homomorphism $\varphi : G \longrightarrow H$ is continuous.*

Proof. By 3.5.8, there is a meager set I in G such that $\varphi|(G \setminus I)$ is continuous. Now take any sequence (g_k) in G converging to an element g. Let

$$J = (g^{-1} \cdot I) \bigcup \bigcup_k (g_k^{-1} \cdot I).$$

By 2.4.7, J is meager. Since G is completely metrizable, it is of second category in itself by 2.5.6. In particular, $J \neq G$. Take any $h \in G \setminus J$. Then, $g_k \cdot h$, $g \cdot h$ are all in $G \setminus I$. Further, $g_k \cdot h \to g \cdot h$ as $k \to \infty$. Since $\varphi|(G \setminus I)$ is continuous, $\varphi(g_k \cdot h) \to \varphi(g \cdot h)$; i.e., $\varphi(g_k) \cdot \varphi(h) \to \varphi(g) \cdot \varphi(h)$. Multiplying by $(\varphi(h))^{-1}$ from the right, we have $\varphi(g_k) \to \varphi(g)$. ■

The following example shows that the above result need not be true if G is not completely metrizable.

Example 3.5.10 Let $\mathbb{Q}^+$ be the multiplicative group of positive rational numbers and $\varphi : \mathbb{Q}^+ \longrightarrow \mathbb{Z}$ the homomorphism satisfying $\varphi(p) = 0$ for primes $p > 2$ and $\varphi(2) = 1$. Since $\mathbb{Q}^+$ is countable, φ is trivially Borel. It is not continuous. To see this, take $q_n = 1 - 2^{-n}$. Then $\varphi(q_n) = -n$. As q_n converges and $\varphi(q_n)$ does not, φ is not continuous.

Exercise 3.5.11 Show that for every Baire measurable homomorphism $f : (\mathbb{R}, +) \longrightarrow (\mathbb{R}, +)$ there is a constant a such that $f(x) = ax$. Also, show that there is a discontinuous homomorphism $f : (\mathbb{R}, +) \longrightarrow (\mathbb{R}, +)$.

Theorem 3.5.12 *(Pettis theorem) Let G be a Baire topological group and H a nonmeager subset with BP. Then there is a neighborhood V of the identity contained in $H^{-1}H$.*

Proof. Since H is nonmeager with BP, there is a nonempty open set U such that $H\Delta U$ is meager. Let $g \in U$. Choose a neighborhood V of the identity such that $gVV^{-1} \subseteq U$. We show that for every $h \in V$, $H\bigcap Hh$ is nonmeager, in particular, nonempty. It will then follow that $V \subseteq H^{-1}H$, and the proof will be complete.

Let $h \in H$. Note that

$$(U\bigcap Uh)\Delta(H\bigcap Hh) \subseteq (U\Delta H)\bigcup((U\Delta H)h). \qquad (*)$$

So, $(U\bigcap Uh)\Delta(H\bigcap Hh)$ is meager. As $gV \subseteq U\bigcap Uh$ and G is Baire, $U\bigcap Uh$ is nonmeager. Therefore, $H\bigcap Hh$ is nonmeager by ($\star$). ■

Corollary 3.5.13 *Every nonmeager Borel subgroup H of a Polish group G is clopen.*

Proof. Let H be a Borel subgroup of G that is not meager. By 3.5.12, H contains a neighborhood of the identity. Hence, H is open. Since H^c is the union of the remaining cosets of H, which are all open, it is open too. ■

We now present a very useful result known as the Kuratowski – Ulam theorem.

For $E \subset X \times Y$, $x \in X$, and $y \in Y$, we set

$$E_x = \{y \in Y : (x, y) \in E\}$$

and

$$E^y = \{x \in X : (x, y) \in E\}.$$

Lemma 3.5.14 *Let X be a Baire space and Y second countable. Suppose $A \subseteq X \times Y$ is a closed, nowhere dense set. Then*

$$\{x \in X : A_x \textit{ is nowhere dense}\}$$

is a dense G_δ set.

Proof. Take any $A \subseteq X \times Y$, closed and nowhere dense. Fix a countable base (V_n) for Y. Let $U = A^c$. Then U is dense and open. Let

$$W_n = \{x \in X : U_x \bigcap V_n \neq \emptyset\}.$$

As

$$W_n = \pi_X(U \bigcap(X \times V_n)),$$

it is open. Also, W_n is dense. Suppose not. Then $(X \setminus \text{cl}(W_n)) \times V_n$ is a nonempty open set disjoint from U. As U is dense, this is a contradiction.

Since for any $x \in X$,

$$A_x \text{ is nowhere dense} \Longleftrightarrow U_x \text{ is dense},$$

it follows that

$$\{x \in X : A_x \text{ is nowhere dense}\} = \bigcap_n W_n.$$

Since X is a Baire space, the result follows. ■

Let X be a nonempty set, Y a topological space; $A \subset X \times Y$; and U nonempty, open in Y. We set

$$A^{\Delta U} = \{x \in X : A_x \text{ is nonmeager in } U\},$$

and

$$A^{*U} = \{x \in X : A_x \text{ is comeager in } U\}.$$

Lemma 3.5.15 *Let X be a Baire space, Y second countable, and suppose $A \subseteq X \times Y$ has BP. The following statements are equivalent.*

(i) *A is meager.*

(ii) *$\{x \in X : A_x$ is meager$\}$ is comeager.*

Proof. (ii) follows from (i) by 3.5.14. Now assume that A is nonmeager. Since A has BP, there exist nonempty open sets U and V in X and Y respectively such that A is comeager in $U \times V$. Therefore, from what we have just proved, A^{*V} is comeager in U. Since U is nonmeager, A^{*V} is nonmeager. In particular, $A^{\Delta X}$ is not meager; i.e., (ii) is false. ■

The following result follows from 3.5.15.

Theorem 3.5.16 *(Kuratowski - Ulam theorem) Let X, Y be second countable Baire spaces and suppose $A \subseteq X \times Y$ has the Baire property. The following are equivalent .*

(i) *A is meager (comeager).*

(ii) *$\{x \in X : A_x$ is meager (comeager)$\}$ is comeager.*

(iii) *$\{y \in Y : A^y$ is meager (comeager)$\}$ is comeager.*

Exercise 3.5.17 Let X, Y, and A be as above. Show that the sets

$$\{x : A_x \text{ has BP}\}$$

and

$$\{y : A^y \text{ has BP}\}$$

are comeager.

Proposition 3.5.18 *Let $(X, \mathcal{A})$ be a measurable space and Y a Polish space. For every $A \in \mathcal{A} \bigotimes \mathcal{B}_Y$ and U open in Y, the sets $A^{\Delta U}$, A^{*U}, and $\{x \in X : A_x$ is meager in $U\}$ are in $\mathcal{A}$.*

Proof. Fix a countable base (U_n) for Y.
Step 1. Let

$$\mathcal{B} = \{A \subseteq X \times Y : A^{\Delta U} \in \mathcal{A} \text{ for all open } U\}.$$

We show that $\mathcal{A} \bigotimes \mathcal{B}_Y \subseteq \mathcal{B}$.

Let $A = B \times V$, $B \in \mathcal{A}$, and V open in Y. Then $A^{\Delta U}$ equals B if $U \bigcap V \neq \emptyset$. Otherwise it is empty. Hence, $A \in \mathcal{B}$.

Our proof will be complete if we show that $\mathcal{B}$ is closed under countable unions and complementation.

For every sequence (A_n) of subsets $X \times Y$,

$$(\bigcup_n A_n)^{\Delta U} = \bigcup_n A_n^{\Delta U}.$$

So, $\mathcal{B}$ is closed under countable unions.

Let $A \in \mathcal{B}$ and U open in Y. Let $x \in X$. We have

$$\begin{aligned}(A^c)_x \text{ is meager in } U &\iff A_x \text{ is comeager in } U\\ &\iff \forall U_n \subseteq U (A_x \text{ is nonmeager in } U_n).\end{aligned}$$

Therefore,

$$(A^c)^{\Delta U} = (\bigcap_{U_n \subseteq U} A^{\Delta U_n})^c.$$

Hence, $A^c \in \mathcal{B}$.
Step 2. Let $A \in \mathcal{A} \bigotimes \mathcal{B}_Y$ and U be open in Y. Then

$$A^{*U} = \bigcap_{U_n \subseteq U} A^{\Delta U_n}.$$

Therefore, $A^{*U} \in \mathcal{A}$ by step 1. The remaining part of the result follows easily. ∎

Exercise 3.5.19 Let $(G, \cdot)$ be a group, X a set, and $a : G \times X \longrightarrow X$ any map. For notational convenience we shall write $g \cdot x$ for $a(g, x)$. We call the map $g \cdot x$ an **action of G on X** if

(i) $e \cdot x = x$, and

(ii) $g \cdot (h \cdot x) = (g \cdot h) \cdot x$,

where e denotes the identity element, $g, h \in G$, and $x \in X$.

Let G be a Polish group acting continuously on a Polish space X. For any $W \subseteq X$ and any nonempty open $U \subseteq G$, define the **Vaught transforms**

$$W^{\Delta U} = \{x \in X : \{g \in U : g \cdot x \in W\} \text{ is nonmeager}\}$$

and

$$W^{*U} = \{x \in X : \{g \in U : g \cdot x \in W\} \text{ is comeager in } U\}.$$

(We shall write simply W^{Δ} and W^* instead of $W^{\Delta G}$ and W^{*G} respectively.) Show the following:

(i) W^{Δ} is invariant.

(ii) W is invariant implies $W = W^{\Delta}$.

(iii) $(\bigcup_n W_n)^{\Delta} = \bigcup_n (W_n^{\Delta})$.

(iv) If $W \subseteq X$ is Borel and $U \subseteq G$ open, then $W^{\Delta U}$ and W^{*U} are Borel.

(Hint: Consider

$$\tilde{W} = \{(x, g) \in X \times G : g \cdot x \in W\}$$

and apply 3.5.18.)

We close this section by showing that the Baire σ-algebra and the Lebesgue σ-algebra are closed under the Souslin operation. The proof presented here is due to Marczewski[113] and proves a much more general result. Call a σ-algebra $\mathcal{B}$ on X **Marczewski complete** if for every $A \subseteq X$ there exists $\hat{A} \in \mathcal{B}$ containing A such that for every B in $\mathcal{B}$ containing A, every subset of $\hat{A} \setminus B$ is in $\mathcal{B}$. Such a set $\hat{A}$ will be called a **minimal $\mathcal{B}$-cover** of A.

Example 3.5.20 Every σ-finite complete measure space is Marczewski complete. We prove this now. Let $(X, \mathcal{B}, \mu)$ be a σ-finite complete measure space. First assume that $\mu^*(A) < \infty$. Take $\hat{A}$ to be a measurable set containing A with $\mu^*(A) = \mu(\hat{A})$. In the general case, write $A = \bigcup A_n$ such that $\mu^*(A_n) < \infty$. Since μ is σ-finite, this is possible. Take $\hat{A} = \bigcup \hat{A}_n$.

Next we show that the Baire σ-algebra of any topological space is Marczewski complete.

Example 3.5.21 Let X be a topological space and $A \subseteq X$. Take A^* to be the union of all open sets U such that A is comeager in U. We first show that $A^* \setminus A$ is meager. Let $\mathcal{U}$ be a maximal family of pairwise disjoint open sets U such that A is comeager in U. Let $W = \bigcup \mathcal{U}$. By the maximality of $\mathcal{U}$, $A^* \subseteq \text{cl}(W)$. By the Banach category therrem, A is comeager in W. Now note that

$$A^* \setminus A \subseteq (A^* \setminus W) \bigcup (W \setminus A) \subseteq (\text{cl}(W) \setminus W) \bigcup (W \setminus A).$$

This shows that $A^* \setminus A$ is meager. Let B be any meager F_σ set containing $A^* \setminus A$. Take $\hat{A} = A^* \bigcup B$.

Theorem 3.5.22 *(Marczewski) If $(X, \mathcal{B})$ is a measurable space with $\mathcal{B}$ Marczewski complete, then $\mathcal{B}$ is closed under the Souslin operation.*

Proof. Let $\{B_s : s \in \mathbb{N}^{<\mathbb{N}}\}$ be a system of sets in $\mathcal{B}$. We have to show that $B = \mathcal{A}(\{B_s\}) \in \mathcal{B}$. Without loss of generality we assume that the system $\{B_s\}$ is regular. For $s \in \mathbb{N}^{<\mathbb{N}}$, let

$$B^s = \bigcup_{\{\alpha : s \prec \alpha\}} \bigcap_n B_{\alpha|n} \subseteq B_s.$$

Note that $B^e = B$ and $B^s = \bigcup_n B^{s\hat{}n}$ for all s. For each $s \in \mathbb{N}^{<\mathbb{N}}$, choose a minimal $\mathcal{B}$-cover $\hat{B}^s$ of B^s. Since $B^s \subseteq B_s$, by replacing $\hat{B}^s$ by $B_s \bigcap \hat{B}^s$ we may assume that $\hat{B}^s \subseteq B_s$. Further, by replacing $\hat{B}^s$ by $\bigcap_{t \preceq s} \hat{B}^t$, we can assume that $\{\hat{B}^s : s \in \mathbb{N}^{<\mathbb{N}}\}$ is regular. Let

$$C_s = \hat{B}^s \setminus \bigcup_n \hat{B}^{s\hat{}n}.$$

Since $B^s = \bigcup_n B^{s\hat{}n} \subseteq \bigcup_n \hat{B}^{s\hat{}n}$, every subset of C_s is in $\mathcal{B}$. Let $C = \bigcup_s C_s$.

Claim: $\hat{B}^e \setminus C \subseteq B$.

Assuming the claim, we complete the proof as follows. Since $\hat{B}^e \setminus B \subseteq C$ and since every subset of C is in $\mathcal{B}$, it follows that $\hat{B}^e \setminus B \in \mathcal{B}$. As $B = \hat{B}^e \setminus (\hat{B}^e \setminus B)$, it belongs to $\mathcal{B}$.

Proof of the claim. Let $x \in \hat{B}^e \setminus C$. Since $x \notin C$, $x \notin C_e$. Since $x \in \hat{B}^e$, there is $\alpha(0) \in \mathbb{N}$ such that $x \in \hat{B}^{\alpha(0)}$. Suppose $n > 0$ and $\alpha(0), \alpha(1), \ldots, \alpha(n-1)$ have been defined such that $x \in \hat{B}^s$, where $s = (\alpha(0), \alpha(1), \ldots, \alpha(n-1))$. Since $x \notin C_s$, there is $\alpha(n) \in \mathbb{N}$ such that $x \in \hat{B}^{s\hat{}\alpha(n)}$. Since $\hat{B}^{\alpha|n} \subset B_{\alpha|n}$ for all n, we conclude that $x \in B$. ∎

3.6 Borel Pointclasses

We shall call a collection of pointsets—subsets of metrizable spaces—a **pointclass**; e.g., the pointclasses of Borel sets, closed sets, open sets. Let X be a metrizable space. For ordinals α, $1 \le \alpha < \omega_1$, we define the following pointclasses by transfinite induction:

$$\mathbf{\Sigma}^0_1(X) = \{U \subseteq X : U \text{ open}\},$$

$$\mathbf{\Pi}^0_1(X) = \{F \subseteq X : F \text{ closed}\};$$

for $1 < \alpha < \omega_1$,

$$\mathbf{\Sigma}^0_\alpha(X) = (\bigcup_{\beta<\alpha} \mathbf{\Pi}^0_\beta(X))_\sigma$$

and

$$\mathbf{\Pi}^0_\alpha(X) = (\bigcup_{\beta<\alpha} \mathbf{\Sigma}^0_\beta(X))_\delta.$$

Finally, for every $1 \le \alpha < \omega_1$,

$$\mathbf{\Delta}^0_\alpha(X) = \mathbf{\Sigma}^0_\alpha(X) \bigcap \mathbf{\Pi}^0_\alpha(X).$$

Note that $\mathbf{\Delta}_1^0(X)$ is the family of all clopen subsets of X; $\mathbf{\Sigma}_2^0(X)$ is the set of all F_σ subsets of X; and $\mathbf{\Pi}_2^0(X)$ is the set of all G_δ sets in X. Sets in $\mathbf{\Sigma}_3^0(X)$ are also called $G_{\delta\sigma}$ sets; those in $\mathbf{\Pi}_3^0(X)$ are called $F_{\sigma\delta}$ sets; etc. The families $\mathbf{\Sigma}_\alpha^0(X)$, $\mathbf{\Pi}_\alpha^0(X)$ and $\mathbf{\Delta}_\alpha^0(X)$ are called **additive, multiplicative, and ambiguous classes** respectively. If there is no ambiguity, or if a statement is true for all X, we sometimes write $\mathbf{\Sigma}_\alpha^0$, $\mathbf{\Pi}_\alpha^0$, and $\mathbf{\Delta}_\alpha^0$ instead of $\mathbf{\Sigma}_\alpha^0(X)$, $\mathbf{\Pi}_\alpha^0(X)$, and $\mathbf{\Delta}_\alpha^0(X)$. A set $A \in \mathbf{\Sigma}_\alpha^0$ is called an **additive class** α set. **Multiplicative class** α sets and **ambiguous class** α sets are similarly defined.

We record below a few elementary facts.

(i) Additive classes are closed under countable unions, and multiplicative ones under countable intersections.

(ii) All the additive, multiplicative, and ambiguous classes are closed under finite unions and finite intersections.

(iii) For all $1 \leq \alpha < \omega_1$,

$$\mathbf{\Sigma}_\alpha^0 = \neg\mathbf{\Pi}_\alpha^0 \text{ (equivalently, } \mathbf{\Pi}_\alpha^0 = \neg\mathbf{\Sigma}_\alpha^0).$$

(iv) For $\alpha \geq 1$, $\mathbf{\Delta}_\alpha^0$ is an algebra.

Proposition 3.6.1 (i) *For every* $1 \leq \alpha < \omega_1$,

$$\mathbf{\Sigma}_\alpha^0, \mathbf{\Pi}_\alpha^0 \subseteq \mathbf{\Delta}_{\alpha+1}^0.$$

Thus we have the following diagram, in which any pointclass is contained in every pointclass to the right of it:

$$\begin{array}{ccccccc} & \mathbf{\Sigma}_1^0 & & \mathbf{\Sigma}_2^0 & & \mathbf{\Sigma}_3^0 & \cdots \\ \mathbf{\Delta}_1^0 & & \mathbf{\Delta}_2^0 & & \mathbf{\Delta}_3^0 & \cdots & \\ & \mathbf{\Pi}_1^0 & & \mathbf{\Pi}_2^0 & & \mathbf{\Pi}_3^0 & \cdots \end{array}$$

(The Hierarchy of Borel Sets)

(ii) *For* $\alpha > 1$, $\mathbf{\Sigma}_\alpha^0 = (\mathbf{\Delta}_\alpha^0)_\sigma$ *and* $\mathbf{\Pi}_\alpha^0 = (\mathbf{\Delta}_\alpha^0)_\delta$. *For zero-dimensional separable metric spaces, this is also true for* $\alpha = 1$.

(iii) *Let* $\alpha < \omega_1$ *be a limit ordinal and* (α_n) *a sequence of ordinals such that* $\alpha = \sup \alpha_n$. *Then*

$$\mathbf{\Sigma}_\alpha^0 = (\bigcup_n \mathbf{\Pi}_{\alpha_n}^0)_\sigma$$

and

$$\mathbf{\Pi}_\alpha^0 = (\bigcup_n \mathbf{\Sigma}_{\alpha_n}^0)_\delta.$$

(iv) *For any metric space X,*

$$\mathcal{B}_X = \bigcup_{\alpha<\omega_1} \mathbf{\Sigma}^0_\alpha(X) = \bigcup_{\alpha<\omega_1} \mathbf{\Pi}^0_\alpha(X).$$

Proof. Since every closed (open) set in a metrizable space is a G_δ set (respectively an F_σ set), (i) is true for $\alpha = 1$. A simple transfinite induction argument completes the proof of (i) for all α.

(ii) Let $\alpha > 1$. By (i), $\mathbf{\Delta}^0_\alpha \supseteq \bigcup_{\beta<\alpha} \mathbf{\Pi}^0_\beta$. Therefore, $(\mathbf{\Delta}^0_\alpha)_\sigma \supseteq \mathbf{\Sigma}^0_\alpha$. Since $\mathbf{\Delta}^0_\alpha \subseteq \mathbf{\Sigma}^0_\alpha$ and $\mathbf{\Sigma}^0_\alpha$ is closed under countable unions, $(\mathbf{\Delta}^0_\alpha)_\sigma \subseteq \mathbf{\Sigma}^0_\alpha$. Thus, $(\mathbf{\Delta}^0_\alpha)_\sigma = \mathbf{\Sigma}^0_\alpha$. Similarly, we show that $(\mathbf{\Delta}^0_\alpha)_\delta = \mathbf{\Pi}^0_\alpha$.

Let X be zero-dimensional and $\alpha = 1$. Then $\mathbf{\Delta}^0_1$ is a base for X. If, moreover, X is second countable, then $(\mathbf{\Delta}^0_1)_\sigma = \mathbf{\Sigma}^0_1$, the family of all open sets. The remaining part of (ii) is seen easily now.

(iii) follows from (i) and (ii).

(iv) By induction on α, we see that for every $1 \le \alpha < \omega_1$, $\mathbf{\Sigma}^0_\alpha(X)$ and $\mathbf{\Pi}^0_\alpha(X)$ are contained in $\mathcal{B}_X$. To prove the other inclusions, set

$$\mathcal{B} = \bigcup_{\alpha<\omega_1} \mathbf{\Sigma}^0_\alpha(X).$$

Then

(a) $\mathcal{B}$ contains all open sets.

(b) If $B \in \mathbf{\Sigma}^0_\alpha$, $B^c \in \mathbf{\Pi}^0_\alpha \subseteq \mathbf{\Sigma}^0_{\alpha+1}$. So, $\mathcal{B}$ is closed under complementation.

(c) Let (B_n) be a sequence in $\mathcal{B}$. Choose $1 \le \alpha_n < \omega_1$ such that $B_n \in \mathbf{\Sigma}^0_{\alpha_n}$. Let $\alpha = \sup \alpha_n + 1$. Then $\bigcup_n B_n \in \mathbf{\Sigma}^0_{\alpha+1}$. So, $\mathcal{B}$ is closed under countable unions.

From (a) – (c), we get that $\mathcal{B} \subseteq \mathcal{B}_X$. Thus, $\mathcal{B}_X = \bigcup_{\alpha<\omega_1} \mathbf{\Sigma}^0_\alpha(X)$.

Similarly we show that $\mathcal{B}_X = \bigcup_{\alpha<\omega_1} \mathbf{\Pi}^0_\alpha(X)$. ■

In 3.6.8, we shall show that for any uncountable Polish space X, the inclusion in (i) is strict.

Corollary 3.6.2 *Let X be an infinite separable metric space.*

(i) *Show that for every α, $|\mathbf{\Sigma}^0_\alpha(X)| = |\mathbf{\Pi}^0_\alpha(X)| = \mathfrak{c}$.*

(ii) *Show that $|\mathcal{B}_X| = \mathfrak{c}$.*

Proposition 3.6.3 *Every set of additive class $\alpha > 2$ is a countable disjoint union of multiplicative class $< \alpha$ sets.*

Proof. Let A be a set of additive class $\alpha > 2$. Write $A = \bigcup A_n$, where A_n is of multiplicative class less than α. Let $B_n = (\bigcup_{i\le n} A_i)^c$. Then B_n is of additive class $< \alpha$. Write $B_n = \bigcup_k B^n_k$, where the B^n_k's are pairwise disjoint ambiguous class $< \alpha$ sets. This is possible since $\alpha > 2$. We have

$$\begin{aligned} A &= A_0 \bigcup (A_1 \bigcap B_0) \bigcup (A_2 \bigcap B_1) \bigcup \cdots \\ &= A_0 \bigcup \bigcup_{n\ge 1} \bigcup_k (A_n \bigcap B^{n-1}_k), \end{aligned}$$

and the result follows. ■

Exercise 3.6.4 **(i)** Let X, Y be metrizable spaces and $f : X \longrightarrow Y$ continuous. Show that if $A \subseteq Y$ is in $\mathbf{\Sigma}^0_\alpha(Y)(\mathbf{\Pi}^0_\alpha(Y))$, then $f^{-1}(A)$ is in $\mathbf{\Sigma}^0_\alpha(X)(\mathbf{\Pi}^0_\alpha(X))$; i.e., the pointclasses $\mathbf{\Sigma}^0_\alpha$ and $\mathbf{\Pi}^0_\alpha$ are closed under continuous preimages.

(ii) Let Y be a subspace of X, $1 \leq \alpha < \omega_1$, and $\mathbf{\Gamma}_\alpha$ the pointclass of additive or multiplicative class α sets. Show that

$$\mathbf{\Gamma}_\alpha(Y) = \mathbf{\Gamma}_\alpha(X)|Y = \{A \bigcap Y : A \in \mathbf{\Gamma}_\alpha(X)\}.$$

(iii) Let $1 \leq \alpha < \omega_1$ and $\mathbf{\Gamma}_\alpha$ the pointclass of additive or multiplicative or ambiguous class α sets. Suppose $A \in \mathbf{\Gamma}_\alpha(X \times Y)$ and $x \in X$. Show that $A_x \in \mathbf{\Gamma}_\alpha(Y)$.

(iv) Let $\alpha > 1$, X a metrizable space, and $E \in \Sigma^0_\alpha(X)$. Show that there is a sequence (E_n) of pairwise disjoint $\Delta^0_\alpha(X)$ sets such that $E = \bigcup_n E_n$. This is true for $\alpha = 1$ if X is a zero-dimensional separable metric space.

Let X and Y be metrizable spaces, $f : X \longrightarrow Y$ a map, and $1 \leq \alpha < \omega_1$. We say that f is **Borel measurable of class** α, or simply of **class** α, if $f^{-1}(U) \in \mathbf{\Sigma}^0_\alpha$ for every open set U. Thus class 1 functions are precisely the continuous functions. A characteristic function χ_A, $A \subseteq X$, is of class α if and only if A is of ambiguous class α. Every class α function is clearly Borel measurable. Let Y be separable and $\mathcal{B}$ a subbase for X. Then f is of class α if and only if $f^{-1}(U)$ is of additive class α for every $U \in \mathcal{B}$. This in particular implies that if Y is separable and f Borel measurable, then f is of class α for some α. To see this, fix a countable base (U_n) for Y. Choose α_n such that $f^{-1}(U_n) \in \mathbf{\Sigma}^0_{\alpha_n}$ and take $\alpha = \sup_n \alpha_n$.

Exercise 3.6.5 **(i)** Let $1 \leq \alpha, \beta < \omega_1$, $f : X \longrightarrow Y$ of class α, and $g : Y \longrightarrow Z$ of class β. Show that $g \circ f$ is of class $\alpha + \beta'$, where $\beta' = \beta$ if β is infinite and is the immediate predecessor of β otherwise.

(ii) Let (f_n) be a sequence of functions of class α converging to f pointwise. Show that f is of class $\alpha + 1$. Also show that if the convergence is uniform, then f is of class α itself.

(iii) Let $X_0, X_1, X_2, \ldots$ be second countable metrizable spaces. Show that $f : X \longrightarrow \prod_{i \in \mathbb{N}} X_i$ is of class α if and only if each $\pi_i \circ f$ is of class α.

We now show that for every uncountable Polish space X and for every $\alpha < \omega_1$, $\mathbf{\Sigma}^0_\alpha(X) \neq \mathbf{\Pi}^0_\alpha(X)$. We shall use universal sets—a very useful notion—to prove our result.

Theorem 3.6.6 *Let* $1 \leq \alpha < \omega_1$ *and* $\mathbf{\Gamma}_\alpha$ *the pointclass of* $\mathbf{\Pi}^0_\alpha$ *or* $\mathbf{\Sigma}^0_\alpha$ *sets. For every second countable metrizable space* Y, *there exists a* $U \in$

$\Gamma_\alpha(\mathbb{N}^\mathbb{N} \times Y)$ *such that*

$$A \in \Gamma_\alpha(Y) \Longleftrightarrow (\exists x \in \mathbb{N}^\mathbb{N})(A = U_x).$$

We call such a set U **universal** *for* $\Gamma_\alpha(Y)$.

Proof. We proceed by induction on α.

Let (V_n) be a countable base for the topology of Y with at least one V_n empty. Define $U \subseteq \mathbb{N}^\mathbb{N} \times Y$ by

$$(x, y) \in U \Longleftrightarrow y \in \bigcup_n V_{x(n)}.$$

Evidently, A is open in Y if and only if $A = U_x$ for some x. It remains to show that U is open. Let $(x_0, y_0) \in U$. Then there is an n such that $y_0 \in V_{x_0(n)}$. Then

$$(x_0, y_0) \in \{x \in \mathbb{N}^\mathbb{N} : x(n) = x_0(n)\} \times V_{x_0(n)} \subseteq U.$$

Thus U is open.

Let $W = U^c$, where $U \subseteq \mathbb{N}^\mathbb{N} \times Y$ is universal for open sets. Clearly, W is universal for closed sets. The result for $\alpha = 1$ is proved.

Suppose $\alpha > 1$ and the result has been proved for all $\beta < \alpha$.

Case 1: α is a limit ordinal

Fix a sequence of countable ordinals (α_n), $1 < \alpha_n < \alpha$, such that $\alpha = \sup \alpha_n$. Let U_n be universal for multiplicative class α_n, $n \in \mathbb{N}$. For $x \in \mathbb{N}^\mathbb{N}$ and $n \in \mathbb{N}$, define $x_n \in \mathbb{N}^\mathbb{N}$ by

$$x_n(m) = x(2^n(2m+1) - 1). \qquad (\star)$$

For each n, $x \longrightarrow x_n$ is a continuous function. Define $U \subseteq \mathbb{N}^\mathbb{N} \times Y$ by

$$(x, y) \in U \Longleftrightarrow (\exists n)((x_n, y) \in U_n).$$

It is routine to check that U is universal for $\Sigma^0_\alpha(Y)$.

Case 2: $\alpha = \beta + 1$, a successor ordinal

Fix a universal Π^0_β set $P \subseteq \mathbb{N}^\mathbb{N} \times Y$. Define $U \subseteq \mathbb{N}^\mathbb{N} \times Y$ by

$$(x, y) \in U \Longleftrightarrow (\exists n)((x_n, y) \in P),$$

where x_n is as defined in $(\star)$. Clearly, U is universal for $\Sigma^0_\alpha(Y)$.

Having defined a universal Σ^0_α set $U \subseteq \mathbb{N}^\mathbb{N} \times Y$, note that U^c is universal for $\Pi^0_\alpha(Y)$. ■

Theorem 3.6.7 *Let* $1 \leq \alpha < \omega_1$ *and* Γ_α *the pointclass of additive or multiplicative class* α *sets. Then for every uncountable Polish space* X, *there is a* $U \in \Gamma_\alpha(X \times X)$ *universal for* $\Gamma_\alpha(X)$.

Proof. Since X is uncountable Polish, it has a subset, say Y, homeomorphic to $\mathbb{N}^{\mathbb{N}}$. By 3.6.6, there is $U \subseteq Y \times X$ universal for $\mathbf{\Gamma}_\alpha(X)$. By 3.6.4(iii), $V \bigcap (Y \times X) = U$ for some $V \in \mathbf{\Gamma}_\alpha(X \times X)$. The set V is universal for $\mathbf{\Gamma}_\alpha(X)$. ■

Corollary 3.6.8 *Let X be any uncountable Polish space and $1 \leq \alpha < \omega_1$. Then there exists an additive class α set that is not of multiplicative class α.*

Proof. Let $U \subseteq X \times X$ be universal for $\mathbf{\Sigma}^0_\alpha(X)$. Take

$$A = \{x \in X : (x, x) \in U\}.$$

Since $\mathbf{\Sigma}^0_\alpha$ is closed under continuous preimages, A is of additive class α. We claim that A is not of multiplicative class α. To the contrary, suppose A is of multiplicative class α. Choose $x_0 \in X$ such that $A^c = U_{x_0}$. Then

$$x_0 \in A^c \iff (x_0, x_0) \in U \iff x_0 \in A.$$

This is a contradiction. ■

This corollary shows that for every uncountable Polish space X and for any α, $\mathbf{\Sigma}^0_\alpha(X) \neq \mathbf{\Sigma}^0_{\alpha+1}(X)$. Is this true for all uncountable separable metric spaces X? For an answer to this question see [83]. The above argument also shows that there does not exist a Borel set $U \subseteq X \times X$ universal for Borel subsets of X for any Polish space X. In fact, we can draw a fairly general conclusion.

Proposition 3.6.9 *Let a pointclass $\mathbf{\Delta}$ be closed under complementation and continuous preimages. Then for no Polish space X is there a set in $\mathbf{\Delta}(X \times X)$ universal for $\mathbf{\Delta}(X)$.*

Proof. Suppose there is a Polish space X and a $U \in \mathbf{\Delta}(X \times X)$ universal for $\mathbf{\Delta}(X)$. Take

$$A = \{x \in X : (x, x) \in U\}.$$

Since $\mathbf{\Delta}$ is closed under continuous preimages, $A \in \mathbf{\Delta}$. As $\mathbf{\Delta}$ is closed under complementation, $A^c \in \mathbf{\Delta}$. Let $A^c = U_{x_0}$ for some $x_0 \in X$. Then

$$x_0 \in A^c \iff (x_0, x_0) \in U \iff x_0 \in A.$$

This is a contradiction. ■

Theorem 3.6.10 *(Reduction theorem for additive classes) Let X be a metrizable space and $1 < \alpha < \omega_1$. Suppose (A_n) is a sequence of additive class α sets in X. Then there exist $B_n \subseteq A_n$ such that*

(a) *The B_n's are pairwise disjoint sets of additive class α, and*

(b) $\bigcup_n A_n = \bigcup_n B_n$.

(See Figure 3.1.) Consequently the B_n's are of ambiguous class α if $\bigcup_n A_n$ is so.

The result is also true for $\alpha = 1$ if X is zero-dimensional and second countable.

Proof. Write

$$A_n = \bigcup_m C_{nm}, \qquad (*)$$

where the C_{nm}'s are of ambiguous class α. If $\alpha > 1$, this is always possible. If $\alpha = 1$, it is possible if X is zero-dimensional and second countable (3.6.1). Enumerate $\{C_{nm} : n, m \in \mathbb{N}\}$ in a single sequence, say (D_i). Let

$$E_i = D_i \setminus \bigcup_{j<i} D_j.$$

Take

$$B_n = \bigcup \{E_i : E_i \subseteq A_n \& (\forall m < n)(E_i \not\subseteq A_m)\}.$$

■

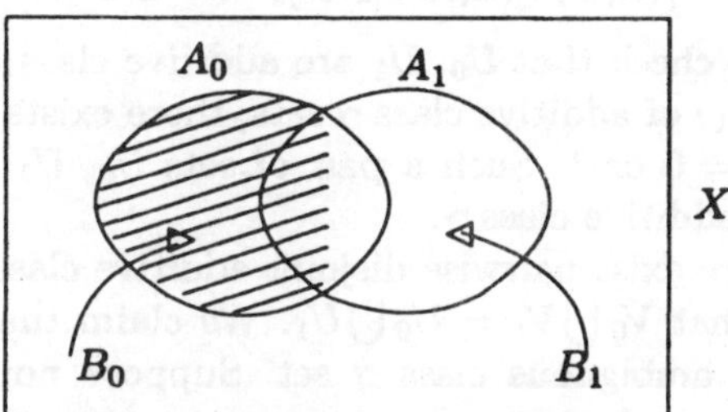

Figure 3.1. Reduction

Theorem 3.6.11 *(Separation theorem for multiplicative classes) Let X be metrizable and $1 < \alpha < \omega_1$. Then for every sequence (A_n) of multiplicative class α sets with $\bigcap A_n = \emptyset$, there exist ambiguous class α sets $B_n \supseteq A_n$ with $\bigcap B_n = \emptyset$.*

In particular, if A and B are two disjoint subsets of X of multiplicative class α, then there is an ambiguous class α set C such that

$$A \subseteq C \;\&\; B \bigcap C = \emptyset.$$

(See Figure 3.2.) This is also true for $\alpha = 1$ if X is zero-dimensional and second countable.

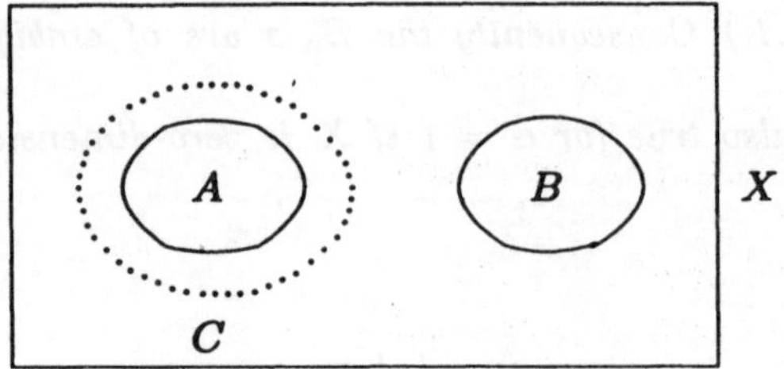

Figure 3.2. Separation

Proof. By 3.6.10, there exist pairwise disjoint additive class α sets $C_n \subseteq A_n^c$ such that $\bigcup_n C_n = \bigcup_n A_n^c = X$. Obviously, the C_n's are of ambiguous class α. Take $B_n = C_n^c$. ■

The next example shows that the separation theorem does not hold for additive classes. Consequently, the reduction theorem does not hold for multiplicative classes.

Example 3.6.12 (a) Fix a homeomorphism $\alpha \longrightarrow (\alpha_0, \alpha_1)$ from $\mathbf{N}^{\mathbf{N}}$ onto $\mathbf{N}^{\mathbf{N}} \times \mathbf{N}^{\mathbf{N}}$. Let γ be any countable ordinal and $U \subseteq \mathbf{N}^{\mathbf{N}} \times \mathbf{N}^{\mathbf{N}}$ a universal $\mathbf{\Sigma}_\gamma^0$ set. Define

$$U_i = \{(\alpha, \beta) : (\alpha_i, \beta) \in U\}, \quad i = 0 \text{ or } 1.$$

It easy quite easy to check that U_0, U_1 are additive class γ sets such that for every pair (A_0, A_1) of additive class γ sets, there exists an $\alpha \in \mathbf{N}^{\mathbf{N}}$ such that $(U_i)_\alpha = A_i$, $i = 0$ or 1. Such a pair of sets U_0, U_1 will be called a **universal pair** for additive class γ.

(b) By 3.6.10, there exist pairwise disjoint additive class γ sets $V_0 \subseteq U_0$ and $V_1 \subseteq U_1$ such that $V_0 \bigcup V_1 = U_0 \bigcup U_1$. We claim that V_0, V_1 cannot be separated by an ambiguous class γ set. Suppose not. Let W be an ambiguous class γ set such that

$$V_0 \subseteq W \text{ and } W \bigcap V_1 = \emptyset.$$

We claim that W is a universal $\mathbf{\Delta}_\gamma^0$ set, which contradicts 3.6.9. To prove our claim, take any $A_0 \in \mathbf{\Delta}_\gamma^0(\mathbf{N}^{\mathbf{N}})$. Let $A_1 = A_0^c$. Then there exists an $\alpha \in \mathbf{N}^{\mathbf{N}}$ such that $(U_i)_\alpha = A_i$, $i = 0$ or 1. Plainly, $A_0 = W_\alpha$.

The next proposition is a very useful one. A sequence (A_n) of sets is called **convergent** if $\liminf_n A_n = \limsup_n A_n = B$, say. In this case we say that (A_n) converges to B and write $\lim A_n = B$. Note that the following two statements are equivalent:

(i) (A_n) is convergent.

(ii) For every $x \in X$, $x \in A_n$ for infinitely many n if and only if $x \in A_n$ for all but finitely many n.

Proposition 3.6.13 *Let X be metrizable and $2 < \alpha < \omega_1$. Suppose $A \in \mathbf{\Delta}^0_\alpha(X)$. Then there is a sequence (A_n) of ambiguous class $< \alpha$ sets such that $A = \lim A_n$.*

The result is also true for $\alpha = 2$, provided that X is separable and zero dimensional.

Proof. We write

$$A = \bigcup_n C_n = \bigcap_n D_n,$$

where the C_n's are multiplicative class $< \alpha$ sets, the D_n's are additive class $< \alpha$ sets, $C_n \subseteq C_{n+1}$, and $D_{n+1} \subseteq D_n$. By 3.6.11, there is a set A_n of ambiguous class$< \alpha$ such that

$$C_n \subseteq A_n \subseteq D_n.$$

Then $A = \lim A_n$ as we now show. Let $x \in \limsup A_n$. Thus, $x \in A_n$ for infinitely many n. Then $x \in D_n$ for infinitely many n and hence for all n. Therefore,

$$\limsup A_n \subseteq A. \tag{1}$$

Now let $x \in A$. Then $x \in C_n$ for all but finitely many n. Since $C_n \subseteq A_n$ for all n,

$$A \subseteq \liminf A_n. \tag{2}$$

The result follows from (1) and (2). ■

We prove the next result for future applications.

Proposition 3.6.14 *Let $2 < \alpha < \omega_1$ and X an uncountable Polish space. There exists a sequence A_n in $\mathbf{\Pi}^0_\alpha(X)$ with $\limsup A_n = \emptyset$ such that there does not exist $B_n \supseteq A_n$ in $\mathbf{\Sigma}^0_\alpha(X)$ with $\limsup B_n = \emptyset$.*

Proof. Take $A \in \mathbf{\Sigma}^0_{\alpha+1}(X) \setminus \mathbf{\Pi}^0_{\alpha+1}(X)$. Such a set exists by 3.6.8. By 3.6.3, we can find disjoint sets $A_n \in \mathbf{\Pi}^0_\alpha(X)$ with union A. Quite trivially, $\limsup A_n = \emptyset$. Suppose there exist $B_n \supseteq A_n$ in $\mathbf{\Sigma}^0_\alpha(X)$ with $\limsup B_n = \emptyset$. We shall get a contradiction.

By 3.6.11, there is a set $C_n \in \mathbf{\Delta}^0_\alpha(X)$ such that $A_n \subseteq C_n \subseteq B_n$. Note that $\limsup C_n = \emptyset$. As the sets A_n are in $\mathbf{\Delta}^0_{\alpha+1}(X)$, by 3.6.13 there are sets $A^k_n \in \mathbf{\Delta}^0_\alpha(X)$ such that $A_n = \lim_k A^k_n$. Now define

$$D_k = (A^k_1 \bigcap C_1) \bigcup (A^k_2 \bigcap C_2) \bigcup \cdots \bigcup (A^k_k \bigcap C_k).$$

Then $D_k \in \mathbf{\Delta}^0_\alpha(X)$. It is now fairly easy to check that $\limsup D_k \subseteq A \subseteq \liminf D_k$, so $A = \lim D_k$. This implies that $A \in \mathbf{\Delta}^0_{\alpha+1}(X)$, and we have arrived at a contradiction. ■

The above observation is due to A. Maitra, C. A. Rogers, and J. E. Jayne.

We close this chapter with another very useful result on Borel functions of class α. In particular, it gives us an analogue of the Lebesgue – Hausdorff theorem (3.1.36) for class α functions.

Theorem 3.6.15 *Suppose X, Y are metrizable spaces with Y second countable and $2 < \alpha < \omega_1$. Then for every Borel function $f : X \longrightarrow Y$ of class α, there is a sequence (f_n) of Borel maps from X to Y of class $< \alpha$ such that $f_n \to f$ pointwise.*

We need some lemmas to prove this result. In what follows, X, Y are metrizable and d is a compatible metric on Y.

Lemma 3.6.16 *Suppose Y is totally bounded. Then every $f : X \longrightarrow Y$ of class α, $\alpha > 1$, is the limit of a uniformly convergent sequence of class α functions $f_n : X \longrightarrow Y$ of finite range.*

Proof. Take any $\epsilon > 0$. We shall obtain a function $g : X \longrightarrow Y$ of class α such that the range of g is finite and $d(g(x), f(x)) < \epsilon$ for all x. Let $\{y_1, y_2, \ldots, y_n\}$ be an ϵ-net in Y. Set

$$A_i = f^{-1}(B(y_i, \epsilon)).$$

The sets $A_1, A_2, \ldots, A_n$ are of additive class α with union X. By 3.6.10, there are pairwise disjoint ambiguous class α sets $B_1, B_2, \ldots, B_n$ such that

$$B_1 \subseteq A_1, B_2 \subseteq A_2, \ldots, B_n \subseteq A_n$$

and

$$\bigcup B_i = \bigcup A_i = X.$$

Define $g : X \longrightarrow Y$ by

$$g(x) = y_i \text{ if } x \in B_i.$$

Then $d(f(x), g(x)) < \epsilon$ for all x. ■

Lemma 3.6.17 *Let $f : X \longrightarrow Y$ be of class $\alpha > 2$ with range contained in a finite set $E = \{y_1, y_2, \ldots, y_n\}$. Then f is the limit of a sequence of functions of class $< \alpha$ with values in E.*

Proof. Let $A_i = f^{-1}(y_i)$, $i = 1, 2, \ldots, n$. Then A_1, A_2, ..., A_n are pairwise disjoint, ambiguous class α sets with union X. By 3.6.13, for each i there is a sequence (A_{im}) of sets of ambiguous class $< \alpha$ such that $A_i = \lim_m A_{im}$. Fix m. Let

$$B_1^m = A_{1m}, B_2^m = A_{2m} \setminus A_{1m}, \ldots, B_n^m = A_{nm} \setminus \bigcup_{j<n} A_{jm}$$

and

$$B_{n+1}^m = X \setminus \bigcup_{j \leq n} A_{jm}.$$

Evidently, the sets B_1^m, B_2^m, ..., B_{n+1}^m are pairwise disjoint and of ambiguous class less than α with union X. So there is a function $f_m : X \longrightarrow Y$ of class $< \alpha$ satisfying

$$f_m(x) = y_i, \text{ if } x \in B_i^m, \quad 1 \leq i \leq n.$$

We claim that $f_m(x_0) \to f(x_0)$ for all $x_0 \in X$. Assume that $x_0 \in A_i$. So, $f(x_0) = y_i$. Since $x_0 \notin \limsup_m A_{jm}$ for all $j \neq i$, there is an integer M such that $x_0 \notin A_{jm}$ for $m > M$ and $j \neq i$. Since $x_0 \in \liminf_m A_{im}$, we can further assume that $x_0 \in A_{im}$ for all $m > M$. Thus, $f_m(x_0) = y_i$ for all $m > M$. Hence, $f_m \to f$ pointwise. ∎

Proof of 3.6.15. Let d be a totally bounded compatible metric on Y. Such a metric exists by 2.1.32 and 2.3.12. By 3.6.16, there is a sequence (g_m) of class α functions, with range finite, converging to f uniformly. Without any loss of generality, we assume that for all x and all m,

$$d(g_m(x), g_{m+1}(x)) < 2^{-m}.$$

By induction on m, we define a sequence (g_{mn}) of functions of class $< \alpha$ of finite range such that for all m and all k,

$$\lim_n g_{mn}(x) = g_m(x) \text{ and } d(g_{m+1,k}(x), g_{m,k}(x)) \leq 2^{-m}. \qquad (*)$$

By 3.6.17, there is a sequence (g_{0n}) of functions of class $< \alpha$, each with range finite, converging pointwise to g_0. Suppose that for some m a sequence (g_{mn}) of class $< \alpha$ functions of finite range converging pointwise to g_m has been defined. We define $(g_{m+1,n})$ such that $(\star)$ is satisfied. By 3.6.17, there is a sequence (h_n) of functions of class $< \alpha$ with finite range converging pointwise to g_{m+1}. Define

$$u_n(x) = d(g_{mn}(x), h_n(x)), \quad x \in X.$$

The map u_n is of class $< \alpha$ taking finitely many values. The set

$$A_n = \{x \in X : u_n(x) \leq 2^{-m}\}$$

is of ambiguous class $< \alpha$. Define $g_{m+1,n}$ by

$$g_{m+1,n}(x) = \begin{cases} h_n(x) & \text{if } x \in A_n, \\ g_{m,n}(x) & \text{otherwise.} \end{cases}$$

It is easily seen that $(\star)$ is satisfied. Define $f_m : X \longrightarrow Y$ by

$$f_m(x) = g_{mm}(x), \quad x \in X.$$

We show that (f_m) converges to f pointwise. Take any $x_0 \in X$. Fix $\epsilon > 0$. Let m be such that $2^{-m+1} < \epsilon/3$ and $d(f(x), g_m(x)) < \epsilon/3$ for all x. Choose $M > m$ such that $d(g_{mi}(x_0), g_m(x_0)) < \epsilon/3$ for all $i > M$. For $i > M$, we have the following.

$$\begin{aligned} d(f_i(x_0), f(x_0)) &= d(g_{ii}(x_0), f(x_0)) \\ &\leq d(g_{ii}(x_0), g_{i-1,i}(x_0)) + \cdots + d(g_{m+1,i}(x_0), g_{mi}(x_0)) \\ &\quad + d(g_{mi}(x_0), g_m(x_0)) + d(g_m(x_0), f(x_0)) \\ &< (2^{-i} + \cdots + 2^{-m}) + \epsilon/3 + \epsilon/3 \\ &< \epsilon. \end{aligned}$$

Our result is proved. ∎

4
Analytic and Coanalytic Sets

In this chapter we present the theory of analytic and coanalytic sets. The theory of analytic and coanalytic sets is of fundamental importance to the theory of Borel sets and Borel functions. It gives the theory of Borel sets its power. Thus the results proved in this chapter are the central results of these notes.

4.1 Projective Sets

Let $B \subseteq X \times Y$. For notational convenience, we denote the projection $\pi_X(B)$ of B to X by $\exists^Y B$; i.e.,

$$\exists^Y B = \{x \in X : (x, y) \in B \text{ for some } y \in Y\}.$$

The **coprojection** of B is defined by

$$\forall^Y B = \{x \in X : (x, y) \in B \text{ for all } y \in Y\}.$$

Clearly,

$$\forall^Y B = (\exists^Y B^c)^c.$$

For any pointclass $\mathbf{\Gamma}$ and any Polish space Y, we set

$$\exists^Y \mathbf{\Gamma} = \{\exists^Y B : B \in \mathbf{\Gamma}(X \times Y), X \text{ a Polish space}\};$$

i.e., $\exists^Y \mathbf{\Gamma}$ is the family of sets of the form $\exists^Y B$ where B is in $\mathbf{\Gamma}(X \times Y)$, X Polish. The pointclass $\forall^Y \mathbf{\Gamma}$ is similarly defined.

Let X be a Polish space. **From now on, a Borel subset of a Polish space will be called a standard Borel set.** A subset A of X is called **analytic** if it is the projection of a Borel subset B of $X \times X$. The pointclass of analytic sets is denoted by $\mathbf{\Sigma}^1_1$. A subset C of X is called **coanalytic** if $X \setminus C$ is analytic.

Note that *a subset A of X is coanalytic if and only if it is the coprojection of a Borel subset of $X \times X$.*

$\mathbf{\Pi}^1_1$ will denote the pointclass of coanalytic sets. Thus $\mathbf{\Pi}^1_1 = \neg\mathbf{\Sigma}^1_1$. Finally, we define

$$\mathbf{\Delta}^1_1 = \mathbf{\Pi}^1_1 \bigcap \mathbf{\Sigma}^1_1.$$

Let X be a Polish space, $C = B \times X$. Then

$$B = \exists^X C = \forall^X C. \tag{$*$}$$

Thus every standard Borel set is analytic as well as coanalytic; i.e., they are $\mathbf{\Delta}^1_1$ sets. The converse of this fact is also true; i.e., every $\mathbf{\Delta}^1_1$ set is Borel (4.4.3). This is one of the most remarkable results on Borel sets. It was proved by Souslin[111] and marked the beginning of descriptive set theory as an independent subject.

Proposition 4.1.1 *Let X be a Polish space and $A \subseteq X$. The following statements are equivalent.*

(i) *A is analytic.*

(ii) *There is a Polish space Y and a Borel set $B \subseteq X \times Y$ whose projection is A.*

(iii) *There is a continuous map $f : \mathbf{N}^{\mathbf{N}} \longrightarrow X$ whose range is A.*

(iv) *There is a closed subset C of $X \times \mathbf{N}^{\mathbf{N}}$ whose projection is A.*

(v) *For every uncountable Polish space Y there is a G_δ set B in $X \times Y$ whose projection is A.*

Proof. (i) trivially implies (ii).

Let Y be a Polish space and B a Borel subset of $X \times Y$ such that $\pi_X(B) = A$, where $\pi_X : X \times Y \longrightarrow X$ is the projection map. By 3.3.17, there is a continuous map g from $\mathbf{N}^{\mathbf{N}}$ onto B. Take $f = \pi_X \circ g$. Since the range of f is A, (ii) implies (iii).

Since the graph of a continuous map $f : \mathbf{N}^{\mathbf{N}} \longrightarrow X$ is a closed subset of $\mathbf{N}^{\mathbf{N}} \times X$ with projection A, (iii) implies (iv).

By 2.6.5, every uncountable Polish space Y contains a homeomorph of $\mathbf{N}^{\mathbf{N}}$, which is necessarily a G_δ set in Y. Therefore, (iv) implies (v).

(i) trivially follows from (v). ■

Proposition 4.1.2 **(i)** *The pointclass* $\mathbf{\Sigma}^1_1$ *is closed under countable unions, countable intersections and Borel preimages. Consequently,* $\mathbf{\Pi}^1_1$ *is closed under these operations.*

(ii) *The pointclass* $\mathbf{\Sigma}^1_1$ *is closed under projection* $\exists^Y$*, and* $\mathbf{\Pi}^1_1$ *is closed under coprojection* $\forall^Y$ *for all Polish* Y*.*

Proof. We first prove (i).

Closure under Borel preimages: Let X and Z be Polish spaces, $A \subseteq X$ analytic, and $f : Z \longrightarrow X$ a Borel map. Choose a Borel subset B of $X \times X$ whose projection is A. Let

$$C = \{(z, x) \in Z \times X : (f(z), x) \in B\}.$$

The set C is Borel, and $\pi_X(C) = f^{-1}(A)$. So $f^{-1}(A)$ is analytic.

Closure under countable unions and countable intersections: Let $A_0, A_1, A_2, \ldots$ be analytic subsets of X. By 4.1.1, there are Borel subsets $B_0, B_1, B_2, \ldots$ of $X \times \mathbb{N}^{\mathbb{N}}$ whose projections are $A_0, A_1, A_2, \ldots$ respectively. Take

$$C = \{(x, \alpha) \in X \times \mathbb{N}^{\mathbb{N}} : (x, \alpha^*) \in B_{\alpha(0)}\}$$

and

$$D = \{(x, \alpha) \in X \times \mathbb{N}^{\mathbb{N}} : (x, f_i(\alpha)) \in B_i \text{ for every } i\},$$

where $\alpha^*(i) = \alpha(i+1)$ and $(f_0, f_1, f_2, \ldots) : \mathbb{N}^{\mathbb{N}} \longrightarrow (\mathbb{N}^{\mathbb{N}})^{\mathbb{N}}$ is a continuous surjection. Note that the map $\alpha \longrightarrow \alpha^*$ is also continuous. Hence, the sets C and D are Borel with projections $\bigcup_i A_i$ and $\bigcap_i A_i$ respectively. We have shown that $\mathbf{\Sigma}^1_1$ is closed under countable unions and countable intersections. The closure properties of $\mathbf{\Pi}^1_1$ follow.

(ii) is trivially seen from the identity $(\star)$ and the fact that the product of two Polish spaces is Polish. ∎

Exercise 4.1.3 Let $B \subseteq X$ be analytic (in particular Borel) and $f : B \longrightarrow Y$ a Borel map. Show that $f(B)$ is analytic.

Is there an analytic set that is not Borel? Recall that in Chapter 3 we used universal sets to show that for any uncountable Polish space X and for any $1 \le \alpha < \omega_1$, $\mathbf{\Sigma}^0_\alpha(X) \neq \mathbf{\Pi}^0_\alpha(X)$. We follow the same ideas to show that there are analytic sets that are not Borel.

Theorem 4.1.4 *For every Polish space* X*, there is an analytic set* $U \subseteq \mathbb{N}^{\mathbb{N}} \times X$ *such that* $A \subseteq X$ *is analytic if and only if* $A = U_\alpha$ *for some* α*; i.e.,* U *is* **universal** *for* $\mathbf{\Sigma}^1_1(X)$.

Proof. Let $C \subseteq \mathbb{N}^{\mathbb{N}} \times (X \times \mathbb{N}^{\mathbb{N}})$ be a universal closed set. The existence of such a set is shown in 3.6.6. Let

$$U = \{(\alpha, x) \in \mathbb{N}^{\mathbb{N}} \times X : (\alpha, x, \beta) \in C \text{ for some } \beta\}.$$

As $U = \exists^{\mathbb{N}^{\mathbb{N}}} C$, it follows that $U \in \mathbf{\Sigma}^1_1$. Let $A \subseteq X$ be $\mathbf{\Sigma}^1_1$. Choose a closed set $F \subseteq X \times \mathbb{N}^{\mathbb{N}}$ whose projection is A (4.1.1). Let $\alpha \in \mathbb{N}^{\mathbb{N}}$ be such that $F = C_\alpha$. Then $A = U_\alpha$. ∎

Theorem 4.1.5 *Let X be an uncountable Polish space.*

(i) *There is an analytic set $U \subseteq X \times X$ such that for every analytic set $A \subseteq X$, there is an $x \in X$ with $A = U_x$.*

(ii) *There is a subset of X that is analytic but not Borel.*

Proof. (i) Since X is uncountable Polish, it contains a homeomorph of $\mathbb{N}^{\mathbb{N}}$, say Y (2.6.5). The set Y is a G_δ set in X (2.2.7). Take $U \subseteq Y \times X$ as in 4.1.4.

(ii) Let

$$A = \{x \in X : (x, x) \in U\}.$$

Since $\mathbf{\Sigma}^1_1$ is closed under continuous preimages, $A \in \mathbf{\Sigma}^1_1$. We claim that A is not coanalytic and hence not Borel. Suppose not. Then A^c analytic. Take an $x_0 \in X$ such that $A^c = U_{x_0}$. Then

$$x_0 \in A \Longleftrightarrow (x_0, x_0) \in U \Longleftrightarrow x_0 \in A^c.$$

We have arrived at a contradiction. ∎

Remark 4.1.6 From the Borel isomorphism and the above theorem we see that every uncountable standard Borel set contains an analytic set that is not Borel.

Just as we defined analytic and coanalytic sets from Borel sets, we can continue with sets that are projections of coanalytic sets, complements of these sets, and so on. More precisely, for each $n \geq 1$, we define pointclasses $\mathbf{\Sigma}^1_n$, $\mathbf{\Pi}^1_n$, and $\mathbf{\Delta}^1_n$ by induction on n as follows: Let X be any Polish space. We have already defined $\mathbf{\Sigma}^1_1(X)$, $\mathbf{\Pi}^1_1(X)$, and $\mathbf{\Delta}^1_1(X)$. Let n be any positive integer. We take

$$\begin{aligned} \mathbf{\Sigma}^1_{n+1}(X) &= \exists^X \mathbf{\Pi}^1_n(X \times X), \\ \mathbf{\Pi}^1_{n+1}(X) &= \neg\mathbf{\Sigma}^1_{n+1}(X), \end{aligned}$$

and

$$\mathbf{\Delta}^1_{n+1}(X) = \mathbf{\Sigma}^1_{n+1}(X) \bigcap \mathbf{\Pi}^1_{n+1}(X).$$

Sets thus obtained are called **projective sets.**

Proposition 4.1.7 *Let n be a positive integer.*

(i) *The pointclasses $\mathbf{\Sigma}^1_n$ and $\mathbf{\Pi}^1_n$ are closed under countable unions, countable intersections and Borel preimages.*

(ii) $\mathbf{\Delta}^1_n$ *is a σ-algebra.*

(iii) *The pointclass $\mathbf{\Sigma}^1_n$ is closed under projections $\exists^Y$, and $\mathbf{\Pi}^1_n$ is closed under coprojections $\forall^Y$, Y Polish.*

Proof. Clearly, (ii) follows from (i). So, we prove (i) and (iii) only. We proceed by induction on n. Let $n > 1$ and $\mathbf{\Pi}^1_{n-1}$ and $\mathbf{\Sigma}^1_{n-1}$ have all the closure properties stated in (i) and (iii). The arguments contained in the proof of 4.1.2 show that $\mathbf{\Sigma}^1_n$ also has the stated closure properties. Since $\mathbf{\Pi}^1_n = \neg\mathbf{\Sigma}^1_n$, the remaining part of the result follows. ■

Exercise 4.1.8 Let $B \subseteq X$ be $\mathbf{\Sigma}^1_n$ and $f : B \longrightarrow Y$ a Borel map. Show that $f(B) \in \mathbf{\Sigma}^1_n$.

Proposition 4.1.9 *For every $n \geq 1$,*

$$\mathbf{\Sigma}^1_n \bigcup \mathbf{\Pi}^1_n \subseteq \mathbf{\Delta}^1_{n+1}.$$

Thus we have the following diagram, in which any pointclass is contained in every pointclass to the right of it:

$$\begin{array}{ccccccc} & \mathbf{\Sigma}^1_1 & & \mathbf{\Sigma}^1_2 & & \mathbf{\Sigma}^1_3 & \cdots \\ \mathbf{\Delta}^1_1 & & \mathbf{\Delta}^1_2 & & \mathbf{\Delta}^1_3 & \cdots & \\ & \mathbf{\Pi}^1_1 & & \mathbf{\Pi}^1_2 & & \mathbf{\Pi}^1_3 & \cdots \end{array}$$

(The Hierarchy of Projective Sets)

Proof. We prove the result by induction on n. Let X be a Polish space and $A \subseteq X$ analytic. As $\mathbf{\Delta}^1_1 \subseteq \mathbf{\Pi}^1_1$, it follows that $\mathbf{\Sigma}^1_1 \subseteq \mathbf{\Sigma}^1_2$. Since $\mathbf{\Sigma}^1_1$ is closed under continuous preimages, the set $C = A \times X$ is analytic. Since

$$A = \forall^X C,$$

A is in $\mathbf{\Pi}^1_2$. Hence $A \in \mathbf{\Delta}^1_2$. The rest of the result now follows fairly easily by induction. ■

Lemma 4.1.10 *Let $n \geq 1$, $\mathbf{\Gamma}$ either $\mathbf{\Sigma}^1_n$ or $\mathbf{\Pi}^1_n$, and X a Polish space. There is a $U \subseteq \mathbb{N}^\mathbb{N} \times X$ in $\mathbf{\Gamma}$ such that $A \subseteq X$ is in $\mathbf{\Gamma}$ if and only if $A = U_\alpha$ for some α; i.e., U is* **universal** *for $\mathbf{\Gamma}(X)$.*

Proof. The result is proved by induction. Suppose $U \subseteq \mathbb{N}^\mathbb{N} \times X$ is universal for $\mathbf{\Sigma}^1_1(X)$. Then U^c is universal for $\mathbf{\Pi}^1_1(X)$. Let $C \subseteq \mathbb{N}^\mathbb{N} \times (X \times \mathbb{N}^\mathbb{N})$ be universal for $\mathbf{\Pi}^1_n(X \times \mathbb{N}^\mathbb{N})$. As in 4.1.4, we see that $\exists^{\mathbb{N}^\mathbb{N}} C$ is universal for $\mathbf{\Sigma}^1_{n+1}(X)$, and its complement is universal for $\mathbf{\Pi}^1_{n+1}(X)$. ■

Theorem 4.1.11 *Let X be an uncountable Polish space and $n \geq 1$.*

(i) *There is a set* $U \in \Sigma^1_n(X \times X)$ *such that for every* $A \in \Sigma^1_n(X)$, *there is an* x *with* $A = U_x$.

(ii) *There is a subset of* X *that is in* $\Sigma^1_n(X)$ *but not in* $\Pi^1_n(X)$.

Proof. The result is proved in exactly the same way as 4.1.5. ■

Exercise 4.1.12 Show that for any Polish space X and for any $n \geq 1$, there is no set $U \in \Delta^1_n(X \times X)$ that is universal for $\Delta^1_n(X)$.

We shall not be much interested in higher projective classes, as they are not of much importance to the theory of Borel sets. Further, regularity properties of projective sets, e.g., questions regarding their cardinalities, measurability, etc., cannot be established without further set-theoretic assumptions. This is beyond the scope of these notes.

The next result gives a very useful connection between the Souslin operation and analytic sets.

Theorem 4.1.13 *Let* X *be a Polish space,* d *a compatible complete metric on* X, *and* $A \subseteq X$. *The following statements are equivalent.*

(i) A *is analytic.*

(ii) *There is a regular system* $\{F_s : s \in \mathbf{N}^{<\mathbf{N}}\}$ *of closed subsets of* X *such that for every* $\alpha \in \mathbf{N}^{\mathbf{N}}$ diameter$(F_{\alpha|n}) \to 0$ *and* $A = \mathcal{A}(\{F_s\})$.

(iii) *There is a system* $\{F_s : s \in \mathbf{N}^{<\mathbf{N}}\}$ *of closed subsets of* X *such that* $A = \mathcal{A}(\{F_s\})$.

Proof. (ii) implies (iii) is obvious.

(iii) $\Longrightarrow$ (i): Let $\{F_s\}$ be a system of closed sets in X such that

$$A = \mathcal{A}(\{F_s\});$$

i.e.,

$$x \in A \Longleftrightarrow \exists\alpha\forall n(x \in F_{\alpha|n}).$$

Let

$$C = \{(x,\alpha) \in X \times \mathbf{N}^{\mathbf{N}} : \forall n(x \in F_{\alpha|n})\}.$$

As

$$C = \bigcap_n \bigcup_{\{s:|s|=n\}} (F_s \times \Sigma(s)),$$

C is closed. Since A is the projection of C, it is analytic.

(i) $\Longrightarrow$ (ii): Let $A \subseteq X$ be analytic. By 4.1.1, there is a continuous map $f : \mathbf{N}^{\mathbf{N}} \longrightarrow X$ whose range is A. Take

$$F_s = \text{cl}(f(\Sigma(s))).$$

Clearly, the system of closed sets $\{F_s : s \in \mathbf{N}^{<\mathbf{N}}\}$ is regular. Since f is continuous, diameter$(F_{\alpha|n})$ converges to 0 as $n \to \infty$.

Let $x = f(\alpha) \in A$. Then for all n, $x \in F_{\alpha|n}$. Thus $A \subseteq \mathcal{A}(\{F_s\})$.

To show the reverse inclusion, take any $x \in \mathcal{A}(\{F_s\})$. Let

$$x \in F_{\alpha|n} = \text{cl}(f(\Sigma(\alpha|n)))$$

for all n. Choose $\alpha_n \in \Sigma(\alpha|n)$ such that $d(x, f(\alpha_n)) < 2^{-n}$. So, $f(\alpha_n) \to x$. Since $\alpha_n \to \alpha$ and f is continuous, $f(\alpha_n) \to f(\alpha)$. Hence, $x \in A$, and the result follows. ■

Theorem 4.1.14 *The pointclass $\mathbf{\Sigma}^1_1$ is closed under the Souslin operation.*

Proof. By 1.13.1, the Souslin operation is idempotent; i.e., for any family $\mathcal{F}$ of sets $\mathcal{A}(\mathcal{A}(\mathcal{F})) = \mathcal{A}(\mathcal{F})$. Since $\mathbf{\Sigma}^1_1 = \mathcal{A}(\mathcal{F})$, where $\mathcal{F}$ is the family of closed sets, the result follows. ■

Remark 4.1.15 Since there are analytic sets that are not coanalytic, $\mathbf{\Pi}^1_1$ is not closed under the Souslin operation.

Exercise 4.1.16 Let X be an uncountable Polish space and $n \geq 2$. Show that $\mathbf{\Sigma}^1_n$, $\mathbf{\Pi}^1_n$, and $\mathbf{\Delta}^1_n$ are closed under the Souslin operation.

Remark 4.1.17 For every Polish space X, there is a pair of analytic sets $U_0, U_1 \subseteq \mathbf{N}^\mathbf{N} \times X$ such that for any pair A_0, A_1 of analytic subsets of X there is an α satisfying $A_i = (U_i)_\alpha$, $i = 0, 1$. To show the existence of such a pair, fix an analytic set $U \subseteq \mathbf{N}^\mathbf{N} \times X$ universal for analytic subsets of X. Let $f(\alpha) = (\alpha_0, \alpha_1)$ be a homeomorphism from $\mathbf{N}^\mathbf{N}$ onto $\mathbf{N}^\mathbf{N} \times \mathbf{N}^\mathbf{N}$. Take

$$U_i = \{(\alpha, x) \in \mathbf{N}^\mathbf{N} \times X : (\alpha_i, x) \in U\}.$$

Since $(\mathbf{N}^\mathbf{N})^\mathbf{N}$ and $\mathbf{N}^\mathbf{N}$ are also homeomorphic, we can say more. There exists a sequence $U_0, U_1, U_2, \ldots$ of analytic subsets of $\mathbf{N}^\mathbf{N} \times X$ such that for any sequence $A_0, A_1, A_2, \ldots$ of analytic subsets of X, there is an $\alpha \in \mathbf{N}^\mathbf{N}$ with $(U_i)_\alpha = A_i$ for all i.

Exercise 4.1.18 Let X be an uncountable Polish space.

(i) Show that there is a sequence (U_n) of analytic subsets of $X \times X$ such that for every sequence (A_n) of analytic subsets of X there is an $x \in X$ with $A_n = (U_n)_x$ for all n.

(ii) Show that there is a set $U \in \mathcal{A}(\mathbf{\Pi}^1_1(\mathbf{N}^\mathbf{N} \times X))$ universal for $\mathcal{A}(\mathbf{\Pi}^1_1(X))$.

Exercise 4.1.19 Show that for any uncountable Polish space X, $\sigma(\mathbf{\Sigma}^1_1(X))$ is not closed under the Souslin operation.

In 2.2.13, we proved that a subset of $\mathbf{N}^\mathbf{N} \times \mathbf{N}^\mathbf{N}$ is closed if and only if it is the body of a tree T on $\mathbf{N} \times \mathbf{N}$. This gives us the following connection between trees and coanalytic sets, which will be used often in the sequel.

Proposition 4.1.20 *Let $A \subseteq \mathbf{N}^{\mathbf{N}}$. The following statements are equivalent.*

(i) *A is coanalytic.*

(ii) *There is a tree T on $\mathbf{N} \times \mathbf{N}$ such that*

$$\begin{aligned} \alpha \in A &\iff T[\alpha] \text{ is well-founded} \\ &\iff T[\alpha] \text{ is well-ordered with respect to } \leq_{KB}. \end{aligned}$$

Proof. Let $A \subseteq \mathbf{N}^{\mathbf{N}}$ be a coanalytic set. Then A^c is analytic. Let C be a closed set in $\mathbf{N}^{\mathbf{N}} \times \mathbf{N}^{\mathbf{N}}$ such that $\pi_1(C) = A^c$, where $\pi_1 : \mathbf{N}^{\mathbf{N}} \times \mathbf{N}^{\mathbf{N}} \longrightarrow \mathbf{N}^{\mathbf{N}}$ is the projection onto the first coordinate space. The existence of such a set follows from 4.1.1. By 2.2.13, there is a tree T on $\mathbf{N} \times \mathbf{N}$ such that $[T] = C$. Now note that

$$\begin{aligned} \alpha \in A^c &\iff \exists\beta((\alpha, \beta) \in [T]) \\ &\iff \exists\beta(\beta \in [T(\alpha)]) \\ &\iff T[\alpha] \text{ is not well-founded} \end{aligned}$$

Thus (ii) follows from (i).

(ii) $\Longrightarrow$ (i): Let $A \subseteq \mathbf{N}^{\mathbf{N}}$ satisfy (ii). Then A^c is the projection of $[T]$, and so A is coanalytic. ■

We close this section by giving a beautiful application of the Borel isomorphism theorem.

Example 4.1.21 Let $g : \mathbb{R} \times \mathbb{R} \longrightarrow \mathbb{R}$ be a Borel function. Define

$$f(x) = \sup_y g(x, y), \quad x \in X.$$

Assume that $f(x) < \infty$ for all x. The function f need not be Borel. To see this, take an analytic set $A \subseteq \mathbb{R}$ that is not Borel. Suppose $B \subseteq \mathbb{R} \times \mathbb{R}$ is a Borel set whose projection is A. Take $g = \chi_B$.

It is interesting to note that we can characterize functions $f : \mathbb{R} \longrightarrow \mathbb{R}$ of the form $f(x) = \sup_y g(x, y)$, g Borel. Call a function $f : \mathbb{R} \longrightarrow \mathbb{R}$ an **A-function** if $\{x : f(x) > t\}$ is analytic for every real number t.

Let $f(x) = \sup_y g(x, y)$, $g : \mathbb{R} \times \mathbb{R} \longrightarrow \mathbb{R}$ Borel. (Assume $f(x) < \infty$.) Then for every real t,

$$f(x) > t \iff (\exists y \in \mathbb{R})(g(x, y) > t).$$

So, f is an A-function. Further, f dominates a Borel function. (A function $u : E \longrightarrow \mathbb{R}$ is said to **dominate** $v : E \longrightarrow \mathbb{R}$ if $v(e) \leq u(e)$ for all $e \in E$.) We show that the converse is true.

Proposition 4.1.22 *(H. Sarbadhikari [99]) For every A-function f : $\mathbb{R} \longrightarrow \mathbb{R}$ dominating a Borel function there is a Borel $g : \mathbb{R} \times \mathbb{R} \longrightarrow \mathbb{R}$ such that $f(x) = \sup_y g(x, y)$.*

Proof. Let $v : \mathbb{R} \longrightarrow \mathbb{R}$ be a Borel function such that $v(x) \leq f(x)$ for all x. For $n \in \mathbb{Z}$, let

$$B_n = \{x \in \mathbb{R} : n \leq v(x) < n + 1\}.$$

Fix an enumeration $\{r_m : m \in \mathbb{N}\}$ of the set of all rational numbers. Let

$$A = \{(x, y) : f(x) > y\}.$$

Since

$$A = \bigcup_m \{(x, y) \in \mathbb{R} \times \mathbb{R} : f(x) > r_m > y\}$$

and f is an A-function, A is analytic. By 4.1.1, there is a Borel set $B \subseteq (\mathbb{R} \times \mathbb{R}) \times \mathbb{R}$ whose projection is A. Define $h : \mathbb{R}^3 \longrightarrow \mathbb{R}$ by

$$h(x, y, z) = \begin{cases} y & \text{if } (x, y, z) \in B, \\ n & \text{if } x \in B_n \ \& \ (x, y, z) \in \mathbb{R}^3 \setminus B. \end{cases}$$

The function h is Borel, and

$$f(x) = \sup_{(y,z)} h(x, y, z).$$

Let $u : \mathbb{R} \longrightarrow \mathbb{R}^2$ be a Borel isomorphism. Such a map exists by the Borel isomorphism theorem. Define g by

$$g(x, y) = h(x, u(y)).$$

■

Remark 4.1.23 Later (4.11.6) we shall give an example of an A-function $f : \mathbb{R} \longrightarrow \mathbb{R}$ that does not dominate a Borel function.

4.2 Σ_1^1 and Π_1^1 Complete Sets

In this section we present a commonly used method to show that a set is analytic or coanalytic but not Borel. Most often if a set is, say, Σ_1^1, then it has a suitable description to show that it is so. However, showing that it is not Borel (say) is generally hard.

Let X be a Polish space and $A \subseteq X$. We say that A is **Σ_1^1-complete** if A is analytic and for every Polish space Y and every analytic $B \subseteq Y$, there is a Borel map $f : Y \longrightarrow X$ such that $f^{-1}(A) = B$. Since there are analytic sets that are not Borel, and since the class of Borel sets is closed under Borel preimages, no Σ_1^1-complete set is Borel. This gives us a technique to show that an analytic set is non-Borel: We simply show that the set under consideration is Σ_1^1-complete. It may appear that we have made the problem more difficult. This is not the case. It has been shown that the statement "every analytic non-Borel set is $\mathbf{\Sigma}_1^1$-complete" is consistent with **ZFC**. Further, whether it is possible to prove the existence of such a set in **ZFC** is still open.

Let X, Y be Polish spaces and $A \subseteq X$, $B \subseteq Y$. We say that A is **Borel reducible** to B if there is a Borel map $f : X \longrightarrow Y$ such that $f^{-1}(B) = A$. Note that if an analytic set A is Borel reducible to B and A is a $\mathbf{\Sigma}_1^1$-complete set, then B is $\mathbf{\Sigma}_1^1$-complete. We define $\mathbf{\Pi}_1^1$-complete sets analogously. All the above remarks clearly hold for $\mathbf{\Pi}_1^1$-complete sets.

We now give a few illustrations of our method.

Example 4.2.1 We identify a tree T on $\mathbb{N}$ with its characteristic function $\chi_T \in 2^{\mathbb{N}^{<\mathbb{N}}}$. So, we put

$$Tr = \{T \in 2^{\mathbb{N}^{<\mathbb{N}}} : T \text{ is a tree on } \mathbb{N}\}.$$

Note that for any $T \in 2^{\mathbb{N}^{<\mathbb{N}}}$,

$$T \in Tr \Longleftrightarrow (\forall s \in \mathbb{N}^{<\mathbb{N}})(\forall t \in \mathbb{N}^{<\mathbb{N}})(s \in T \ \&\ t \prec s \Longrightarrow t \in T).$$

Hence, Tr is a G_δ set in $2^{\mathbb{N}^{<\mathbb{N}}}$, where $2^{\mathbb{N}^{<\mathbb{N}}}$ is equipped with the product of discrete topologies on $2 = \{0, 1\}$, and hence is a Polish space. Let

$$WF = \{T \in Tr : T \text{ is well-founded}\}.$$

We show that WF is $\mathbf{\Pi}_1^1$-complete.

Observe that

$$T \in WF \Longleftrightarrow T \in Tr \ \&\ \forall\beta\exists n(T(\beta|n) = 0).$$

Therefore, $WF = \forall^{\mathbb{N}^{\mathbb{N}}} E$, where

$$E = \{(T, \beta) \in 2^{\mathbb{N}^{<\mathbb{N}}} \times \mathbb{N}^{\mathbb{N}} : T \in Tr \ \&\ \exists n(T(\beta|n) = 0)\}.$$

It is quite easy to see that the set E is Borel. Hence, WF is coanalytic.

Now take any coanalytic set C in $\mathbb{N}^{\mathbb{N}}$. By 4.1.20, there is a tree T on $\mathbb{N} \times \mathbb{N}$ such that

$$\alpha \in C \Longleftrightarrow T[\alpha] \text{ is well-founded.}$$

Define $f : \mathbb{N}^{\mathbb{N}} \longrightarrow Tr$ by

$$f(\alpha) = T[\alpha],$$

the section of T at α. The map f is continuous: Take any $s \in \mathbb{N}^{<\mathbb{N}}$ and note that

$$f(\alpha)(s) = 1 \Longleftrightarrow T(\alpha||s|, s) = 1.$$

Thus $\pi_s \circ f$ is continuous for all s, and so f is continuous.

As $C = f^{-1}(WF)$, by the Borel isomorphism theorem it follows that WF is $\mathbf{\Pi}_1^1$-complete.

Example 4.2.2 We identify binary relations on $\mathbf{N}$ with points of $2^{\mathbf{N}\times\mathbf{N}}$. As before, we equip $2^{\mathbf{N}\times\mathbf{N}}$ with the product of discrete topologies on $2 = \{0,1\}$. Let

$$LO = \{\alpha \in 2^{\mathbf{N}\times\mathbf{N}} : \alpha \text{ is a linear order}\}.$$

It is easy to check that LO is Borel. Define

$$WO = \{\alpha \in 2^{\mathbf{N}\times\mathbf{N}} : \alpha \text{ is a well-order}\}.$$

Arguing as in 4.2.1, we see that WO is coanalytic. We now show that WO is Π_1^1-complete. It is sufficient to show that there is a continuous map $R : Tr \longrightarrow 2^{\mathbf{N}\times\mathbf{N}}$ such that $WF = R^{-1}(WO)$.

Fix a bijection $u : \mathbf{N} \longrightarrow \mathbf{N}^{<\mathbf{N}}$. To each $T \in Tr$, associate a binary relation $R(T)$ on $\mathbf{N}$ as follows:

$$\begin{aligned} k\ R(T)\ l \iff & (u(k), u(l) \notin T \ \&\ k \le l) \\ & \vee (u(k) \in T \ \&\ u(l) \notin T) \\ & \vee (u(k), u(l) \in T \ \&\ \ u(k) \le_{KB} u(l)) \end{aligned}$$

It is easy to check that $T \longrightarrow R(T)$ is a continuous map from Tr to $2^{\mathbf{N}\times\mathbf{N}}$. Since a tree T on $\mathbf{N}$ is well-founded if and only if $\le_{KB}$ is a well-order on T (1.10.10.), $WF = R^{-1}(WO)$.

Exercise 4.2.3 Let

$$N = \{\alpha \in \mathbf{N}^{\mathbf{N}} : \alpha(i) > 0 \text{ for infinitely many } i\}.$$

Show the following

(i) N is Polish.

(ii) The set

$$IF^* = \{K \in K(\mathbf{N}^{\mathbf{N}}) : N \bigcap K \neq \emptyset\}$$

is Σ_1^1-complete.

Exercise 4.2.4 Show that the set

$$\{K \in K(\mathbb{R}) : K \subseteq \mathbb{Q}\}$$

is Π_1^1-complete, where $K(\mathbb{R})$ is the space of all compact subsets of $\mathbb{R}$ equipped with the Vietoris topology.

Proposition 4.2.5 *Let X be an uncountable Polish space. Then*

$$U(X) = \{K \in K(X) : K \text{ is uncountable}\}$$

is Σ_1^1-complete.

Proof. We first show that $U(X) \in \mathbf{\Sigma}^1_1$. Let $P(X)$ denote the set of all nonempty perfect subsets of X. Then $P(X)$ is Borel in $K(X)$. To see this, take a countable base (V_n) for X. We have

$$\begin{aligned} K \text{ is perfect} \iff & \forall n(K \bigcap V_n \neq \emptyset \\ & \Longrightarrow \exists k \exists l(V_k, V_l \subseteq V_n \\ & \& \ V_k \bigcap V_l = \emptyset \ \& \ K \bigcap V_k, K \bigcap V_l \neq \emptyset)). \end{aligned}$$

So,

$$P(X) = \bigcap_n [A_n^c \bigcup \bigcup_{(k,l) \in S_n} (A_k \bigcap A_l)],$$

where

$$A_n = \{K \in K(X) : K \bigcap V_n \neq \emptyset\}$$

and

$$S_n = \{(k,l) : V_k \subseteq V_n \ \& \ V_l \subseteq V_n \ \& \ V_k \bigcap V_l = \emptyset\}.$$

Hence, $P(X)$ is Borel. Let $K \in K(X)$. By 2.6.3,

$$K \text{ is uncountable} \iff (\exists P \in K(X))(P \in P(X) \ \& \ P \subseteq K).$$

By 2.4.11, the set

$$\{(K,L) \in K(X) \times K(X) : K \subseteq L\}$$

is closed. Hence, $U(X) \in \mathbf{\Sigma}^1_1$.

It remains to show that $U(X)$ is $\mathbf{\Sigma}^1_1$-complete. Since every uncountable Polish space contains a G_δ set homeomorphic to $\mathbb{N}^\mathbb{N}$, it is sufficient to prove the result for $X = \mathbb{N}^\mathbb{N}$. Let N be as in 4.2.3. Define $f : \mathbb{N}^\mathbb{N} \longrightarrow K(\mathbb{N}^\mathbb{N})$ by

$$f(\alpha) = \{\beta \in \mathbb{N}^\mathbb{N} : \beta \leq \alpha \text{ pointwise}\}.$$

Then f is continuous. Further,

$$\alpha \in N \iff f(\alpha) \text{ is uncountable}.$$

Now consider the map $g : K(K(\mathbb{N}^\mathbb{N})) \longrightarrow K(\mathbb{N}^\mathbb{N})$ defined by

$$g(\mathcal{K}) = \bigcup \mathcal{K}, \quad \mathcal{K} \in K(K(\mathbb{N}^\mathbb{N})).$$

The map g is continuous (2.4.11). Define

$$h(K) = g(f(K)), \quad K \in K(\mathbb{N}^\mathbb{N}).$$

The map h is continuous, and

$$IF^* = h^{-1}(\{K \in K(\mathbb{N}^\mathbb{N}) : K \text{ is uncountable}\}).$$

The result follows from 4.2.3. ■

Corollary 4.2.6 *Let X be an uncountable Polish space. Then*

$$\{K \in K(X) : K \text{ is countable}\}$$

is Π_1^1-complete.

Proposition 4.2.7 *(Mazurkiewicz) The set $DIFF$ of everywhere differentiable functions $f : [0,1] \longrightarrow \mathbb{R}$ is Π_1^1-complete. In particular, it is a coanalytic, non-Borel subset of $C[0,1]$.*

Proof. We know that the map $(f, x) \longrightarrow f(x)$ is continuous on $C[0,1] \times X$. From this it easily follows that $DIFF$ is Π_1^1. We now show that WF is Borel reducible to $DIFF$. This will complete the proof.

Let $s \longrightarrow \langle s \rangle$ be a bijection from $\mathbb{N}^{<\mathbb{N}}$ onto $\mathbb{N}$. For each $s \in \mathbb{N}^{<\mathbb{N}}$, define an open interval $J_s \subseteq [0,1]$ and a nonempty closed interval K_s satisfying the following conditions.

(i) K_s and J_s are concentric.

(ii) $|K_s| \le 2^{-\langle s \rangle}(|J_s| - |K_s|)$.

(iii) $J_{s^\frown n} \subseteq K_s^{(L)}$, where $K_s^{(L)}$ is the left half of K_s.

(iv) $J_{s^\frown n} \bigcap J_{s^\frown m} = \emptyset$, if $n \neq m$.

Let $K_s^{(R)}$ denote the right half of K_s. So the $K_s^{(R)}$'s are pairwise disjoint. Also, for every $\alpha \in \mathbb{N}^{\mathbb{N}}$, $\bigcap_k J_{\alpha|k} = \bigcap_k K_{\alpha|k} = \bigcap_k K_{\alpha|k}^{(L)}$ is a sigleton. For any tree T on $\mathbb{N}$, set

$$G_T = \bigcup_{\alpha \in [T]} \bigcap_k J_{\alpha|k}.$$

Clearly,

$$T \in WF \Longleftrightarrow G_T = \emptyset. \qquad (*)$$

Further,

$$G_T = \bigcup_{\alpha \in [T]} \bigcap_k K_{\alpha|k}^{(L)} = \bigcap_k \bigcup_{s \in T \bigcap \mathbb{N}^k} J_s.$$

For each closed interval $I = [a, b] \subseteq [0,1]$, let $\varphi_I : [0,1] \longrightarrow [0, |I|]$ be a function in $DIFF$ that is positive precisely on (a, b), and $\varphi_I(\frac{a+b}{2}) = b - a$.

Let T be a tree on $\mathbb{N}$ and $x \in [0,1]$. Define

$$F_T(x) = \sum_{s \in T} \varphi_{K_s^{(R)}}(x).$$

Since $0 \le \varphi_{K_s^{(R)}}(x) \le |K_s^{(R)}| \le 2^{-\langle s \rangle}$, F_T is a continuous function.

$T \longrightarrow F_T$ is a continuous map from Tr to $C[0,1]$: Let S and T be two trees on $\mathbb{N}$ such that

$$T \bigcap \{s \in Tr : \langle s \rangle < N\} = S \bigcap \{s \in Tr : \langle s \rangle < N\}.$$

Then, for any $x \in [0,1]$,

$$|F_T(x) - F_S(x)| \le \sum_{\langle s\rangle \ge N} (\varphi_{K_S^R}(x) - \varphi_{K_T^R}(x)) \le 2^{-N}.$$

Hence, $T \longrightarrow F_T$ is continuous.

Our proof will be complete if we show that

$$T \in WF \Longleftrightarrow F_T \in DIFF.$$

By (*) it is sufficient to show that for every $x \in [0,1]$,

$$x \notin G_T \Longleftrightarrow F_T \text{ is not differentiable at } x.$$

Let $x \in G_T$. Choose $\alpha \in [T]$ such that $x \in K_{\alpha|k}^{(L)}$ for every k. Let $l_k = |K_{\alpha|k}|$, and let c_k be the midpoint of $K_{\alpha|k}^{(R)}$. Since $x \notin K_s^{(R)}$ for any s, $F_T(x) = 0$. Also $F_T(c_k + l_k/4) = 0$. So $\frac{F_T(c_k+l_k/4)-F_T(x)}{c_k+l_k/4-x} = 0$. On the other hand, $|\frac{F_T(c_k)-F_T(x)}{c_k-x}| = |\frac{F_T(c_k)}{c_k-x}| \ge \frac{2}{3}$. Since $c_k, c_k + l_k \to x$, it follows that f is not differentiable at x.

Now assume that $x \notin G_T$. Then there exists a positive integer N such that for no $s \in T$ with $\langle s\rangle \ge N$, $x \in J_s$. Let $s \in T$ with $\langle s\rangle \ge N$. Then for any $h \ne 0$,

$$\begin{aligned} \left|\frac{\varphi_{K_s^{(R)}}(x+h)-\varphi_{K_s^{(R)}}(x)}{h}\right| &= \frac{\varphi_{K_s^{(R)}}(x+h)}{h} \\ &\le \frac{|K_s^{(R)}|}{|J_s|-|K_s|} \\ &\le 2^{-\langle s\rangle}. \end{aligned}$$

For any $n \ge N$, set

$$F_T^n(x) = \sum_{s\in T, \langle s\rangle \le n} \varphi_{K_s^{(R)}}(x).$$

We have

$$\frac{F_T(x+h) - F_T(x)}{h} - \frac{F_T^n(x+h) - F_T^n(x)}{h} \le 2^{-n}.$$

Since F_T^n is differentiable at x, it follows that

$$\limsup_{h\to 0} \frac{F_T(x+h) - F_T(x)}{h} - \liminf_{h\to 0} \frac{F_T(x+h) - F_T(x)}{h} \le 2^{-n+1}.$$

Letting $n \to \infty$, we see that F_T is differentiable at x. ∎

4.3 Regularity Properties

In this section we show that analytic sets have nice structural properties; e.g., they are measurable with respect to all finite measures, they have the Baire property, and they satisfy the continuum hypothesis. We also discuss the possible cardinalities of coanalytic sets. These are very useful facts, and subsequently we give several applications of these.

In 3.5.22, we proved that if $(X, \mathcal{B}, \mu)$ is a complete σ-finite measure space, then $\mathcal{B}$ is closed under the Souslin operation. We also proved that the σ-algebra of sets with the Baire property is closed under the Souslin operation. Using these and 4.1.14, we get the following two theorems.

Theorem 4.3.1 *Let μ be a σ-finite measure on $(X, \mathcal{B}_X)$, X Polish. Then every analytic subset of X is μ-measurable.*

Theorem 4.3.2 *Every analytic subset of a Polish space has the Baire property .*

Exercise 4.3.3 Let X be an uncountable Polish space and $\mathcal{B}$ either the σ-algebra of subsets of X having the Baire property or the completion $\overline{\mathcal{B}_X}^{\mu}$, where μ is a continuous probability on $\mathcal{B}_X$. Show that no σ-algebra $\mathcal{A}$ satisfying

$$\sigma(\mathbf{\Sigma}^1_1) \subseteq \mathcal{A} \subseteq \mathcal{B}$$

is countably generated.

As mentioned earlier, we shall give several applications of these results in the sequel. At present we use it to give a solution to a problem of Ulam[121]. Recall that in Chapter 3 we considered the following problem: Is

$$\mathcal{P}(\mathbb{R}) \bigotimes \mathcal{P}(\mathbb{R}) = \mathcal{P}(\mathbb{R} \times \mathbb{R})?$$

We showed that under **CH** the answer to this question is yes. In the same spirit, Ulam[121] asked the following question: Is

$$\sigma(\mathbf{\Sigma}^1_1(\mathbb{R})) \bigotimes \sigma(\mathbf{\Sigma}^1_1(\mathbb{R})) = \sigma(\mathbf{\Sigma}^1_1(\mathbb{R} \times \mathbb{R}))?$$

The answer to this question is no.

Theorem 4.3.4 *(B. V. Rao[95]) Let X be an uncountable Polish space and $U \subseteq X \times X$ universal analytic. Then*

$$U \notin \mathcal{P}(X) \bigotimes \mathcal{B},$$

where $\mathcal{B}$ is as in 4.3.3.

Proof. Suppose $U \in \mathcal{P}(X) \bigotimes \mathcal{B}$. We shall get a contradiction. From 3.1.7, there are $C_0, C_1, C_2, \ldots \subseteq X$ and $D_0, D_1, D_2, \ldots$ in $\mathcal{B}$ such that

$U \in \sigma(\{C_i \times D_i : i \in \mathbb{N}\})$. Let Y be an uncountable Borel subset of X such that each $D_i \bigcap Y$ is Borel. In particular, every section $(U \bigcap (X \times Y))_x$, $x \in X$, is Borel. Let E be an analytic non-Borel set contained in Y. Since U is universal,

$$E = U_{x_0} = (U \bigcap (X \times Y))_{x_0}$$

for some $x_0 \in X$. We have arrived at a contradiction. ■

Next we show that analytic sets satisfy the continuum hypothesis.

Theorem 4.3.5 *Every uncountable analytic set contains a homeomorph of the Cantor set and hence is of cardinality* $\mathfrak{c}$.

Proof. Let X be a Polish space and $f : \mathbb{N}^{\mathbb{N}} \longrightarrow X$ a continuous map whose range is uncountable. We first show that there is a Cantor scheme $\{F_s : s \in 2^{<\mathbb{N}}\}$ of closed subsets of $\mathbb{N}^{\mathbb{N}}$ such that whenever $|s| = |t|$ and $s \neq t$, $f(F_s) \bigcap f(F_t) = \emptyset$.

Since the range of f is uncountable, we get an uncountable $Z \subseteq \mathbb{N}^{\mathbb{N}}$ such that $f|Z$ is one-to-one. By the Cantor – Bendixson theorem (2.6.2), we can further assume that Z is dense-in-itself. Take a compatible complete metric $d < 1$ on $\mathbb{N}^{\mathbb{N}}$. We define a system $\{U_s : s \in 2^{<\mathbb{N}}\}$ of nonempty open subsets of $\mathbb{N}^{\mathbb{N}}$ satisfying the following conditions:

(i) diameter$(U_s) < 2^{-|s|}$;

(ii) $U_s \bigcap Z \neq \emptyset$;

(iii) $\text{cl}(U_{s^\frown\epsilon}) \subseteq U_s$, $\epsilon = 0, 1$; and

(iv) whenever $|s| = |t|$ and $s \neq t$, $f(\text{cl}(U_s)) \bigcap f(\text{cl}(U_t)) = \emptyset$. In particular, $\text{cl}(U_s) \bigcap \text{cl}(U_t) = \emptyset$.

We define such a system by induction on $|s|$. Take $U_e = X$. Suppose U_s has been defined for some s. Since Z is dense-in-itself and U_s open, $U_s \bigcap Z$ has at least two distinct points, say x_0, x_1. Then $f(x_0) \neq f(x_1)$. Let W_0 and W_1 be disjoint open sets containing $f(x_0)$ and $f(x_1)$ respectively. Since f is continuous, there are open sets $U_{s^\frown 0}$ and $U_{s^\frown 1}$ satisfying the following conditions:

(a) $x_\epsilon \in U_{s^\frown\epsilon} \subseteq \text{cl}(U_{s^\frown\epsilon}) \subseteq U_s$, $\epsilon = 0$ or 1;

(b) diameter$(U_{s^\frown\epsilon}) < \frac{1}{2^{|s|+1}}$; and

(c) $f(\text{cl}(U_{s^\frown\epsilon})) \subseteq W_\epsilon$, $\epsilon = 0$ or 1.

(d) In particular, $f(\text{cl}(U_{s^\frown 0})) \bigcap f(\text{cl}(U_{s^\frown 1})) = \emptyset$.

Put $F_s = \text{cl}(U_s)$. Let $C = \mathcal{A}(\{F_s\})$. Then C is homeomorphic to the Cantor set, and $f|C$, being one-to-one and continuous, is an embedding. ■

Remark 4.3.6 The above proof shows more: *Let X, Y be Polish spaces and $f : X \longrightarrow Y$ a continuous map with range uncountable. Then there is a homeomorph of the Cantor set $C \subseteq X$ such that $f|C$ is one-to-one.*

We now give some consequences of 4.3.5 (and 4.3.6).

Proposition 4.3.7 *Let X be a Polish space and $A \subseteq X$. The following statements are equivalent.*

(i) *A is analytic.*

(ii) *There is a closed set $C \subseteq X \times \mathbb{N}^{\mathbb{N}}$ such that*

$$A = \{x \in X : C_x \text{ is uncountable}\}.$$

(iii) *There is a Polish space Y and an analytic set $B \subseteq X \times Y$ such that*

$$A = \{x \in X : B_x \text{ is uncountable}\}.$$

Proof. (i) $\Longrightarrow$ (ii): Let $f : \mathbb{N}^{\mathbb{N}} \longrightarrow X$ be a continuous map with range A and $\pi_1 : \mathbb{N}^{\mathbb{N}} \times \mathbb{N}^{\mathbb{N}} \longrightarrow \mathbb{N}^{\mathbb{N}}$ the projection map. Note that π_1 is continuous and $\pi_1^{-1}(\alpha)$ uncountable for all α. Since $\mathbb{N}^{\mathbb{N}} \times \mathbb{N}^{\mathbb{N}}$ is homeomorphic to $\mathbb{N}^{\mathbb{N}}$, this shows that there is a continuous map $h : \mathbb{N}^{\mathbb{N}} \longrightarrow \mathbb{N}^{\mathbb{N}}$ such that $h^{-1}(\alpha)$ is uncountable for all α. Take $C = \text{graph}(f \circ h)$.

(iii) is a special case of (ii).

(iii) $\Longrightarrow$ (i): By (4.3.6), we have the following: Let P, Q be Polish spaces and $f : P \longrightarrow Q$ a continuous map. The range of f is uncountable if and only if there is a countable dense-in-itself subset Z of P such that $f|Z$ is one-to-one.

Note also that the set

$$D = \{(x_n) \in (\mathbb{N}^{\mathbb{N}})^{\mathbb{N}} : \{x_n : n \in \mathbb{N}\} \text{ is dense-in-itself}\}$$

is a G_δ set in $(\mathbb{N}^{\mathbb{N}})^{\mathbb{N}}$.

Now let X, Y and B be as in (iii). Let $f : \mathbb{N}^{\mathbb{N}} \longrightarrow X \times Y$ be a continuous map with range B. By (a),

$$\begin{aligned} B_x \text{ is uncountable} \iff\ & (\exists (z_n) \in D)(\forall i \forall j (i \neq j \Longrightarrow f(z_i) \neq f(z_j)), \\ & \& \forall k(\pi_X(f(z_k)) = x)), \end{aligned}$$

where $\pi_X : X \times Y \longrightarrow X$ is the projection map. The result follows from (b). ■

We know that if X is a separable metric space, Y a metrizable space, and $f : X \longrightarrow Y$ a continuous map, then $f(X)$ is separable. Using 4.3.5, we now show that this result is true even for Borel f when X is analytic. The beautiful proof given below is due to S. Simpson.

Theorem 4.3.8 *(S. Simpson [79]) Let X be an analytic subset of a Polish space, Y a metrizable space, and $f : X \longrightarrow Y$ a Borel map. Then $f(X)$ is separable.*

Proof. Without any loss of generality, we assume that X is Polish and $Y = f(X)$. Suppose Y is not separable. Then there is an uncountable closed discrete subspace Z of Y. As $|X| = \mathfrak{c}$, $|Y| \leq \mathfrak{c}$, and hence $|Z| \leq \mathfrak{c}$. Let $X' = f^{-1}(Z)$. Note that X' is Borel. Now take any $A \subseteq \mathbb{R}$ of the same cardinality as Z that does not contain any uncountable closed set. We have proved the existence of such a set in 3.2.8. Let g be any one-to-one map from Z onto A. Since Z is discrete, g is continuous. Clearly, $g \circ f$ is Borel. As $A = g(f(X'))$, A is an uncountable analytic set not containing a perfect set. This contradicts 4.3.5. ■

Corollary 4.3.9 *Every Borel homomorphism $\varphi : G \longrightarrow H$ from a completely metrizable group G to a metrizable group H is continuous.*

Proof. Let (g_n) be a sequence in G converging to g. Replacing G by the closed subgroup generated by $\{g_n : n \in \mathbb{N}\}$, we assume that G is Polish. By 4.3.8, $\varphi(G)$ is separable. The result follows from 3.5.9. ■

As another application of 4.3.8, we give a partial answer to a question raised by A. H. Stone [120]: Let X, Y be metrizable spaces and $f : X \longrightarrow Y$ a Borel map. Is there an ordinal $\alpha < \omega_1$ such that f is of class α? The answer to this question is clearly yes if Y is second countable. By 4.3.8, Y is separable if X is analytic. So, Stone's question has a positive answer if X is analytic. *This problem is open even for coanalytic X!*

Finally, we apply 4.3.5 to give a partial solution to a well-known problem in set theory. A set A of reals has **strong measure zero** if for every sequence (a_n) of positive real numbers, there exists a sequence (I_n) of open intervals such that $|I_n| \leq a_n$ and $A \subseteq \bigcup_n I_n$.

Proposition 4.3.10 **(i)** *Every countable set of reals has strong measure zero.*

(ii) *Every strong measure zero set is of (Lebesgue) measure zero.*

(iii) *The family of all strong measure zero sets forms a σ-ideal.*

Proof. (i) and (ii) are immediate consequences of the definition. We prove (iii) now. Let (A_n) be a sequence of strong measure zero sets. Take any sequence (a_n) of positive real numbers. Choose pairwise disjoint infinite subsets $I_0, I_1, I_2, \ldots$ of $\mathbb{N}$ whose union is $\mathbb{N}$. For each n choose open intervals I_m^n, $m \in I_n$, such that $|I_m^n| \leq a_m$ and $A_n \subseteq \bigcup_{m \in I_n} I_m^n$. Note that

$$\bigcup_n A_n \subseteq \bigcup_{n \in \mathbb{N}} \bigcup_{m \in I_n} I_m^n.$$

The proof of (iii) is clearly seen now. ■

Here is another simple but useful fact about strong measure zero sets.

Proposition 4.3.11 *Let $A \subseteq [0,1]$ be a strong measure zero set and $f : [0,1] \longrightarrow \mathbb{R}$ a continuous map. Then the set $f(A)$ has strong measure zero.*

Proof. Let (a_n) be any sequence of positive real numbers. We have to show that there exist open intervals J_n, $n \in \mathbb{N}$, such that $|J_n| \leq a_n$ and $f(A) \subseteq \bigcup_n J_n$. Since f is uniformly continuous, for each n there is a positive real number b_n such that whenever $X \subseteq [0,1]$ is of diameter at most b_n, the diameter of $f(X)$ is at most a_n. Since A has strong measure zero, there are open intervals I_n, $n \in \mathbb{N}$, such that $|I_n| \leq b_n$ and $A \subseteq \bigcup_n I_n$. Take $J_n = f(I_n)$. ■

Here are some interesting questions on strong measure zero sets. Is there an uncountable set of reals that is not a strong measure zero set? Do all measure zero sets have strong measure zero? We consider the second question first.

Example 4.3.12 It is easy to see that there is no sequence (I_n) of open intervals such that the length of I_n is at most $3^{-(n+1)}$ and (I_n) cover the Cantor ternary set C. Hence, C is not a strong measure zero set. It follows that not all measure zero sets have strong measure zero.

From 4.3.12 and 4.3.11 we get the following interesting result.

Proposition 4.3.13 *No set of reals containing a perfect set has strong measure zero.*

The Borel conjecture [20]: *No uncountable set of reals is a strong measure zero set.* ■

From 4.3.13 and 4.3.5, we now have the following.

Proposition 4.3.14 *No uncountable analytic $A \subseteq \mathbb{R}$ has strong measure zero.*

Thus, no analytic set can be a counterexample to the Borel conjecture. It has been shown that the Borel conjecture is independent of **ZFC**. The proof of this is obviously beyond the scope of this book. We refer the interested reader to [9]. Here, under the continuum hypothesis, we give an example of an uncountable strong measure zero set.

Exercise 4.3.15 **(i)** Show that there is a set A of reals of cardinality $\mathfrak{c}$ such that $A \bigcap C$ is countable for every closed, nowhere dense set. (Such a set A is called a **Lusin set.**)

(ii) Show that every Lusin set is a strong measure zero set.

Does **CH** hold for coanalytic sets? This cannot be decided in **ZFC**. However, in **ZFC** we can say something about the cardinalities of coanalytic sets—*a coanalytic set is either countable or is of cardinality* $\aleph_1$ *or* $\mathfrak{c}$. We prove these facts now.

Let T be a well-founded tree on $\mathbf{N}$. Recall the definition of the rank function $\rho_T : T \longrightarrow \mathbf{ON}$ given in Chapter 1:

$$\rho_T(u) = \sup\{\rho_T(v) + 1 : u \prec v, v \in T\},\ u \in T.$$

(We take $\sup(\emptyset) = 0$.) Note that $\rho_T(u) = 0$ if u is terminal in T.

We extend this notion for ill-founded trees too. Let T be an ill-founded tree and $s \in \mathbf{N}^{<\mathbf{N}}$. Define

$$\rho_T(s) = \begin{cases} 0 & \text{if } s \notin T, \\ \rho_{T_s}(e) & \text{if } s \in T \ \&\ T_s \text{ is well-founded}, \\ \omega_1 & \text{otherwise.} \end{cases}$$

Note that T is well-founded if and only if $\rho_T(e) < \omega_1$.

Lemma 4.3.16 *Let T be a tree on $\mathbf{N} \times \mathbf{N}$ and $\xi < \omega_1$. For every $s \in \mathbf{N}^{<\mathbf{N}}$,*

$$C_s^\xi = \{\alpha \in \mathbf{N}^\mathbf{N} : \rho_{T[\alpha]}(s) \le \xi\}$$

is Borel.

Proof. We prove the result by induction on ξ. Note that

$$C_s^0 = \{\alpha \in \mathbf{N}^\mathbf{N} : \forall i((\alpha|(|s|+1), s\hat{\ }i) \notin T)\}.$$

So, C_s^0 is Borel (in fact closed) for all s. Since for any countable ordinal $\xi > 0$,

$$C_s^\xi = \bigcap_i \bigcup_{\eta<\xi} C_{s\hat{\ }i}^\eta,$$

the proof is easily completed by transfinite induction. ■

Theorem 4.3.17 *Every coanalytic set is a union of* $\aleph_1$ *Borel sets.*

Proof. Let X be Polish and $C \subseteq X$ coanalytic. By the Borel isomorphism theorem (3.3.13), without any loss of generality we may assume that $X = \mathbf{N}^\mathbf{N}$. By 4.1.20, there is a tree T on $\mathbf{N} \times \mathbf{N}$ such that

$$\alpha \in C \Longleftrightarrow T[\alpha] \text{ is well-founded.}$$

So,

$$\alpha \in C \Longleftrightarrow \rho_{T[\alpha]}(e) < \omega_1.$$

Therefore,

$$C = \bigcup_{\xi<\omega_1} C_e^\xi,$$

where the C_e^ξ are as in 4.3.16. ■

The sets $C_e^\xi, \xi < \omega_1$, defined in the above proof are called the **constituents** of C. Since **CH** holds for Borel sets, we now have the following result.

Theorem 4.3.18 *A coanalytic set is either countable or of cardinality* $\aleph_1$ *or* $\mathfrak{c}$.

The following question remains: Does **CH** hold for coanalytic sets? Another related question is, Is there an uncountable coanalytic set that does not contain a perfect set (equivalently, an uncountable Borel set)? Gödel[45] showed that in the universe L of constructible sets, which is a model of **ZFC**, there is an uncountable coanalytic set that does not contain a perfect set. (See also [49], p. 529.) On the other hand, under **"analytic determinacy"** ([53], p. 206) every uncountable coanalytic set contains a perfect set. Hence under this hypothesis every uncountable coanalytic set is of cardinality $\mathfrak{c}$. **"Analytic determinacy"** can be proved from the existence of large cardinals. Thus, the statement "there is an uncountable coanalytic set not containing a perfect set" cannot be decided in **ZFC**. Any further discussion on this topic is beyond the scope of these notes.

4.4 The First Separation Theorem

The separartion theorems and the dual results—the reduction theorems—are among the most important results on analytic and coanalytic sets, with far-reaching consequences on Borel sets.

Theorem 4.4.1 *(The first separation theorem for analytic sets) Let A and B be disjoint analytic subsets of a Polish space X. Then there is a Borel set C such that*

$$A \subseteq C \text{ and } B \bigcap C = \emptyset. \qquad (*)$$

(If $(\star)$ is satisfied, we say that C separates A from B.)

The proof of this theorem is based on the following combinatorial lemma.

Lemma 4.4.2 *Suppose $E = \bigcup_n E_n$ cannot be separated from $F = \bigcup_m F_m$ by a Borel set. Then there exist m, n such that E_n cannot be separated from F_m by a Borel set.*

Proof. Suppose for every m, n there is a Borel set C_{mn} such that

$$E_n \subseteq C_{mn} \text{ and } F_m \bigcap C_{mn} = \emptyset.$$

It is fairly easy to check that the Borel set

$$C = \bigcup_n \bigcap_m C_{mn}$$

separates E from F. ∎

Proof of 4.4.1. Let A and B be two disjoint analytic subsets of X. Suppose there is no Borel set C such that

$$A \subseteq C \text{ and } B \bigcap C = \emptyset.$$

We shall get a contradiction. Let $f : \mathbf{N}^{\mathbf{N}} \longrightarrow A$ and $g : \mathbf{N}^{\mathbf{N}} \longrightarrow B$ be continuous surjections. We shall get $\alpha, \beta \in \mathbf{N}^{\mathbf{N}}$ such that $f(\Sigma(\alpha|n))$ cannot be separated from $g(\Sigma(\beta|n))$ by a Borel set for any $n \in \mathbf{N}$.

We first complete the proof assuming that α, β satisfying the above properties have been defined. Since A and B are disjoint, $f(\alpha) \neq g(\beta)$. Since f and g are continuous, there exist disjoint open sets U and V containing $f(\alpha)$ and $g(\beta)$ respectively. By the continuity of f and g, there exists an $n \in \mathbf{N}$ such that $f(\Sigma(\alpha|n)) \subseteq U$ and $g(\Sigma(\beta|n)) \subseteq V$. In particular, $f(\Sigma(\alpha|n))$ is separated from $g(\Sigma(\beta|n))$ by a Borel set. This is a contradiction.

Definition of α, β: We proceed by induction.

Since $A = \bigcup f(\Sigma(n))$ and $B = \bigcup g(\Sigma(m))$, by 4.4.2 there exist $\alpha(0)$ and $\beta(0)$ such that $f(\Sigma(\alpha(0)))$ cannot be separated from $g(\Sigma(\beta(0)))$ by a Borel set. Suppose $\alpha(0), \alpha(1), \ldots, \alpha(k)$ and $\beta(0), \beta(1), \ldots, \beta(k)$ satisfying the above conditions have been defined. Since

$$f(\Sigma(\alpha(0), \alpha(1), \ldots, \alpha(k))) = \bigcup_n f(\Sigma(\alpha(0), \alpha(1), \ldots, \alpha(k), n))$$

and

$$g(\Sigma(\beta(0), \beta(1), \ldots, \beta(k))) = \bigcup_m g(\Sigma(\beta(0), \beta(1), \ldots, \beta(k), m)),$$

by 4.4.2 again we get $\alpha(k+1)$ and $\beta(k+1)$ with the desired properties. ■

Theorem 4.4.3 *(Souslin) A subset A of a Polish space X is Borel if and only if it is both analytic and coanalytic; i.e., $\Delta^1_1(X) = \mathcal{B}_X$.*

Proof. The "only if" part is trivial. Suppose both A and A^c are analytic. Since A is the only set separating A from A^c, the "if part" immediately follows from 4.4.1. ■

Proposition 4.4.4 *Suppose $A_0, A_1, \ldots$ are pairwise disjoint analytic subsets of a Polish space X. Then there exist pairwise disjoint Borel sets $B_0, B_1, \ldots$ such that $B_n \supseteq A_n$ for all n.*

Proof. By 4.4.1, for each n there is a Borel set C_n such that

$$A_n \subseteq C_n \text{ and } C_n \bigcap \bigcup_{m \neq n} A_m = \emptyset.$$

Take

$$B_n = C_n \bigcap \bigcap_{m \neq n} (X \setminus C_m).$$

■

Theorem 4.4.5 *Let $E \subseteq X \times X$ be an analytic equivalence relation on a Polish space X. Suppose A and B are disjoint analytic subsets of X. Assume that B is invariant with respect to E (i.e., B is a union of E-equivalence classes). Then there is an E-invariant Borel set C separating A from B.*

Proof. First we note the following. Let D be an analytic subset of X and D^* the smallest invariant set containing D. Since

$$D^* = \pi_X(E\bigcap(D \times X)),$$

where $\pi_X : X \times X \longrightarrow X$ is the projection to the second coordinate space, D^* is analytic.

We show that there is a sequence (A_n) of invariant analytic sets and a sequence (B_n) of Borel sets such that

(i) $A \subseteq A_0$,

(ii) $A_n \subseteq B_n \subseteq A_{n+1}$, and

(iii) $B \bigcap B_n = \emptyset$.

Take $A_0 = A^*$. Since B is invariant, $A_0 \bigcap B = \emptyset$. By 4.4.1, let B_0 be a Borel set containing A_0 and disjoint from B. Suppose A_i, B_i, $0 \leq i \leq n$, satisfying (i), (ii), and iii) have been defined. Put $A_{n+1} = B_n^*$. Since B is invariant, $A_{n+1} \bigcap B = \emptyset$. By 4.4.1, let B_{n+1} be a Borel set containing A_{n+1} and disjoint from B.

Having defined (A_n), (B_n), let $C = \bigcup_n B_n$. Clearly, C is a Borel set containing A and disjoint from B. Since $C = \bigcup_n A_n$, it is also invariant. ■

Exercise 4.4.6 (Preiss [92]) Fix a positive integer ℓ. Let $\mathcal{CB}(\ell)$ be the smallest family of subsets of $\mathbb{R}^\ell$ satisfying the following conditions.

(a) $\mathcal{CB}(\ell)$ contains all open (closed) convex subsets of $\mathbb{R}^\ell$.

(b) $\mathcal{CB}(\ell)$ is closed under countable intersection.

(c) For every nondecreasing sequence (B_n) in $\mathcal{CB}(\ell)$, $\bigcup_n B_n \in \mathcal{CB}(\ell)$.

Let A and B be any two subsets of $\mathbb{R}^\ell$. Say that A is separated from B by a set in $\mathcal{CB}(\ell)$ if $A \subseteq C \subseteq B^c$ for some $C \in \mathcal{CB}(\ell)$.

(i) Suppose $A = \bigcup_m A_m$, $A_m \subseteq A_{m+1}$, and $B = \bigcup_n B_n$. Assume that A is not separated from B by a set in $\mathcal{CB}(\ell)$. Show that there exist integers m and n such that A_m is not separated from B_n by a set in $\mathcal{CB}(\ell)$.

In the rest of this exercise we assume that A and B are analytic.

(ii) Let $f : \mathbb{N}^{\mathbb{N}} \longrightarrow A$ and $g : \mathbb{N}^{\mathbb{N}} \longrightarrow B$ be continuous surjections. Suppose A is not separated from B by a set in $\mathcal{CB}(\ell)$. Show that there exist $\alpha, \beta \in \mathbb{N}^{\mathbb{N}}$ such that for every k, $f(\Sigma^*(\alpha|k))$ is not separated from $g(\Sigma(\beta|k))$ by a set in $\mathcal{CB}(\ell)$, where $\Sigma^*(\alpha|k) = \{\gamma \in \mathbb{N}^{\mathbb{N}} : \forall i < k(\gamma(i) \leq \alpha(i))\}$.

(iii) Now assume that A is convex and disjoint from B. Show that A is separated from B by a set in $\mathcal{CB}(\ell)$.

(Hint: The convex hull of any compact set in $\mathbb{R}^\ell$ is compact.)

(iv) Show that $\mathcal{CB}(\ell)$ equals the set of all convex Borel subsets of $\mathbb{R}^\ell$.

4.5 One-to-One Borel Functions

In this section we give some consequences of the results proved in the last section.

Proposition 4.5.1 *Let A be an analytic subset of a Polish space, Y a Polish space, and $f : A \longrightarrow Y$ a one-to-one Borel map. Then $f : A \longrightarrow f(A)$ is a Borel isomorphism.*

Proof. Let $B \subseteq A$ be Borel in A. We need to show that $f(B)$ is Borel in $f(A)$. As both B and $C = A \setminus B$ are analytic and f Borel, $f(B)$ and $f(C)$ are analytic. Since f is one-to-one, these two sets are disjoint. So, by 4.4.1, there is a Borel set $D \subseteq Y$ such that $f(B) \subseteq D$ and $f(C) \bigcap D = \emptyset$. Since $f(B) = D \bigcap f(A)$, the result follows. ■

Theorem 4.5.2 *Let X, Y be Polish spaces, $A \subseteq X$ analytic, and $f : A \longrightarrow Y$ any map. The following statements are equivalent*

(i) *f is Borel measurable.*

(ii) graph(f) *is Borel in $A \times Y$.*

(iii) graph(f) *is analytic.*

Proof. We only need to show that (iii) implies (i). The other implications are quite easy to see. Let U be an open set in Y. As

$$f^{-1}(U) = \pi_X(\text{graph}(f) \bigcap (X \times U)),$$

where $\pi_X : X \times Y \longrightarrow X$ is the projection map, it is analytic. Similarly, $f^{-1}(U^c)$ is analytic. By 4.4.1, there is a Borel set $B \subseteq X$ such that

$$f^{-1}(U) \subseteq B \text{ and } B \bigcap f^{-1}(U^c) = \emptyset.$$

Since $f^{-1}(U) = B \bigcap A$, it is Borel in A, and the result follows. ■

Exercise 4.5.3 Let X be a separable Banach space and X_1 a Borel subspace of X. Suppose there is a Borel subspace X_2 of X such that

(i) $X_1 \bigcap X_2 = \{0\}$, and

(ii) every $x \in X$ can be expressed in the form $x_1 + x_2$, where $x_1 \in X_1$ and $x_2 \in X_2$.

Show that X_1 is closed in X.
(Hint: Using 4.5.2 show that the map $x \longrightarrow x_1$ is Borel measurable. Now argue as in 3.5.9 and conclude that the map $x \longrightarrow x_1$ is, in fact, continuous.)

Solovay [110] gave an example of a coanalytic set $C \subseteq \mathbb{N}^{\mathbb{N}}$ and a non-Borel measurable function $f : C \times \mathbb{R} \longrightarrow 2^{\mathbb{N}}$ whose graph is Borel in $C \times \mathbb{R} \times 2^{\mathbb{N}}$. This example is based on a coding of Borel subsets of $\mathbb{R}$ that we describe now in some detail.

Solovay's Coding of Borel Sets

Let (r_i) be an enumeration of the rationals and let J be the pairing function on $\mathbb{N} \times \mathbb{N}$ defined by

$$J(m, n) = 2^m(2n + 1).$$

We define the coding recursively as follows:

1. $\alpha \in \mathbb{N}^{\mathbb{N}}$ codes $[r_i, r_j]$ if $\alpha(0) \equiv 0 \pmod 3$, $\alpha(1) = i$, and $\alpha(2) = j$.

2. Suppose $\alpha_i \in \mathbb{N}^{\mathbb{N}}$ codes $B_i \subseteq \mathbb{R}$, $i = 0, 1, 2, \ldots$; then $\alpha \in \mathbb{N}^{\mathbb{N}}$ codes $\bigcup_i B_i$ if $\alpha(0) \equiv 1 \pmod 3$ and $\alpha(J(m, n)) = \alpha_m(n)$.

3. Suppose $\beta \in \mathbb{N}^{\mathbb{N}}$ codes B, $\alpha(0) \equiv 2 \pmod 3$, and $\alpha(n+1) = \beta(n)$. Then α codes B^c.

4. α codes $B \subseteq \mathbb{R}$ only as required by 1 – 3.

Note the following.

a. Every $\alpha \in \mathbb{N}^{\mathbb{N}}$ codes at most one subset of $\mathbb{R}$.

b. Every Borel subset of $\mathbb{R}$ is coded by some $\alpha \in \mathbb{N}^{\mathbb{N}}$. (One shows this by showing that the class of all sets having a code contains all $[r_i, r_j]$ and is closed under countable unions and complementation.)

c. If a subset of $\mathbb{R}$ is coded by α, it is Borel. (This is true because the class of all $\alpha \in \mathbb{N}^{\mathbb{N}}$ that code a Borel set $B \subseteq \mathbb{R}$ is closed under 1 – 3.)

Next, we define a function $\Phi : \mathbb{N}^{\mathbb{N}} \times \mathbb{N} \longrightarrow \mathbb{N}^{\mathbb{N}}$ with the property that if α codes a Borel set B, then $\Phi(\alpha, \cdot)$ recovers the Borel sets from which B is constructed. For this definition, we fix an enumeration (s_n), without repetitions, of $\mathbb{N}^{<\mathbb{N}}$ such that $s_n \prec s_m \Longrightarrow n \leq m$. So s_0 is the empty sequence. The definition of $\Phi(\alpha, n)$ will proceed by induction on n.

Set

$$\Phi(\alpha, 0) = \alpha, \quad \alpha \in \mathbb{N}^{\mathbb{N}}.$$

Let $n > 0$ and suppose that $\Phi(\alpha, m)$ has been defined for all $\alpha \in \mathbb{N}^{\mathbb{N}}$ and all $m < n$. Since $n > 0$, s_n is of positive length. Let $m < n$ and u be such that $s_n = s_m \hat{\ } u$. Now define for $i \in \mathbb{N}$

$$\Phi(\alpha, n)(i) = \begin{cases} 0 & \text{if } \Phi(\alpha, m)(0) \equiv 0 \pmod 3, \\ \Phi(\alpha, m)(J(u, i)) & \text{if } \Phi(\alpha, m)(0) \equiv 1 \pmod 3, \\ \Phi(\alpha, m)(i+1) & \text{if } \Phi(\alpha, m)(0) \equiv 2 \pmod 3. \end{cases}$$

It is easy to see that the graph of Φ is Borel. Hence, Φ is Borel measurable by 4.5.2. Also, by induction on n, we see that if α codes a Borel set, then for all n, $\Phi(\alpha, n)$ codes a Borel set.

For $\beta \in \mathbb{N}^{\mathbb{N}}$, define $\overline{\beta} \in \mathbb{N}^{\mathbb{N}}$ such that for every $n \in \mathbb{N}$,

$$s_{\overline{\beta}(n)} = (\beta(0), \beta(1), \ldots, \beta(n-1)).$$

Plainly, the map $\beta \longrightarrow \overline{\beta}$ is continuous. Now define a coanalytic set

$$C = \{\alpha \in \mathbb{N}^{\mathbb{N}} : (\forall \beta)(\exists n)\Phi(\alpha, \overline{\beta}(n)) = 0\}.$$

It is easily seen that C is closed under 1 – 3. Hence, if $\alpha \in \mathbb{N}^{\mathbb{N}}$ codes a Borel set, then $\alpha \in C$. Conversely, if α fails to code a Borel set, then by induction, one can construct a function $\beta : \mathbb{N} \longrightarrow \mathbb{N}$ such that for all n, $\Phi(\alpha, \overline{\beta}(n))$ fails to code a Borel set. But then, for all n, $\Phi(\alpha, \overline{\beta}(n)) \neq 0$.

We now proceed to give an example of a function with domain coanalytic whose graph is Borel and that is not Borel measurable.

Let $\varphi : \mathbb{N} \times \mathbb{N} \longrightarrow \mathbb{N}$ be the function satisfying $s_{\varphi(n,i)} = s_n \hat{\ } i$. Let $E \subseteq \mathbb{R} \times 2^{\mathbb{N}}$ be defined as follows:

$$\begin{aligned} (\alpha, x, \gamma) \in E \iff & (\forall n)[\{\Phi(\alpha, n)(0) \equiv 0 (\text{mod } 3) \\ & \Longrightarrow \{\gamma(n) = 1 \iff (\exists i)(\exists j)(\Phi(\alpha, n)(1) = i \\ & \& \Phi(\alpha, n)(2) = j \& x \in [r_i, r_j])\}] \\ & \& (\forall n)[\{\Phi(\alpha, n)(0) \equiv 1 (\text{mod } 3) \\ & \Longrightarrow \{\gamma(n) = 1 \iff (\exists i)(\gamma(\varphi(n, i)) = 1)\}] \\ & \& (\forall n)[\{\Phi(\alpha, n)(0) \equiv 2 (\text{mod } 3) \\ & \Longrightarrow \{\gamma(n) = 1 \iff \gamma(\varphi(n, 0)) = 0)\}]. \end{aligned}$$

Since Φ is Borel, E is Borel. Further, for every $\alpha \in C$ and every $x \in \mathbb{R}$, there is a unique $\gamma \in 2^{\mathbb{N}}$ such that $(\alpha, x, \gamma) \in E$. Thus $E \bigcap (C \times \mathbb{R} \times 2^{\mathbb{N}})$ is the graph of a function $f : C \times \mathbb{R} \longrightarrow 2^{\mathbb{N}}$, say, that is Borel in $C \times \mathbb{R} \times 2^{\mathbb{N}}$.

We show that f is not Borel measurable. Towards a contradiction, assume that f is Borel measurable on $C \times \mathbb{R}$. Consider the set

$$F = \{(\alpha, x) \in C \times \mathbb{R} : f(\alpha, x)(0) = 0\}.$$

According to the observations made in preceeding paragraphs, the condition "$\alpha \in C$ and $f(\alpha, x)(0) = 0$" states that x does not belong to the Borel set coded by α. Since f is Borel measurable, F is Borel in $C \times \mathbb{R}$, so there must exist a Borel subset D of $\mathbb{N}^{\mathbb{N}} \times \mathbb{R}$ such that $F = D \bigcap (C \times \mathbb{R})$. Fix a Borel isomorphism h from $\mathbb{R}$ onto $\mathbb{N}^{\mathbb{N}}$. Let

$$B = \{x \in \mathbb{R} : (h(x), x) \in D\}.$$

Plainly, B is a Borel subset of $\mathbb{R}$. So, there is $\alpha^* \in C$ such that α^* codes B. Set $x^* = h^{-1}(\alpha^*)$. Then

$$\begin{aligned} x^* \in B &\iff (\alpha^*, x^*) \in D \\ &\iff (\alpha^*, x^*) \in F \\ &\iff f(\alpha^*, x^*)(0) = 0 \\ &\iff x^* \notin \text{the Borel set coded by } \alpha^* \\ &\iff x^* \notin B, \end{aligned}$$

a contradiction.

Theorem 4.5.4 *Let X, Y be Polish spaces, A a Borel subset of X, and $f : A \longrightarrow Y$ a one-to-one Borel map. Then $f(A)$ is Borel.*

Proof. Replacing X by $X \times Y$, A by graph(f), and f by $\pi_Y|\text{graph}(f)$, without any loss of generality, we assume that f is continuous. Since every Borel set is a one-to-one continuous image of a closed subset of $\mathbb{N}^{\mathbb{N}}$ (3.3.17), we further assume that $X = \mathbb{N}^{\mathbb{N}}$ and that A is a closed set.

For every $s \in \mathbb{N}^{<\mathbb{N}}$, we get a Borel subset B_s of Y such that for every $s, t \in \mathbb{N}^{<\mathbb{N}}$,

(i) $f(\Sigma(s) \bigcap A) \subseteq B_s \subseteq \text{cl}(f(\Sigma(s) \bigcap A))$,

(ii) $s \succ t \Longrightarrow B_s \subseteq B_t$, and

(iii) whenever $s \neq t$ and $|s| = |t|$, $B_s \bigcap B_t = \emptyset$.

We first complete the proof assuming that such a system $\{B_s : s \in \mathbb{N}^{<\mathbb{N}}\}$ of Borel sets exists. Let

$$D = \bigcap_n \bigcup_{|s|=n} B_s.$$

Then D is Borel. We show that

$$f(A) = D.$$

Let $\alpha \in A$. Then $f(\alpha) \in B_{\alpha|n}$ for all n. Thus, $f(A) \subseteq D$. For the reverse inclusion, let $y \in D$. By (ii) and (iii), there is an α such that $y \in B_{\alpha|n}$ for every n. Since $B_{\alpha|n} \subseteq \text{cl}(f(\Sigma(\alpha|n) \bigcap A))$, we get an $\alpha_n \in \Sigma(\alpha|n) \bigcap A$ such that $d(y, f(\alpha_n)) < 2^{-n}$. Clearly, $\alpha_n \to \alpha$. As A is closed, $\alpha \in A$. Since f is continuous, $f(\alpha) = \lim_n f(\alpha_n) = y$. Hence, $y \in f(A)$.

It remains to show that a system of Borel sets $\{B_s : s \in \mathbb{N}^{<\mathbb{N}}\}$ satisfying (i) – (iii) exists. We proceed by induction on the length of s.

Take $B_e = \text{cl}(f(A))$. Suppose B_s has been defined. Since $f|A$ is one-to-one, $f(\Sigma(s\,\hat{}\,0) \bigcap A), f(\Sigma(s\,\hat{}\,1) \bigcap A), f(\Sigma(s\,\hat{}\,2) \bigcap A), \ldots$ are pairwise disjoint. Further, they are analytic. By 4.4.4, there exist pairwise disjoint Borel sets $B'_{s\hat{}n} \supseteq f(\Sigma(s\hat{}n) \bigcap A)$. Take

$$B_{s\hat{}n} = B_s \bigcap B'_{s\hat{}n} \bigcap \text{cl}(f(\Sigma(s\hat{}n) \bigcap A)).$$

■

Corollary 4.5.5 *Let X be a standard Borel space and Y a metrizable space. Suppose there is a one-to-one Borel map f from X onto Y. Then Y is standard Borel and f a Borel isomorphism.*

Proof. By 4.3.8, Y is separable. The result follows from 4.5.4.

Exercise 4.5.6 Let $\mathcal{T}$ and $\mathcal{T}'$ be two Polish topologies on X such that $\mathcal{T}' \subseteq \sigma(\mathcal{T})$. Show that $\sigma(\mathcal{T}) = \sigma(\mathcal{T}')$.

Theorem 4.5.7 *(Blackwell – Mackey theorem, [13]) Let X be an analytic subset of a Polish space and $\mathcal{A}$ a countably generated sub σ-algebra of the Borel σ-algebra $\mathcal{B}_X$. Let $B \subseteq X$ be a Borel set that is a union of atoms of $\mathcal{A}$. Then $B \in \mathcal{A}$.*

Proof. Let $\{B_n : n \in \mathbb{N}\}$ be a countable generator of $\mathcal{A}$. Consider the map $f : X \longrightarrow 2^{\mathbb{N}}$ defined by

$$f(x) = (\chi_{B_0}(x), \chi_{B_1}(x), \ldots), \quad x \in X.$$

Then $\mathcal{A} = f^{-1}(\mathcal{B}_{2^{\mathbb{N}}})$. In particular, $f : X \longrightarrow 2^{\mathbb{N}}$ is Borel measurable. So, $f(B)$ and $f(B^c)$ are disjoint analytic subsets of $2^{\mathbb{N}}$. By 4.4.1, there is a Borel set C containing $f(B)$ and disjoint from $f(B^c)$. Clearly, $B = f^{-1}(C)$, and so it belongs to $\mathcal{A}$. ■

Remark 4.5.8 The condition that $\mathcal{A}$ is countably generated cannot be dropped from the above result. To see this, let $\mathcal{A}$ be the countable – co-countable σ-algebra on $\mathbb{R}$. By 3.1.16, $\mathcal{A}$ is not countably generated. As any Borel set is a union of atoms of $\mathcal{A}$, the above theorem does not hold for $\mathcal{A}$.

Remark 4.5.9 In the next chapter we shall show that 4.5.7 is not true for coanalytic X.

Corollary 4.5.10 *Let X be an analytic subset of a Polish space and $\mathcal{A}_1$, $\mathcal{A}_2$ two countably generated sub σ-algebras of $\mathcal{B}_X$ with the same set of atoms. Then $\mathcal{A}_1 = \mathcal{A}_2$. In particular, if $\mathcal{A}$ is a countably generated sub σ-algebra containing all the singletons, then $\mathcal{A} = \mathcal{B}_X$.*

4.6 The Generalized First Separation Theorem

Theorem 4.6.1 *(The generalized first separation theorem, Novikov[90]) Let (A_n) be a sequence of analytic subsets of a Polish space X such that $\bigcap A_n = \emptyset$. Then there exist Borel sets $B_n \supseteq A_n$ such that $\bigcap B_n = \emptyset$.*

(If (A_n) satisfies the conclusion of this result, we call it *Borel separated.*)

As in the proof of the first separation theorem, the proof of this result is also based on a combinatorial lemma.

Lemma 4.6.2 *Let (E_n) be a sequence of subsets of X, $k \in \mathbb{N}$, and $E_i = \bigcup_n E_{in}$ for $i \leq k$. Suppose (E_n) is not Borel separated. Then there exist $n_0, n_1, \ldots, n_k$ such that the sequence $E_{0n_0}, E_{1n_1}, \ldots, E_{kn_k}, E_{k+1}, E_{k+2}, \ldots$ is not Borel separated.*

Proof. We prove the result by induction on k.

Initial step: $k = 0$. Suppose the result is not true. Hence, for every n, there is a sequence $(B_{in})_{i\in\mathbb{N}}$ of Borel sets such that

(i) $\bigcap_i B_{in} = \emptyset$,

(ii) $B_{0n} \supseteq E_{0n}$, and

(iii) $B_{in} \supseteq E_i$ for all i.

Let

$$\begin{aligned} B_i &= \bigcup_n B_{in} \quad \text{if} \quad i = 0, \\ &= \bigcap_n B_{in} \quad \text{if} \quad i > 0. \end{aligned}$$

Then $B_i \supseteq E_i$, the B_i's are Borel and $\bigcap B_i = \emptyset$. This contradicts the hypothesis that (E_n) is not Borel separated, and we have proved the result for $k = 0$.

Inductive step. Suppose $k > 0$ and the result is true for all integers less than k. By the induction hypothesis, there are integers $n_0, n_1, \ldots, n_{k-1}$ such that $E_{0n_0}, E_{1n_1}, \ldots, E_{k-1n_{k-1}}, E_k, E_{k+1}, \ldots$ is not Borel separated. By the initial step, there is an n_k such that $E_{0n_0}, E_{1n_1}, \ldots, E_{kn_k}, E_{k+1}, E_{k+2}, \ldots$ is not Borel separated. ■

Proof of 4.6.1. (Mokobodzki [86]) Let (A_n) be a sequence of analytic sets that is not Borel separated and such that $\bigcap_n A_n = \emptyset$. For each n, fix a continuous surjection $f_n : \mathbb{N}^\mathbb{N} \longrightarrow A_n$. We get a sequence $\alpha_0, \alpha_1, \ldots$ in $\mathbb{N}^\mathbb{N}$ such that for every $k > 0$ the sequence

$$f_0(\Sigma(\alpha_0|k)), f_1(\Sigma(\alpha_1|(k-1))), \ldots, f_{k-1}(\Sigma(\alpha_{k-1}|1)), A_k, A_{k+1}, \ldots$$

is not Borel separated.

To see that such a sequence exists we proceed by induction. Write $A_0 = \bigcup_n f_0(\Sigma(n))$. By 4.6.2, there exists $\alpha_0(0) \in \mathbb{N}$ such that the sequence $f_0(\Sigma(\alpha_0(0))), A_1, A_2, \ldots$ is not Borel separated. Write $f_0(\Sigma(\alpha_0(0))) = \bigcup_m f_0(\Sigma(\alpha_0(0)m))$ and $A_1 = \bigcup_n f_1(\Sigma(n))$. Apply 4.6.2 again to get $\alpha_0(1), \alpha_1(0) \in \mathbb{N}$ such that the sequence $f_0(\Sigma(\alpha_0(0)\alpha_0(1))), f_1(\Sigma(\alpha_1(0))), A_2, A_3, \ldots$ is not Borel separated. Proceeding similarly we get the sequence $\alpha_0, \alpha_1, \alpha_2, \ldots$ satisfying the desired conditions.

Since $\bigcap A_n = \emptyset$, there exist $i < j$ such that $f_i(\alpha_i) \neq f_j(\alpha_j)$. Since f_i and f_j are continuous, there exist disjoint open sets U_i, U_j in X such that $f_i(\alpha_i) \in U_i$ and $f_j(\alpha_j) \in U_j$. Using the continuity of f_i and f_j again, we get a large enough k such that $f_i(\Sigma(\alpha_i|k-i)) \subseteq U_i$ and $f_j(\Sigma(\alpha_j|k-j)) \subseteq U_j$. Thus the sequence (B_n) of Borel sets, where

$$B_n = \begin{cases} U_n & \text{if } n = i \text{ or } j, \\ X & \text{otherwise,} \end{cases}$$

separates $f_0(\Sigma(\alpha_0|k)), f_1(\Sigma(\alpha_1|(k-1))), \ldots, f_{k-1}(\Sigma(\alpha_{k-1}|1)), A_k, A_{k+1}, \ldots$, which is a contradiction. ■

Corollary 4.6.3 *Let (A_n) be a sequence of analytic subsets of a Polish space X such that* $\limsup A_n = \emptyset$. *Then there exist Borel sets $B_n \supseteq A_n$ such that* $\limsup B_n = \emptyset$.

Remark 4.6.4 Later in this chapter we shall show that 4.6.3 is not true for coanalytic A_n's.

Theorem 4.6.5 *(Weak reduction principle for coanalytic sets) Let $C_0, C_1, C_2, \ldots$ be a sequence of coanalytic subsets of a Polish space such that $\bigcup C_n$ is Borel. Then there exist pairwise disjoint Borel sets $B_n \subseteq C_n$ such that $\bigcup B_n = \bigcup C_n$.*

Proof. Let $A_n = X \setminus C_n$, where $X = \bigcup_n C_n$. Then (A_n) is a sequence of analytic sets such that $\bigcap_n A_n = \emptyset$. By 4.6.1, there exist Borel sets $D_n \supseteq A_n$ such that $\bigcap_n D_n = \emptyset$. Take

$$B_n = B'_n \setminus \bigcup_{m<n} B'_m,$$

where $B'_n = X \setminus D_n$. ■

Exercise 4.6.6 Let E be an analytic equivalence relation on a Polish space X. Suppose $A_0, A_1, A_2, \ldots$ are invariant analytic subsets of X such that $\bigcap A_n = \emptyset$. Show that there exist invariant Borel sets $B_n \supseteq A_n$ with $\bigcap_n B_n = \emptyset$. Conclude that if $C_0, C_1, C_2, \ldots$ is a sequence of invariant coanalytic sets whose union is Borel, then there exist pairwise disjoint invariant Borel sets $B_n \subseteq C_n$ with $\bigcup B_n = \bigcup C_n$.

(Hint: Use 4.4.5 and 4.6.1.)

4.7 Borel Sets with Compact Sections

Throughout this section, X and Y are fixed Polish spaces and (V_n) a countable base for Y.

Theorem 4.7.1 *(Saint Raymond[97]) Let A_0 and A_1 be disjoint analytic subsets of $X \times Y$ with the sections $(A_0)_x$, $x \in X$, closed in Y. Then there is a sequence (B_n) of Borel subsets of X such that*

$$A_1 \subseteq \bigcup_n (B_n \times V_n) \text{ and } A_0 \bigcap \bigcup_n (B_n \times V_n) = \emptyset. \qquad (*)$$

Proof. By 4.4.1, there is a Borel set containing A_1 and disjoint from A_0. So, without any loss of generality, we assume that A_1 is Borel. For each n, let

$$C_n = \{x \in X : V_n \subseteq (A_0)_x^c\}.$$

Then C_n is coanalytic and

$$(A_0)^c = \bigcup_n (C_n \times V_n).$$

Note that $((C_n \times V_n) \bigcap A_1)$ is a sequence of coanalytic sets whose union is Borel. Hence, by 4.6.5, there exist Borel sets $D_n \subseteq (C_n \times V_n) \bigcap A_1$ such that

$$\bigcup D_n = \bigcup_n (A_1 \bigcap (C_n \times V_n)) = A_1.$$

By 4.4.1, there exist Borel sets B_n such that

$$\pi_X(D_n) \subseteq B_n \subseteq C_n,$$

where $\pi_X : X \times Y \longrightarrow X$ is the projection map. It is now fairly easy to see that (B_n) satisfies $(\star)$. ■

As a direct consequence of 4.7.1, we get the following structure theorem for Borel sets with open sections.

Theorem 4.7.2 *(Kunugui, Novikov) Suppose $B \subseteq X \times Y$ is any Borel set with sections B_x open, $x \in X$. Then there is a sequence (B_n) of Borel subsets of X such that*

$$B = \bigcup (B_n \times V_n).$$

Proof. Apply 4.7.1 to $A_0 = B^c$ and $A_1 = B$. ■

Corollary 4.7.3 *Let A_0 and A_1 be disjoint analytic subsets of $X \times Y$ with sections $(A_0)_x$ and $(A_1)_x$ closed for all $x \in X$. Then there exist disjoint Borel sets B_0 and B_1 with closed sections such that $A_0 \subseteq B_0$ and $A_1 \subseteq B_1$.*

Corollary 4.7.4 *Suppose $B \subseteq X \times Y$ is a Borel set with the sections B_x closed. Then there is a Polish topology $\mathcal{T}$ finer than the given topology on X generating the same Borel σ-algebra such that B is closed relative to the product topology on $X \times Y$, X being equipped with the new topology $\mathcal{T}$.*

Proof. By 4.7.2, write

$$B^c = \bigcup_n (B_n \times V_n),$$

the B_n's Borel. By 3.2.5, take a finer Polish topology $\mathcal{T}$ on X generating the same Borel σ-algebra such that B_n is $\mathcal{T}$-open. ■

Exercise 4.7.5 Let A_0 and A_1 be disjoint analytic subsets of $X \times Y$ with sections $(A_0)_x$ compact. Show that there exists a Borel set B_0 in $X \times Y$ with compact sections separating A_0 from A_1.

Exercise 4.7.6 [102] Let X, Y be Polish and $A_0, A_1 \subseteq X \times Y$ disjoint analytic. Assume that the sections $(A_0)_x$, $(A_1)_x$ are closed. Show that there exists a Borel map $u : X \times Y \longrightarrow [0,1]$ such that $y \longrightarrow u(x,y)$ is continuous for all x and

$$u(x) = \begin{cases} 0 & \text{if } x \in A_0, \\ 1 & \text{if } x \in A_1. \end{cases}$$

In the next section we shall show that 4.7.6 does not hold for A_0, A_1 coanalytic.

Exercise 4.7.7 [102] Let X, Y be Polish, $B \subseteq X \times Y$ Borel with sections closed, and $f : B \longrightarrow [0,1]$ a Borel map such that $y \longrightarrow f(x,y)$ is continuous for all x. Show that there is a finer Polish topology $\mathcal{T}$ on X generating the same Borel σ-algebra such that when X is equipped with the topology $\mathcal{T}$, B is closed and f continuous. Conclude that there is a Borel extension $F : X \times Y \longrightarrow [0,1]$ of f such that $y \longrightarrow F(x,y)$ is continuous for all x.

Generalize this with the range space $[0,1]$ replaced by any compact convex subset of $\mathbb{R}^n$.

Remark 4.7.8 We can generalize the concluding part of 4.7.7 for analytic B. This is done by imitating the usual proof of the Tietze extension theorem for normal spaces and using 4.7.6 repeatedly. We invite the reader to carry out the exercise. (See [102].)

We give below an example showing that 4.7.7 does not hold for coanalytic B.

Example 4.7.9 (H. Sarbadhikari) Let $A \subseteq [0,1]$ be an analytic non-Borel set and $E \subseteq [0,1] \times \mathbb{N}^{\mathbb{N}}$ a closed set whose projection is A. Set $B = E \bigcup (([0,1] \setminus A) \times \mathbb{N}^{\mathbb{N}})$ and $f : B \longrightarrow [0,1]$ the characteristic function of E. We claim that there is no Borel extension $F : [0,1] \times \mathbb{N}^{\mathbb{N}} \longrightarrow [0,1]$ of f such that $y \longrightarrow F(x,y)$ is continuous. Suppose not. Consider $C = F^{-1}((0,1])$. Then C is a Borel set with sections C_x open and whose projection is A. Hence A is Borel. (See the paragraph below.) We have arrived at a contradiction.

We have seen that the projection of a Borel set need not be Borel. We give below some conditions on the sections of a Borel set under which its projection is Borel.

Let $B \subseteq X \times Y$ be a Borel set. Assume that the sections B_x are open in Y. Then $\pi_X(B)$ is Borel. To see this, take a countable dense set $\{r_n : n \in \mathbb{N}\}$ in Y. Note that

$$x \in \pi_X(B) \Longleftrightarrow \exists n(x, r_n) \in B,$$

i.e., $\pi_X(B) = \bigcup_n \{x \in X : (x, r_n) \in B\}$. Hence, it is Borel.

We have also seen that $\pi_X(B)$ is Borel if the Borel set $B \subseteq X \times Y$ satisfies any one of the following conditions:

(i) For every $x \in \pi_X(B)$, the section B_x contains exactly one point (4.5.4).

(ii) For every $x \in \pi_X(B)$, B_x is nonmeager (3.5.18).

(iii) For every $x \in \pi_X(B)$, $P(x, B_x) > 0$, where P is any transition probability on $X \times Y$ (3.4.24).

Exercise 4.7.10 Let X be a Polish space and $B \subseteq X \times \mathbb{R}^n$ a Borel set with convex sections. Show that $\pi_X(B)$ is Borel.

Theorem 4.7.11 *(Novikov) Let X and Y be Polish spaces and B a Borel subset of $X \times Y$ with sections B_x compact. Then $\pi_X(B)$ is Borel in X.*

Proof. (Srivastava) Since every Polish space is homeomorphic to a G_δ subset of the Hilbert cube $\mathbb{H}$, without any loss of generality, we assume that Y is a compact metric space. Note that the sections B_x are closed in Y. By 4.7.4, there is a finer Polish topology on X generating the same Borel σ-algebra and making B closed in $X \times Y$. Hence, by 2.3.24, $\pi_X(B)$ is closed in X, X being equipped with the new topology. But the Borel structure of X is the same with respect to both the topologies. The result follows. ∎

Using 4.7.2, we give another elementary proof of this important result.

Alternative Proof of 4.7.11. (Srivastava) As above, we assume that Y is compact. By 4.7.2, write

$$(X \times Y) \setminus B = \bigcup_n (B_n \times V_n),$$

the B_n's Borel, the V_n's open. Now note that

$$X \setminus \pi_X(B) = \bigcup_{\{F \subseteq \mathbb{N} : F \text{ is finite } \& \bigcup_{n\in F} V_n = Y\}} \bigcap_{n \in F} B_n.$$

∎

Corollary 4.7.12 *Let X, Y be Polish spaces with Y σ-compact (equivalently, locally compact). Then the projection of every Borel set B in $X \times Y$ with x-sections closed in Y is Borel.*

Proof. Write $Y = \bigcup_n Y_n$, Y_n compact. Then

$$\pi_X(B) = \bigcup_n \pi_X(B \bigcap (X \times Y_n)).$$

Now apply 4.7.11. ■

4.8 Polish Groups

The theory of Borel sets is very useful in analysis (see [4], [54], [72], [73], [124], etc.). In this section we present some very basic results on Polish groups that are often used in analysis. Some more applications are given in the next chapter.

Theorem 4.8.1 *Let $(G, \cdot)$ be a Polish group and H a closed subgroup. Suppose $E = \{(x, y) : x \cdot y^{-1} \in H\}$; i.e., E is the equivalence relation induced by the right cosets. Then the σ-algebra of invariant Borel sets is countably generated.*

Proof. Let $\{U_n : n \in \mathbb{N}\}$ be a countable base for the topology of G. Put

$$B_n = \bigcup_{y \in H} y \cdot U_n.$$

So, the B_n's are Borel (in fact, open). We show that $\{B_n : n \in \mathbb{N}\}$ generates $\mathcal{B}$.

Let H_1 and H_2 be two distinct cosets. Since H is closed, H_1 and H_2 are closed. Since they are disjoint, there is a basic open set U_n such that $U_n \bigcap H_1 \neq \emptyset$ and $U_n \bigcap H_2 = \emptyset$. Then $H_1 \subset B_n$ and $B_n \bigcap H_2 = \emptyset$. It follows that the right cosets are precisely the atoms of $\sigma(\{B_n : n \in \mathbb{N}\})$. The result now follows from 4.5.7. ■

In the next chapter we shall give a proof of 4.8.1 without using the theory of analytic sets.

It is interesting to note that the converse of 4.8.1 is also true.

Theorem 4.8.2 *(Miller[84]) Let G be a Polish group and H a Borel subgroup. Suppose the σ-algebra of invariant Borel sets is countably generated. Then H is closed.*

We need a few preliminary results to prove the above theorem.

Proposition 4.8.3 *Let X be a Polish space and G a group of homeomorphisms of X such that for every pair U, V of nonempty open sets there is a $g \in G$ with $g(U) \bigcap V \neq \emptyset$. Suppose A is a G-invariant Borel set; i.e., $g(A) = A$ for all $g \in G$. Then either A or A^c is meager in X.*

Proof. Suppose neither A nor A^c is meager in X. Then there exist nonempty open sets U, V such that A and A^c are comeager in U and V respectively. By our hypothesis, there is a $g \in G$ such that $g(U) \bigcap V \neq \emptyset$. Let $W = g(U) \bigcap V$. It follows that W is meager. This contradicts the Baire category theorem. ■

Let $x \in X$. The set

$$G_x = \{g \in G : g \cdot x = x\}$$

is called the **stabilizer** of x. Clearly, G_x is a subgroup of G.

Theorem 4.8.4 *(Miller[84]) Let $(G, \cdot)$ be a Polish group, X a second countable T_1 space, and $(g, x) \longrightarrow g \cdot x$ an action of G on X. Suppose that for a given x, the map $g \longrightarrow g \cdot x$ is Borel. Then the stabilizer G_x is closed.*

Proof. Let $H = \text{cl}(G_x)$. It is fairly easy to see that we can replace G by H. Hence, without loss of generality we assume that G_x is dense in G.

Since X is second countable and T_1, G_x is Borel. Therefore, by 3.5.13, we shall be done if we show that G_x is nonmeager. Suppose not. We shall get a contradiction. Take a countable base (U_n) for X. Let $f(g) = g \cdot x$. As f is Borel, $f^{-1}(U_n) = A_n$, say, is Borel. For every $h \in G_x$, $A_n \cdot h = A_n$. Since X is T_1, for any two g, h we have

$$g \cdot x = h \cdot x \Longleftrightarrow \forall n (g \in A_n \Longleftrightarrow h \in A_n).$$

Hence, for any $g \in G$

$$gG_x = \bigcap \{A_n : g \in A_n\}.$$

Applying 4.8.3 to the group of homeomorphisms of G induced by right multiplication by elements of G_x, we see that A_n is either meager or comeager. Since G_x is meager, there exists n such that $g \in A_n$ and A_n is meager. Hence,

$$G = \bigcup \{A_n : A_n \text{ meager}\}.$$

This contradicts the Baire category theorem, and our result is proved. ■

Remark 4.8.5 A close examination of the proof of 4.8.4 shows that it holds when X is a countably generated measurable space with singletons as atoms.

Proof of 4.8.2. Let $X = G/H$, the set of right cosets, and $q : G \longrightarrow G/H$ the quotient map. Equip G/H with the largest σ-algebra making q Borel measurable. By our hypothesis, X is a countably generated measurable space with singletons as atoms. Consider the action $(g, g'H) \longrightarrow g \cdot g'H$ of G on X. Let $x = H$. Then the stabilizer

$$G_x = \{g \in G : g \cdot x = x\} = H.$$

Since $g \longrightarrow g \cdot x$ is Borel, the result follows from 4.8.5. ∎

Theorem 4.8.6 *Let G be a Polish group, X a Polish space, and $a(g, x) = g \cdot x$ an action of G on X. Assume that $g \cdot x$ is continuous in x for all g and Borel in g for all x. Then the action is continuous.*

Proof. By 3.1.30, the action $a : G \times X \longrightarrow X$ is Borel. Let (V_n) be a countable base for X. Put $C_n = a^{-1}(V_n)$. Then C_n is Borel with open sections. By 4.7.2, write

$$C_n = \bigcup_m (B_{nm} \times W_{nm}),$$

the B_{nm}'s Borel, the W_{nm}'s open. By 3.5.1, B_{nm} has the Baire property. Let I_{nm} be a meager set in G such that $B_{nm} \Delta I_{nm}$ is open. Put $I = \bigcup_{nm} I_{nm}$. Then I is meager in G and $a|(G \setminus I) \times X$ is continuous.

Now take a sequence (g_k, x_k) in $G \times X$ converging to (g, x), say. We need to show that $g_k \cdot x_k \to g \cdot x$. Let

$$J = \bigcup_k I \cdot g_k^{-1} \bigcup I \cdot g^{-1}.$$

Since G is a topological group, J is meager in G. By the Baire category theorem, $G \neq J$. Take any $h \in G \setminus J$. Then $h \cdot g, h \cdot g_k \in G \setminus I$. As $g_k \to g$, $h \cdot g_k \to h \cdot g$. Since $a|(G \setminus I) \times X$ is continuous, $(h \cdot g_k) \cdot x_k \to (h \cdot g) \cdot x$. Since the action is continuous in the second variable,

$$g_k \cdot x_k = h^{-1} \cdot ((h \cdot g_k) \cdot x_k) \to h^{-1} \cdot ((h \cdot g) \cdot x) = g \cdot x.$$

∎

Exercise 4.8.7 Generalise 4.8.6 for completely metrizable groups G and completely metrizable X that are not necessarily separable.

It is worth noting that in the above proof we used only the following: G has a Polish topology such that the multiplication is separately continuous in each variable. Now observe the following result.

Lemma 4.8.8 *If $(G, \cdot)$ is a group with a Polish topology such that the group operation $(g, h) \longrightarrow g \cdot h$ is Borel, then $g \longrightarrow g^{-1}$ is continuous.*

Proof. Since $(g,h) \longrightarrow g \cdot h$ is Borel, the graph

$$\{(g,h) : g \cdot h = e\}$$

of $g \longrightarrow g^{-1}$ is Borel. Hence, by 4.5.2, $g \longrightarrow g^{-1}$ is Borel measurable. An imitation of the proof of 3.5.9 shows that $g \longrightarrow g^{-1}$ is continuous. ■

From these observations we get the following result.

Proposition 4.8.9 *If $(G, \cdot)$ is a group with a Polish topology such that the group operation is separately continuous in each variable, then G is a topological group.*

Proof. In view of 4.8.8, we have only to show that the group operation is jointly continuous. This we get immediately by applying 4.8.6 to $X = G$ and action $g \cdot x$ the group operation. ■

This result is substantially generalized as follows.

Theorem 4.8.10 *(S. Solecki and S. M. Srivastava[109]) Let $(G, \cdot)$ be a group with a Polish topology such that $h \longrightarrow g \cdot h$ is continuous for every $g \in G$, and $g \longrightarrow g \cdot h$ Borel for all h. Then G is a topological group.*

Proof. By 4.8.9, we only have to show that the group operation $g \cdot h$ is jointly continuous. A close examination of the proof of 4.8.6 shows that this follows from the following result. ■

Lemma 4.8.11 *Let G satisfy the hypothesis of our theorem. Then for every meager set I and every g,*

$$Ig = \{h \cdot g : h \in I\}$$

is meager.

Proof.

Claim. If I is meager in G, so is $I^{-1} = \{h \in G : h^{-1} \in I\}$.

Assuming the claim, we prove the lemma as follows. Let I be meager in G and $g \in G$. By the claim, I^{-1} is meager. Since the group operation is continuous in the second varible, $J = g^{-1} \cdot I^{-1}$ is meager. As $I \cdot g = J^{-1}$, it is meager by our claim.

Proof of the claim. Let I be meager. Since every meager set is contained in a meager F_σ, without any loss of generality we assume that I is Borel. By 3.1.30, the group operation $(g,h) \longrightarrow g \cdot h$ is a Borel map. Since the graph of $g \longrightarrow g^{-1}$ is Borel, $g \longrightarrow g^{-1}$ is Borel measurable (4.5.2). Hence, $(g,h) \longrightarrow g^{-1} \cdot h$ is Borel measurable. Let

$$\hat{I} = \{(h,g) : g^{-1} \cdot h \in I\}.$$

Since $\hat{I}$ is a Borel set, it has the Baire property. Now, for every $g \in G$,

$$\hat{I}^g = \{h \in G : g^{-1} \cdot h \in I\} = g \cdot I.$$

Hence, by our hypothesis, $\hat{I}^g$ is meager for every g. Therefore, by the Kuratowski - Ulam theorem (3.5.16), the set $\{h : \hat{I}_h \text{ is meager}\}$ is comeager and hence nonempty by the Baire category theorem. In particular, there exists $h \in G$ such that $\hat{I}_h = h \cdot I^{-1}$ is meager. It follows that $I^{-1} = h^{-1}(hI^{-1})$ is meager. ■

Remark 4.8.12 S. Solecki and S. M. Srivastava have shown that 4.8.10 can be generalized as follows: *Let $(G, \cdot)$ be a group with a topology that is metrizable, separable, and Baire. Suppose the multiplication $g \cdot h$ is continuous in h for all g and Baire measurable in g for all h. Then G is a topological group.* (See [109] for details and applications of this result.)

The following example shows that 4.8.10 is not necessarily true if the group operation $g \cdot h$ is Borel but not continuous in any one of the variables.

Example 4.8.13 Consider the additive group $(\mathbb{R}, +)$ of real numbers. Let $(\mathbb{R}, \mathcal{T})$ be the topological sum $(\mathbb{R} \setminus \{0\}$, usual topology$) \bigoplus \{0\}$ So, $\mathcal{T}$ is generated by the usual open sets and $\{0\}$. Clearly, $\mathcal{T}$ is a Polish topology on $\mathbb{R}$ inducing the usual Borel σ-algebra. In particular, the addition $(x, y) \longrightarrow x + y$ is Borel. If $(\mathbb{R}, \mathcal{T})$ were a topological group it would be discrete, which is not the case.

The next example shows that we cannot drop the condition of measurability of the group operation $g \cdot h$ in one of the variables from 4.8.10. Note that if G were, moreover, abelian, the result is trivially true in this generality. Also, in Solovay'smodel of **ZF** every set has the Baire property. So, we cannot refute this statement without **AC**. The next example shows that under **AC**, the measurability condition in one of the variables cannot be dropped.

Example 4.8.14 (G. Hjorth) Under **AC**, there is a discontinuous group isomorphism $\varphi : \mathbb{R} \longrightarrow \mathbb{R}$. Take G to be $\mathbb{R} \times \mathbb{R}$ with the product topology and the group operation defined by

$$(r, s) \cdot (p, q) = (r + 2^{\varphi(s)} p, s + q),$$

i.e., the group is a semidirect product of two copies of $\mathbb{R}$ with respect to the homomorphism $\bar{\varphi} : \mathbb{R} \longrightarrow \text{Aut}(\mathbb{R})$ naturally induced by φ, $\bar{\varphi}(s)(p) = 2^{\varphi(s)} p$.

4.9 Reduction Theorems

In Section 2, we showed that a subset C of a Polish space X is coanalytic if and only if there is a Borel map $f : X \longrightarrow Tr$ such that $x \in C \iff f(x)$ is well-founded. Then, to each x we assigned an ordinal $\alpha < \omega_1$, namely the rank of the tree $f(x)$, and used it to compute the possible cardinalities of coanalytic sets. This assignment satisfies some definability conditions that are of fundamental importance.

A **norm** on a set S is a map $\varphi : S \longrightarrow \mathbf{ON}$. (Recall that **ON** denotes the class of all ordinal numbers (Chapter 1)). Let φ be a norm on S. Let $\leq_\varphi$ be the binary relation on S defined by

$$x \leq_\varphi y \Longleftrightarrow \varphi(x) \leq \varphi(y).$$

Then $\leq_\varphi$ is (i) reflexive, (ii) transitive, (iii) connected; i.e., for every $x, y \in S$, at least one of $x \leq_\varphi y$ or $y \leq_\varphi x$ holds, and (iv) there is no sequence (x_n) of elements in S such that $x_{n+1} <_\varphi x_n$ for all n, where

$$x <_\varphi y \Longleftrightarrow \varphi(x) < \varphi(y) \Longleftrightarrow x \leq_\varphi y \;\&\; \neg y \leq_\varphi x.$$

A binary relation satisfying i) – iv) is called a **prewellordering** on S.

Let X be a Polish space and $A \subseteq X$ coanalytic. A norm φ on A is called a $\mathbf{\Pi}^1_1$-norm if there are binary relations $\leq_\varphi^{\mathbf{\Pi}^1_1} \in \mathbf{\Pi}^1_1$ and $\leq_\varphi^{\mathbf{\Sigma}^1_1} \in \mathbf{\Sigma}^1_1$ on X such that for $y \in A$,

$$x \in A \;\&\; \varphi(x) \leq \varphi(y) \Longleftrightarrow x \leq_\varphi^{\mathbf{\Pi}^1_1} y \Longleftrightarrow x \leq_\varphi^{\mathbf{\Sigma}^1_1} y.$$

The following is the main result of this section.

Theorem 4.9.1 *(Moschovakis) Every $\mathbf{\Pi}^1_1$ set A in a Polish space X admits a $\mathbf{\Pi}^1_1$-norm $\varphi : A \longrightarrow \omega_1$.*

Its proof is given later in the section.

The following two lemmas are very useful. We give only sketches of their proofs as they are straghtforward verifications.

Lemma 4.9.2 *Let X be a Polish space and $A \subseteq X$ coanalytic. A norm $\varphi : A \longrightarrow \mathbf{ON}$ is a $\mathbf{\Pi}^1_1$-norm if and only if there are binary relations $\leq_\varphi^{\mathbf{\Sigma}^1_1}$, and $<_\varphi^{\mathbf{\Sigma}^1_1}$ on X, both in $\mathbf{\Sigma}^1_1$, such that for every $y \in A$,*

$$x \in A \;\&\; \varphi(x) \leq \varphi(y) \Longleftrightarrow x \leq_\varphi^{\mathbf{\Sigma}^1_1} y$$

and

$$x \in A \;\&\; \varphi(x) < \varphi(y) \Longleftrightarrow x <_\varphi^{\mathbf{\Sigma}^1_1} y.$$

Proof. We prove the "only if" part first. Let $\leq_\varphi^{\mathbf{\Pi}^1_1}$ and $\leq_\varphi^{\mathbf{\Sigma}^1_1}$ witness that φ is a $\mathbf{\Pi}^1_1$-norm on A. Define $<_\varphi^{\mathbf{\Sigma}^1_1}$ by

$$x <_\varphi^{\mathbf{\Sigma}^1_1} y \Longleftrightarrow \neg y \leq_\varphi^{\mathbf{\Pi}^1_1} x \;\&\; x \leq_\varphi^{\mathbf{\Sigma}^1_1} y.$$

To prove the converse, assume that $\leq_\varphi^{\mathbf{\Sigma}^1_1}$ and $<_\varphi^{\mathbf{\Sigma}^1_1}$ are given as above. Define

$$x \leq_\varphi^{\mathbf{\Pi}^1_1} y \Longleftrightarrow x \in A \;\&\; \neg y <_\varphi^{\mathbf{\Sigma}^1_1} x.$$

■

Let $A \subseteq X$ and φ be a norm on A. Define $\leq_\varphi^*$ and $<_\varphi^*$ on X by

$$x \leq_\varphi^* y \Longleftrightarrow x \in A \ \&\ (y \notin A \text{ or } (y \in A \ \&\ \varphi(x) \leq \varphi(y)))$$

and

$$x <_\varphi^* y \Longleftrightarrow x \in A \ \&\ (y \notin A \text{ or } (y \in A \ \&\ \varphi(x) < \varphi(y))).$$

Lemma 4.9.3 *Let X be a Polish space, $A \subseteq X$ coanalytic, and φ a norm on A. Then φ is a $\mathbf{\Pi}_1^1$-norm if and only if both $\leq_\varphi^*, <_\varphi^*$ are coanalytic.*

Proof. We first prove the "only if" part. Let φ be a $\mathbf{\Pi}_1^1$-norm on A and $\leq_\varphi^{\mathbf{\Pi}_1^1}$ and $\leq_\varphi^{\mathbf{\Sigma}_1^1}$ witness this. For x, y in X, note that

$$x \leq_\varphi^* y \Longleftrightarrow x \in A \ \&\ [x \leq_\varphi^{\mathbf{\Pi}_1^1} y \text{ or } \neg y \leq_\varphi^{\mathbf{\Sigma}_1^1} x]$$

and

$$x <_\varphi^* y \Longleftrightarrow x \in A \ \&\ \neg y \leq_\varphi^{\mathbf{\Sigma}_1^1} x.$$

Thus $\leq_\varphi^*$ and $<_\varphi^*$ are in $\mathbf{\Pi}_1^1$.

Conversely, assume that $\leq_\varphi^*$ and $<_\varphi^*$ are in $\mathbf{\Pi}_1^1$. Take $\leq_\varphi^{\mathbf{\Pi}_1^1}$ to be $\leq_\varphi^*$ itself and define $\leq_\varphi^{\mathbf{\Sigma}_1^1}$ by

$$x \leq_\varphi^{\mathbf{\Sigma}_1^1} y \Longleftrightarrow \neg(y <_\varphi^* x).$$

■

Example 4.9.4 Let $X = 2^{\mathbf{N}\times\mathbf{N}}$ and $A = WO$. For $x \in WO$, Let $|x| < \omega_1$ be the order type of x.

For $x \in 2^{\mathbf{N}\times\mathbf{N}}$, define

$$m <_x n \Longleftrightarrow x(m,n) = 1 \ \&\ x(n,m) = 0.$$

For x, y in $2^{\mathbf{N}\times\mathbf{N}}$, set

$$x \leq_{|\cdot|}^{\mathbf{\Sigma}_1^1} y \Longleftrightarrow \exists z \in \mathbf{N}^{\mathbf{N}} \forall m \forall n [m <_x n \Longleftrightarrow z(m) <_y z(n)]$$

and

$$\begin{aligned} x <_{|\cdot|}^{\mathbf{\Sigma}_1^1} y \quad \Longleftrightarrow \quad & \exists k \exists z \in \mathbf{N}^{\mathbf{N}} \forall m \forall n [z(m) <_y k \\ & \&(m <_x n \Longleftrightarrow z(m) <_y z(n))]. \end{aligned}$$

Thus, $x \leq_{|\cdot|}^{\mathbf{\Sigma}_1^1} y$ if and only if there is an order-preserving map from x to y, and $x <_{|\cdot|}^{\mathbf{\Sigma}_1^1} y$ if and only if there is an order-preserving map from x into an initial segment of y. The sets $\leq_{|\cdot|}^{\mathbf{\Sigma}_1^1}$ and $<_{|\cdot|}^{\mathbf{\Sigma}_1^1}$ are clearly $\mathbf{\Sigma}_1^1$. Further, for $y \in WO$,

$$x \leq_{|\cdot|}^{\mathbf{\Sigma}_1^1} y \Longleftrightarrow x \in WO \ \&\ |x| \leq |y|,$$

and

$$x <^{\Sigma^1_1}_{|\cdot|} y \iff x \in WO \;\&\; |x| < |y|.$$

Therefore, by 4.9.2, $|\cdot|$ is a $\mathbf{\Pi}^1_1$-norm on WO, which we shall call the **canonical norm** on WO.

Exercise 4.9.5 Let $X = 2^{\mathbb{N}^{<\mathbb{N}}}$ and $A = WF$. Show that $T \longrightarrow \rho_T$, the rank of T, is a $\mathbf{\Pi}^1_1$-norm on WF.

Example 4.9.6 Let X be a Polish space and $\mathbf{C}$ the set of all nonempty countable compact subsets of X. In 2.5.13, for each $K \in \mathbf{C}$ we defined $\rho(K)$ to be the first ordinal α such that the α th Cantor – Bendixson derivative K^α of K is empty. We show that $K \longrightarrow \rho(K)$ is a $\mathbf{\Pi}^1_1$-norm on $\mathbf{C}$, where $C \subseteq K(X)$, $K(X)$ being equipped with the Vietoris topology.

For any $\alpha \in 2^{\mathbb{N}\times\mathbb{N}}$, let

$$D(\alpha) = \{m \in \mathbb{N} : \alpha(m,m) = 1\},$$

$$LO^* = \{\alpha \in LO : \alpha(0,m) = 1 \text{ for every } m \in D(\alpha)\},$$

and

$$WO^* = \{\alpha \in WO : \alpha(0,m) = 1 \text{ for every } m \in D(\alpha)\}.$$

Thus, LO^* is the set of all α that encode linear orders on subsets of $\mathbb{N}$ for which 0 is the least element. This is Borel. Similarly, WO^* is the set of all α that encode well-orders on subsets of $\mathbb{N}$ having 0 as the first element. WO^* is a coanalytic set. Using the fact that WO is not analytic, one shows easily that WO^* is not analytic.

Now define two analytic sets $R, S \subseteq LO^* \times K(X)$ by

$$\begin{aligned} R(\alpha,K) \iff\; & \alpha \in LO^* \;\&\; [\exists f \in K(X)^{\mathbb{N}}(f(0) = K \\ & \&\forall m \in D(\alpha)[f(m) \neq \emptyset \\ & \&\{m \neq 0 \longrightarrow \forall n(n <^*_\alpha m \longrightarrow f(m) \subseteq f(n)')\}])], \end{aligned}$$

and

$$\begin{aligned} S(\alpha,K) \iff\; & \alpha \in LO^* \;\&\; [\exists f \in K(X)^{\mathbb{N}}(f(0) = K \\ & \&\forall m \in D(\alpha)[f(m) \neq \emptyset \\ & \&(m \neq 0 \longrightarrow f(m) = \textstyle\bigcap_{n <^*_\alpha m} f(n)')] \\ & \&\textstyle\bigcap_{m \in D(\alpha)} f(m)' = \emptyset)], \end{aligned}$$

where $n <^*_\alpha m \iff \alpha(n,m) = 1 \;\&\; \alpha(m,n) = 0$.

Using 3.3.9 it is fairly easy to see that R and S are analytic. Further, if $\alpha \in WO^*$ and $K \in K(X)$ is countable, then

$$R(\alpha,K) \iff \rho(K) \geq |\alpha| + 1$$

and

$$S(\alpha,K) \iff \rho(K) = |\alpha| + 1.$$

Now take a Borel map $\alpha \longrightarrow \alpha'$ from LO^* to LO^* such that

(i) $\alpha \in WO^* \iff \alpha' \in WO^*$, and

(ii) $|\alpha'| = |\alpha| + 1$ for $\alpha \in WO^*$.

Define analytic binary relations $\leq_\rho$, $<_\rho$ on $K(X)$ by

$$K \leq_\rho L \iff \exists\alpha(R(\alpha, L) \,\&\, S(\alpha, K))$$

and

$$K <_\rho L \iff \exists\alpha(R(\alpha', L) \,\&\, S(\alpha, K)).$$

It is straightforward to verify that for any nonempty countable compact set L,

$$K \text{ is nonempty countable compact } \& \ \rho(K) \leq \rho(L) \iff K \leq_\rho L$$

and

$$K \text{ is nonempty countable compact } \& \ \rho(K) < \rho(L) \iff K <_\rho L.$$

Hence, by 4.9.2, ρ is a $\mathbf{\Pi}_1^1$-norm on C.

Proof of 4.9.1. By 4.2.2, there exists a Borel measurable function $f : X \longrightarrow 2^{\mathbb{N}\times\mathbb{N}}$ such that

$$x \in A \iff f(x) \in WO.$$

Define a norm $\varphi : A \longrightarrow \mathbf{ON}$ by

$$\varphi(x) = |f(x)|, \ \ x \in A,$$

where $|\cdot|$ is the canonical $\mathbf{\Pi}_1^1$-norm on WO defined in 4.9.4. It is easy to check that φ is a $\mathbf{\Pi}_1^1$-norm on A. ∎

Remark 4.9.7 Let $\varphi : A \longrightarrow \omega_1$ be a $\mathbf{\Pi}_1^1$-norm on A and $y \in A$. Then $\{x \in A : \varphi(x) \leq \varphi(y)\}$ is $\mathbf{\Delta}_1^1$ and so by Souslin's theorem (4.4.3), Borel. Thus every coanalytic set is a union of $\aleph_1$ Borel sets. This is a result we obtained earlier. The present proof is essentially the same as the one given earlier.

Theorem 4.9.8 *(Boundedness theorem for $\mathbf{\Pi}_1^1$-norms) Suppose A is a $\mathbf{\Pi}_1^1$ set in a Polish space X and φ a norm on A as defined in 4.9.1. Then for every $\mathbf{\Sigma}_1^1$ set $B \subseteq A$, $\sup\{\varphi(x) : x \in B\} < \omega_1$.*

Hence, A is Borel if and only if $\sup\{\varphi(x) : x \in A\} < \omega_1$.

Proof. Suppose $\sup\{\varphi(y) : y \in B\} = \omega_1$. Take any $\mathbf{\Pi}_1^1$ set C that is not $\mathbf{\Sigma}_1^1$. Fix a Borel function g such that

$$x \in C \iff g(x) \in WO.$$

Then,

$$\begin{aligned} x \in C &\iff \exists y(y \in B \;\&\; |g(x)| \le \varphi(y)) \\ &\iff \exists y(y \in B \;\&\; g(x) \le^{\Sigma^1_1}_{|.|} f(y)), \end{aligned}$$

where f is as in 4.9.1. This contradicts the fact that C is not $\boldsymbol{\Sigma}^1_1$. Hence,

$$\sup\{\varphi(x) : x \in B\} < \omega_1.$$

If A is Borel, then taking B to be A, we see that $\sup\{\varphi(x) : x \in A\} < \omega_1$. On the other hand, if $\sup\{\varphi(x) : x \in A\} < \omega_1$, then A is a union of countably many Borel sets of the form $\{x \in A : \varphi(x) = \xi\}$, $\xi < \omega_1$. So A is Borel. ∎

Remark 4.9.9 This result gives an alternative proof of the first separation theorem for analytic sets (4.4.1).

The Borel isomorphism theorem says that any two uncountable Borel sets are isomorphic. Are any two analytic non-Borel sets isomorphic? Are any two coanalytic non-Borel sets isomorphic? We discuss these questions now.

Exercise 4.9.10 Let X and Y be uncountable Polish spaces. Suppose $U \subseteq X \times X$ and $V \subseteq Y \times Y$ are universal analytic. Show that U and V are Borel isomorphic.

Example 4.9.11 (A. Maitra and C. Ryll-Nardzewski[76]) Let X, Y be uncountable Polish spaces. Let $U \subseteq X \times X$ be universal analytic and $C \subseteq Y$ an uncountable coanalytic set not containing a perfect set. We mentioned earlier that Gödel's axiom of constructibility implies the existence of such a set. The set C does not contain any uncountable Borel set. Take $A = Y \setminus C$. Then U and A are not Borel isomorphic. Here is a proof. Suppose they are Borel isomorphic. Take a Borel isomorphism $f : U \longrightarrow A$. By 3.3.5, there exist Borel sets $B_1 \supseteq U$, $B_2 \supseteq A$ and a Borel isomorphism $g : B_1 \longrightarrow B_2$ extending f. Let φ be a $\boldsymbol{\Pi}^1_1$ norm on U^c as defined in 4.9.8. It is easy to verify that for uncountably many $\xi < \omega_1$, $\{(x, y) \in U^c : \varphi(x, y) = \xi\}$ is uncountable. By 4.9.8, $\sup\{\varphi(x, y) : (x, y) \in B_1^c\} < \omega_1$. Therefore, $B_1 \setminus U$ contains an uncountable Borel set. It follows that C contains an uncountable Borel set, which is not the case. Hence, U and A are not Borel isomorphic.

Exercise 4.9.12 Show that the statement "any two analytic non-Borel sets are isomorphic" is equivalent to "any two coanalytic non-Borel sets are isomorphic."

Remark 4.9.13 J. Steel has shown that under **"analytic determinacy"** any two analytic non-Borel sets are isomorphic. Hence, the statement "any two analytic non-Borel sets are isomorphic" cannot be decided in **ZFC**.

Theorem 4.9.14 *(The reduction principle for coanalytic sets) (Kuratowski) Let (A_n) be sequence of $\mathbf{\Pi}^1_1$ sets in a Polish space X. Then there is a sequence (A^*_n) of $\mathbf{\Pi}^1_1$ sets such that they are pairwise disjoint, $A^*_n \subseteq A_n$, and $\bigcup_n A^*_n = \bigcup_n A_n$.*

Proof. Consider $A \subseteq X \times \mathbb{N}$ given by

$$(x, n) \in A \iff x \in A_n.$$

Clearly, A is $\mathbf{\Pi}^1_1$ with projection $\bigcup_n A_n$. Let φ be a $\mathbf{\Pi}^1_1$-norm on A. Define $A^* \subseteq X \times \mathbb{N}$ by

$$\begin{aligned}(x, n) \in A^* \quad \iff \quad & (x, n) \in A \;\&\; \forall m[(x, n) \leq^*_\varphi (x, m)] \\ & \&\; \forall m[(x, n) <^*_\varphi (x, m) \text{ or } n \leq m].\end{aligned}$$

Thus, for each x in the projecton of A we first look at the set of integers n with $(x, n) \in A$ such that $\varphi(x, n)$ is the minimum. Then we choose the least among these integers. Note that A^* is $\mathbf{\Pi}^1_1$, $A^* \subseteq A$, and for every $x \in \bigcup_n A_n$ there is exactly one n such that $(x, n) \in A^*$. Let

$$A^*_n = \{x : (x, n) \in A*\}.$$

Clearly, A^*_n is $\mathbf{\Pi}^1_1$. It is easy to check that the A^*_n's are pairwise disjoint and $\bigcup_n A^*_n = \bigcup_n A_n$. ■

Corollary 4.9.15 *Let X be Polish and A_0, A_1 coanalytic subsets of X. Then there exist pairwise disjoint coanalytic sets A^*_0, A^*_1 contained in A_0, A_1 respectively such that $A^*_0 \bigcup A^*_1 = A_0 \bigcup A_1$.*

Proof. In the above theorem, take $A_n = \emptyset$ for $n > 1$.

Remark 4.9.16 Let $B = \bigcup_n A_n$ be Borel. As

$$A^*_n = B \setminus \bigcup_{i \neq n} A^*_i,$$

it is $\mathbf{\Sigma}^1_1$. So each A^*_n is Borel by Souslin's theorem (4.4.3). This gives an alternative proof of 4.6.1.

The following examples show that analytic sets do not satisfy the reduction principle and the separation theorems are not true for coanalytic sets.

Example 4.9.17 Let U_0, U_1 be a universal pair of analytic subsets of $\mathbb{N}^\mathbb{N} \times \mathbb{N}^\mathbb{N}$ (4.1.17). Suppose there exist pairwise disjoint analytic sets $V_0 \subseteq U_0$, $V_1 \subseteq U_1$ such that $V_0 \bigcup V_1 = U_0 \bigcup U_1$. By the first separation theorem for analytic sets, (4.4.1), there is a Borel set B containing V_0 and disjoint from V_1. We claim that B is universal Borel, which contradicts 3.6.9. To prove our claim, take any Borel $E \subseteq \mathbb{N}^\mathbb{N}$. Since U_0, U_1 is a universal pair of analytic sets, there is an α such that $E = (U_0)_\alpha$ and $E^c = (U_1)_\alpha$. Plainly, $E = B_\alpha$.

Exercise 4.9.18 Let C_0, C_1 be a universal pair of coanalytic subsets of $\mathbf{N}^\mathbf{N} \times \mathbf{N}^\mathbf{N}$. By reduction principle for coanalytic sets, there exist disjoint coanalytic sets D_0, D_1 such that $D_i \subseteq C_i$, $i = 0$ or 1 and $D_0 \bigcup D_1 = C_0 \bigcup C_1$. Show that there is no Borel set B containing D_0 and disjoint from D_1.

Using the above idea, we also get a very useful parametrization of Borel sets.

Theorem 4.9.19 *Let X be a Polish space. Then there exist sets $C \in \mathbf{\Pi}_1^1(\mathbf{N}^\mathbf{N})$ and $V \in \mathbf{\Pi}_1^1(\mathbf{N}^\mathbf{N} \times X)$, $U \in \mathbf{\Sigma}_1^1(\mathbf{N}^\mathbf{N} \times X)$ such that for every $\alpha \in C$, $U_\alpha = V_\alpha$ and*

$$\mathbf{\Delta}_1^1(X) = \{U_\alpha : \alpha \in C\}.$$

In particular, there are a coanalytic set and an analytic set contained in $\mathbf{N}^\mathbf{N} \times X$ that are universal for $\mathbf{\Delta}_1^1(X)$.

Proof. Let W_0, W_1 be coanalytic subsets of $\mathbf{N}^\mathbf{N} \times X$ such that for every pair (C_0, C_1) of sets in $\mathbf{\Pi}_1^1(X)$ there is an α with $C_i = (W_i)_\alpha$, $i = 0$ or 1. By the reduction principle for coanalytic sets, (4.9.15), there are pairwise disjoint coanalytic sets $V_i \subseteq W_i$, $i = 0$ or 1, such that $V_0 \bigcup V_1 = W_0 \bigcup W_1$. Define

$$C = \{\alpha : \forall x((\alpha, x) \in V_0 \bigcup V_1)\}.$$

So, C is coanalytic. Take $V = V_0$ and $U = V_1^c$. A routine argument shows that C, U, and V have the desired properties. So, we have proved the first part of the result.

To see the second part, note that $V \bigcap (C \times X)$ is a coanalytic set universal for $\mathbf{\Delta}_1^1(X)$, and its complement is an analytic set universal for $\mathbf{\Delta}_1^1(X)$. ■

Exercise 4.9.20 Show that in 4.9.19 we cannot replace $C \in \mathbf{\Pi}_1^1$ by $C \in \mathbf{\Sigma}_1^1$.

The next example shows that 4.7.6 cannot be generalized for coanalytic sets A_0, A_1.

Example 4.9.21 Let C_0 and C_1 be disjoint coanalytic subsets of $I = [0, 1]$ that are not Borel separated; i.e., there is no Borel set containing C_0 and disjoint from C_1. Let

$$A_0 = (I \times \{0\}) \bigcup (C_0 \times [0, 3/4])$$

and

$$A_1 = (I \times \{1\}) \bigcup (C_1 \times [1/4, 1]).$$

Clearly, A_0, A_1 are disjoint coanalytic sets with sections closed. Suppose there is a Borel map $u : I \times I \longrightarrow I$ such that $u|B$ is the characteristic function of A_1, where $B = A_0 \bigcup A_1$. Then, the set

$$E = \{x \in I : u(x, 1/2) = 0\}$$

is Borel and separates C_0 from C_1.

4.10 Choquet Capacitability Theorem

In this section we introduce the notion of a capacity, and prove the Choquet capacitability theorem. The notion of a capacity was introduced by Choquet [30]. It lies in the heart of the theory of analytic sets [33]. It is used particularly in stochastic process and potential theory [34], [38].

A **capacity** on a Polish space X is a set-map $I : \mathcal{P}(X) \longrightarrow [0, \infty]$ satisfying the following conditions.

(i) I is "monotone"; i.e., $A \subseteq B \Longrightarrow I(A) \leq I(B)$.

(ii) $A_0 \subseteq A_1 \subseteq A_2 \subseteq \cdots \Longrightarrow \lim I(A_n) = I(A)$, where $A = \bigcup_n A_n$. (We express this by saying that "I is going up".)

(iii) $I(K) < \infty$ for every compact $K \subseteq X$.

(iv) For every compact K and every $t > 0$, $I(K) < t$ implies that there is an open set $U \supseteq K$ such that $I(U) < t$. (We express this by saying that "I is right-continuous over compacta".)

Example 4.10.1 Let μ be a finite Borel measure on a Polish space X and μ^* the associated outer measure. Thus, for any $A \subseteq X$,

$$\mu^*(A) = \inf\{\mu(B) : B \supseteq A, B \text{ Borel}\}.$$

It is easy to check that μ^* is a capacity on X.

Example 4.10.2 (Separation capacity) Let X be a polish space. Define $I : \mathcal{P}(X \times X) \longrightarrow \{0, 1\}$ by

$$I(A) = \begin{cases} 0 & \text{if } \pi_1(A) \bigcap \pi_2(A) = \emptyset, \\ 1 & \text{otherwise,} \end{cases}$$

where π_1 and π_2 are the two projection maps on $X \times X$. For $A \subseteq X \times X$, set

$$R[A] = \pi_1(A) \times \pi_2(A);$$

i. e., $R[A]$ is the smallest rectangle containing A. Clearly, $I(A) = 0$ if and only if $R[A]$ is disjoint from the diagonal in $X \times X$. It is easy to verify that I is a capacity on $X \times X$. Later in this section, using this capacity we shall give a rather beautiful proof of the first separation theorem for analytic sets. Because of this, I is called the **separation capacity** on $X \times X$.

Example 4.10.3 Let X and Y be Polish spaces and $f : X \longrightarrow Y$ a continuous map. Suppose that I is a capacity on Y. Define

$$I_f(A) = I(f(A)), \quad A \subseteq X.$$

It easy to see that I_f is a capacity on X.

We can generalize 4.10.1 as follows.

Proposition 4.10.4 *Let I be a capacity on a Polish space X and that $I^* : \mathcal{P}(X) \longrightarrow [0, \infty]$ be defined by*

$$I^*(A) = \inf\{I(B) : B \supseteq A, B \textit{ Borel}\}.$$

Then I^* is a capacity on X.

Proof. Clearly, I^* is monotone. Further, I^* and I coincide on Borel sets. As I is a capacity, it follows that $I^*(K) < \infty$ for every compact K and that I^* is right-continuous over compacta.

To show that I^* is going up, take any nondecreasing sequence (A_n) of subsets of X. Set $A = \bigcup_n A_n$. Note that for every $C \subseteq X$, there is a Borel $D \supseteq C$ such that $I^*(C) = I(D)$. Hence, for every n there is a Borel $B_n \supseteq A_n$ such that $I(B_n) = I^*(A_n)$. Replacing B_n by $\bigcap_{m \geq n} B_m$, we may assume that (B_n) is nondecreasing. Set $B = \bigcup_n B_n$. Clearly,

$$I(B) \geq I^*(A) \geq I^*(A_n) = I(B_n)$$

for every n. Since I is going up, $\lim I(B_n) = I(B)$. It follows that $\lim I^*(A_n) = I^*A)$. ■

Exercise 4.10.5 Let I be a capacity on a Polish space X. Suppose (K_n) is a nonincreasing sequence of compact subsets of X decreasing to, say, K. Show that $I(K_n)$ converges to $I(K)$.

Example 4.10.6 Consider $I : \mathcal{P}(\mathbb{N}^{\mathbb{N}}) \longrightarrow \{0, 1\}$ defined by

$$I(A) = \begin{cases} 0 & \text{if } A \text{ is contained in a } K_\sigma \text{ set,} \\ 1 & \text{otherwise.} \end{cases}$$

Then I satisfies the conditions (i), (ii), and (iii) of the definition of a capacity. Further, if (K_n) is a nonincreasing sequence of compact sets with intersection, say, K, then $I(K_n) \to I(K)$. But since no open set in $\mathbb{N}^{\mathbb{N}}$ is contained in a K_σ, I is not right-continuous over compacta.

Exercise 4.10.7 Suppose X is a compact metric space and $I : \mathcal{P}(X) \longrightarrow [0, \infty]$ satisfies the conditions (i), (ii), and (iii) of the definition of a capacity. Further, assume that whenever (K_n) is a nonincreasing sequence of compact sets with intersection K, $I(K_n) \to I(K)$. Show that I is a capacity.

We now introduce the key notion of this section. Let X be a Polish space, I a capacity on X, and $A \subseteq X$. We say that A is **I-capacitable** if

$$I(A) = \sup\{I(K) : K \subseteq A \text{ compact}\}.$$

The set A is called **universally capacitable** if it is I-capacitable with respect to all capacities I on X.

Exercise 4.10.8 Let X and Y be Polish spaces and $f : X \longrightarrow Y$ a continuous map. Assume that $A \subseteq X$ is universally capacitable. Show that $f(A)$ is universally capacitable.

Remark 4.10.9 This is almost the only known stability property of the class of universally capacitable sets. For instance, later in this section we shall show that the complement of a universally capacitable set need not be universally capacitable.

Proposition 4.10.10 *Let I be a capacity on a Polish space X and $A \subseteq X$ universally capacitable. Then*

$$I(A) = I^*(A),$$

where I^* is as defined in 4.10.4.

Proof. By 4.10.4, I^* is a capacity. Now note the following.

$$\begin{array}{rcll} I^*(A) & = & \sup\{I^*(K) : K \subseteq A \text{ compact}\} & (\text{as } A \text{ is } I^* - \text{capacitable}) \\ & = & \sup\{I(K) : K \subseteq A \text{ compact}\} & \\ & = & I(A) & (\text{as } A \text{ is } I - \text{capacitable}) \end{array}$$

■

Proposition 4.10.11 $\mathbb{N}^{\mathbb{N}}$ *is universally capacitable.*

Proof. For any $s = (n_0, n_1, \ldots, n_{k-1}) \in \mathbb{N}^{<\mathbb{N}}$, set

$$\Sigma^*(s) = \{\alpha \in \mathbb{N}^{\mathbb{N}} : (\forall i < k)(\alpha(i) \leq n_i)\}.$$

Take any capacity I on $\mathbb{N}^{\mathbb{N}}$ and a real number t such that $I(\mathbb{N}^{\mathbb{N}}) > t$. To prove our result, we shall show that there is a compact set K such that $I(K) \geq t$.

Since the sequence $(\Sigma^*(n))$ increases to $\mathbb{N}^{\mathbb{N}}$, there is a natural number n_0 such that $I(\Sigma^*(n_0)) > t$. Again, since $(\Sigma^*(n_0 n))$ increases to $\Sigma^*(n_0)$, there is a natural number n_1 such that $I(\Sigma^*(n_0 n_1)) > t$. Proceeding similarly, we get a sequence $n_0, n_1, n_2, \ldots$ of natural numbers such that

$$I(\Sigma^*(n_0 n_1 \ldots n_{k-1})) > t$$

for every k. Now consider

$$K = \{\alpha \in \mathbb{N}^{\mathbb{N}} : \alpha(i) \leq n_i \text{ for every } i\}.$$

Clearly, K is compact. We claim that $I(K) \geq t$. Suppose not. Since I is right-continuous over compacta, there is an open set $U \supseteq K$ such that $I(U) < t$. It is not very hard to show that $U \supseteq \Sigma^*(n_0 n_1 \ldots n_{k-1}))$ for some k. Since I is monotone, we have arrived at a contradiction. ■

Theorem 4.10.12 *(Choquet capacitability theorem [30], [107]) Every analytic subset of a Polish space is universally capacitable.*

Proof. Let X be a Polish space and $A \subseteq X$ analytic. Let I be any capacity on X. Suppose $I(A) > t$. Let $f : \mathbf{N}^{\mathbf{N}} \longrightarrow X$ be a continuous map with range A. By 4.10.11, there is a compact $K \subseteq \mathbf{N}^{\mathbf{N}}$ such that $I_f(K) > t$. Plainly $I(f(K)) > t$, and our result is proved. ■

The next result will show that the notion of a capacity is quite relevant to the theory of analytic sets.

Proposition 4.10.13 *Let X be a Polish space and I the separation capacity on $X \times X$ as defined in 4.10.2. Assume that a rectangle $A_1 \times A_2$ be universally capacitable. If $I(A_1 \times A_2) = 0$, then there is a Borel rectangle $B = B_1 \times B_2$ containing $A_1 \times A_2$ of I-capacity 0.*

Proof of 4.10.13. Set $C_0 = A_1 \times A_2$. By 4.10.10, there is a Borel $C_1 \supseteq C_0$ such that $I(C_1) = 0$. Set $C_2 = R[C_1]$. (Recall that R[A] denotes the smallest rectangle containing A.) Clearly $I(C_2) = 0$. Since C_2 is analytic, by 4.10.12, it is universally capacitable. By 4.10.10, there is a Borel $C_3 \supseteq C_2$ such that $I(C_3) = 0$. Set $C_4 = R[C_3]$. Proceeding similarly, we get a nondecreasing sequence (C_n) of subsets of $X \times X$ such that C_n is a rectangle for even n and C_n's are Borel for odd n. Further $I(C_n) = 0$ for all n. Take $B = \bigcup_n C_n$. ■

Here are a few applications of the above result. By 4.10.12 and 4.10.13, we immediately get an alternative proof of the first separation theorem for analytic sets. To see another application, let A_1 and A_2 be two disjoint coanalytic subsets of an uncountable Polish space that cannot be separated by disjoint Borel sets(4.9.18). By 4.10.13, the coanalytic set $A_1 \times A_2$ is not universally capacitable. Thus, the complement of a universally capacitable set need not be universally capacitable.

4.11 The Second Separation Theorem

In this section we prove yet another separation theorem for analytic sets. In the next section we apply it and show that every countable-to-one Borel map is bimeasurable.

Theorem 4.11.1 *(Second separation theorem for analytic sets) (Kuratowski) Let X be a Polish space and A, B two analytic subsets. There* *exist disjoint coanalytic sets C and D such that*

$$A \setminus B \subseteq C \text{ and } B \setminus A \subseteq D.$$

Proof. By 4.1.20, there exist Borel maps $f : X \longrightarrow LO$, $g : X \longrightarrow LO$ such that $f^{-1}(WO) = A^c$ and $g^{-1}(WO) = B^c$.

For α, β in LO, define

$$\alpha \preceq \beta \iff \exists f \in \mathbf{N}^{\mathbf{N}}(f|D(\alpha) \text{ is one-to-one} \\ \& \forall m \forall n(\alpha(m,n) = 1 \iff \beta(f(m), f(n)) = 1)).$$

(Recall that for any $\alpha \in \mathbf{N}^{\mathbf{N}}$, $n \in D(\alpha) \iff \alpha(n,n) = 1$.) So $\preceq$ is an analytic subset of $\mathbf{N}^{\mathbf{N}} \times \mathbf{N}^{\mathbf{N}}$.

Let

$$C = B^c \bigcap \{x \in X : f(x) \preceq g(x)\}^c$$

and

$$D = A^c \bigcap \{x \in X : g(x) \preceq f(x)\}^c.$$

Clearly, C and D are coanalytic. We claim that C and D are disjoint. Suppose not. Take any $x \in C \bigcap D$. Then both $f(x)$ and $g(x)$ are in WO. Therefore, either $|f(x)| \le |g(x)|$ or $|g(x)| \le |f(x)|$. Since $x \in C \bigcap D$, this is impossible.

Finally, we show that $A \setminus B \subseteq C$. Let $x \in A \setminus B$. Then, of course, $x \in B^c$. As $f(x) \notin WO$ and $g(x) \in WO$, there is no order-preserving one-to-one map from $D(f(x))$ into $D(g(x))$. So, $x \in C$. Similarly it follows that $B \setminus A \subseteq D$. ■

Exercise 4.11.2 Let A, B be analytic subsets of a Polish space X and $f, g : X \longrightarrow LO$ Borel maps such that $f^{-1}(WO) = A^c$ and $g^{-1}(WO) = B^c$. Define $\beta_A : X \longrightarrow \omega_1 + 1$ by

$$\beta_A(x) = \begin{cases} |f(x)| & \text{if } x \in A^c, \\ \omega_1 & \text{otherwise.} \end{cases}$$

Define $\beta_B : X \longrightarrow \omega_1 + 1$ analogously. Show that

$$\{x \in X : \beta_A(x) \le \beta_B(x)\} \in \mathbf{\Sigma}^1_1.$$

Corollary 4.11.3 *Suppose X is a Polish space and (A_n) a sequence of analytic subsets of X. Then there exists a sequence (C_n) of pairwise disjoint coanalytic sets such that*

$$A_n \setminus \bigcup_{m \ne n} A_m \subseteq C_n.$$

Proof. By the second separation theorem, for each n there exist pairwise disjoint coanalytic sets C'_n and D'_n such that

$$A_n \setminus \bigcup_{m \neq n} A_m \subseteq C'_n \text{ and } \bigcup_{m \neq n} A_m \setminus A_n \subseteq D'_n.$$

Take

$$C_n = C'_n \bigcap \bigcap_{m \neq n} D'_m.$$

■

Proposition 4.11.4 *Suppose X is a Polish space and (A_n) a sequence of analytic subsets of X. Then there exists a sequence (C_n) of coanalytic subsets of X such that*

$$A_n \setminus \limsup A_m \subseteq C_n \tag{1}$$

and

$$\limsup C_n = \emptyset. \tag{2}$$

Proof. For each n, set $\beta_n = \beta_{A_n}$, where β_{A_n} is as defined in 4.11.2. Let

$$Q_{nm} = \{x \in X : \beta_n(x) \leq \beta_m(x)\}.$$

Q_{nm} is analytic by 4.11.2. Take

$$C_n = [\limsup_m \{Q_{nm}\}]^c.$$

Then C_n is coanalytic and

$$x \notin C_n \iff \exists \eta \subseteq \mathbb{N}(\eta \text{ infinite, } \& \; x \in \bigcap_{m \in \eta} Q_{nm}). \tag{$*$}$$

Proof of (1): Let $x \in A_n \setminus \limsup A_m$. Then $\beta_n(x) = \omega_1$. Let η be any infinite subset of $\mathbb{N}$. Find $m \in \eta$ such that $x \notin A_m$. Then $\beta_m(x) < \omega_1 = \beta_n(x)$. So, $x \notin Q_{nm}$. By ($*$) $x \in C_n$.

Proof of (2): Suppose $\limsup C_n \neq \emptyset$. Take any $x \in \limsup C_n$. Choose an infinite subset η of $\mathbb{N}$ such that $x \in C_n$ for all $n \in \eta$. Choose $n_0 \in \eta$ such that $\beta_{n_0}(x) = \min\{\beta_n(x) : n \in \eta\}$. So, $x \notin C_{n_0}$ by ($\star$). This is a contradiction. Hence, $\limsup C_n = \emptyset$. ■

Remark 4.11.5 (J. Jayne, A. Maitra, and C. A. Rogers) The generalized first separation principle (4.6.1) does not hold for coanalytic sets. This follows from the fact that the following statement does not hold in general.

(Q) *Whenever (C_n) is a sequence of coanalytic subsets of an uncountable Polish space X such that* $\limsup C_n = \emptyset$, *there exist Borel sets B_n in X such that $C_n \subseteq B_n$ and* $\limsup B_n = \emptyset$.

Assume (**Q**). Find a sequence (U_n) of analytic subsets of $X \times X$ universal for sequences of analytic subsets of X (4.1.18). Apply 4.11.4 to these analytic sets U_n. We will get coanalytic subsets C_n of $X \times X$ such that

$$U_n \setminus \limsup U_m \subseteq C_n,$$

and

$$\limsup C_n = \emptyset.$$

By (**Q**), there exist Borel sets B_n in $X \times X$ such that $C_n \subseteq B_n$ for all n and $\limsup B_n = \emptyset$. Choose $2 < \alpha < \omega_1$ such that every B_i is of additive class α. To establish our claim we now show that 3.6.14 is false for α. Towards proving this, let (E_n) be a sequence of multiplicative class α sets in X such that $\limsup E_i = \emptyset$. Choose $\sigma \in \mathbf{N}^{\mathbf{N}}$ such that $E_i = (U_i)_\sigma$ for each i. Then

$$(\limsup U_i)_\sigma = \limsup (U_i)_\sigma = \limsup E_i = \emptyset.$$

So,

$$\forall i (E_i = (U_i)_\sigma \subseteq (C_i)_\sigma \subseteq (B_i)_\sigma)$$

and

$$\limsup (B_i)_\sigma = \emptyset.$$

Since the sets $(B_i)_\sigma$ are of additive class α, we have shown that 3.6.14 does not hold. Thus (**Q**) is false.

Exercise 4.11.6 Show that there is an A-function $f : \mathbb{R} \longrightarrow \mathbb{R}$ that does not dominate a Borel function.

4.12 Countable-to-One Borel Functions

Theorem 4.12.1 *Let X be a Borel subset of a Polish space, Y Polish, and $f : X \longrightarrow Y$ Borel. Then*

$$Z_f = \{y \in Y : f^{-1}(y) \textit{ is a singleton}\}$$

is coanalytic.

Proof. We first prove the result in case X is a closed subset of $\mathbf{N}^{\mathbf{N}}$ and f continuous.

For $s \in \mathbf{N}^{<\mathbf{N}}$, let

$$A_s = f(\Sigma(s) \bigcap X) \text{ and } B_s = f(X \setminus \Sigma(s)).$$

Then

$$\begin{aligned} A_s &= \bigcup_n A_{s^\frown n}, \\ B_{s^\frown n} &\supseteq B_s, \end{aligned}$$

and

$$B_s = \bigcup_{\{t:|s|=|t| \ \&\ s\neq t\}} A_t.$$

Note that $\{A_s \setminus B_s : |s| = k\}$ is a pairwise disjoint family for any k. Further, $\{A_s \setminus B_s : s \in \mathbb{N}^{<\mathbb{N}}\}$ is a regular system. Also, as f is continuous, for any $\alpha \in X$,

$$\{f(\alpha)\} = \bigcap_k A_{\alpha|k} = \bigcap_k \mathrm{cl}(A_{\alpha|k}).$$

Now,

$$\begin{aligned} Z_f &= \bigcup_\alpha [\{f(\alpha)\} \setminus f(X \setminus \{\alpha\})] \\ &= \bigcup_\alpha [\bigcap_k f(X \bigcap \Sigma(\alpha|k)) \setminus f(\bigcup_k (X \setminus \Sigma(\alpha|k)))] \\ &= \bigcup_\alpha \bigcap_k (A_{\alpha|k} \setminus B_{\alpha|k}). \end{aligned}$$

By 4.11.3, for each k there is a family $\{C_s : |s| = k\}$ of pairwise disjoint coanalytic sets such that

$$A_s \setminus B_s \subseteq C_s.$$

Replacing C_s by $\bigcap_{t \preceq s} C_t$, we assume that $\{C_s : s \in \mathbb{N}^{<\mathbb{N}}\}$ is regular. Let

$$C^*_s = C_s \bigcap \ (\mathrm{cl}(A_s) \setminus B_s).$$

Then

$$A_s \setminus B_s \subseteq C^*_s \subseteq \mathrm{cl}(A_s) \setminus B_s. \qquad (*)$$

Further, for any s and any m,

$$\begin{aligned} C^*_{s^\frown m} &= [\mathrm{cl}(A_{s^\frown m}) \setminus B_{s^\frown m}] \bigcap C_{s^\frown m} \\ &\subseteq [\mathrm{cl}(A_s) \setminus B_s] \bigcap C_s \\ &= C^*_s \end{aligned}$$

This shows that $\{C^*_s : s \in \mathbb{N}^{<\mathbb{N}}\}$ is regular and

$$(|s| = |t| \ \&\ s \neq t) \Longrightarrow C^*_s \bigcap C^*_t = \emptyset.$$

By $(\star)$,

$$Z_f \subseteq \bigcup_\alpha \bigcap_k C^*_{\alpha|k} \subseteq \bigcup_\alpha \bigcap_k (\mathrm{cl}(A_{\alpha|k}) \setminus B_{\alpha|k}).$$

For every $\alpha \in \mathbb{N}^{\mathbb{N}}$, we have

$$\begin{aligned} \bigcap_k (\mathrm{cl}(A_{\alpha|k}) \setminus B_{\alpha|k}) &= \bigcap_k \mathrm{cl}(A_{\alpha|k}) \bigcap \bigcap_k B^c_{\alpha|k} \\ &= \bigcap_k A_{\alpha|k} \bigcap \bigcap_k B^c_{\alpha|k} \\ &= \bigcap_k (A_{\alpha|k} \setminus B_{\alpha|k}) \\ &\subseteq Z_f \end{aligned}$$

Hence,

$$Z_f = \bigcup_\alpha \bigcap_k C^*_{\alpha|k}.$$

By 1.12.3,

$$\bigcup_\alpha \bigcap_k C^*_{\alpha|k} = \bigcap_k \bigcup_{|s|=k} C^*_s,$$

and the result in the special case follows.

For the general case, note that by 3.3.15 and 2.6.9, graph(f) is a one-to-one continuous image of a closed subset D of $\mathbb{N}^\mathbb{N}$. Now apply the above case to $X = D$ and $f = \pi_Y \circ g$, where $g : D \longrightarrow$ graph(f) is a continuous bijection and $\pi_Y : X \times Y$ is the projection onto Y. ■

Corollary 4.12.2 *Let X, Y be Polish spaces and $B \subseteq X \times Y$ a Borel set. Then the set*

$$Z = \{x \in X : B_x \textit{ is a singleton}\}$$

is coanalytic.

Theorem 4.12.3 *(Lusin[71]) If X, Y are Polish and B a Borel subset of $X \times Y$ such that for every $x \in X$ the section B_x is countable, then $\pi_X(B)$ is Borel.*

Proof. Let $E \subseteq \mathbb{N}^\mathbb{N}$ be a closed set and $f : E \longrightarrow X \times Y$ a one-to-one continuous map from E onto B. Consider $g = \pi_X \circ f$. For every $x \in \pi_X(B)$, $g^{-1}(x)$ is a countable closed subset of E. Hence, by the Baire category theorem, $g^{-1}(x)$ has an isolated point. Let $g_s = g|\Sigma(s)$, $s \in \mathbb{N}^{<\mathbb{N}}$. As

$$\pi_X(B) = \bigcup_s Z_{g_s},$$

it is coanalytic by 4.12.1. The result follows from Souslin's theorem. ■

In the next chapter we shall present the result of Lusin in its full generality: *Every analytic subset of the product of two Polish spaces X and Y with the sections E_x countable can be covered by countably many Borel graphs.*

Theorem 4.12.4 *Suppose X, Y are Polish spaces and $f : X \longrightarrow Y$ is a countable-to-one Borel map. Then $f(B)$ is Borel for every Borel set B in X.*

Proof. The result follows from 4.12.3 and the identity

$$f(B) = \pi_Y(\text{graph}(f) \bigcap (B \times Y)).$$

■

It is interesting to note that the converse of the above theorem is also true.

Theorem 4.12.5 *(Purves [93]) Let X be a standard Borel space, Y Polish, and $f : X \longrightarrow Y$ a bimeasurable map. Then*

$$\{y \in Y : f^{-1}(y) \text{ is uncountable}\}$$

is countable.

We need a lemma.

Lemma 4.12.6 *Let X be a standard Borel space, Y Polish, and $A \subseteq X \times Y$ analytic with $\pi_X(A)$ uncountable. Suppose that for every $x \in \pi_X(A)$, the section A_x is perfect. Then there is a $C \subseteq \pi_X(A)$ homeomorphic to the Cantor set and a one-to-one Borel map $f : C \times 2^{\mathbb{N}} \longrightarrow A$ such that $\pi_X(f(x, \alpha)) = x$ for every x and every α.*

Granting the Lemma, the proof is completed as follows.

Proof of 4.12.5. Assume that $f^{-1}(y)$ is uncountable for uncountably many y. We shall show that there is a Borel $B \subseteq X$ such that $f(B)$ is not Borel.

Case 1: f is continuous.

Fix a countable base (U_n) for the topology of X. Let $G = \text{graph}(f)$. For each n, let

$$E_n = \{y \in Y : U_n \bigcap G^y \text{ is countable}\}$$

and

$$A = G \setminus \bigcup_n (U_n \times E_n).$$

By 4.3.7, E_n is coanalytic. Hence, A is analytic. Further, $\pi_Y(A)$ is uncountable and A^y is perfect for every $y \in \pi_Y(A)$. By 4.12.6, there is a homeomorph of the Cantor set C contained in $\pi_Y(A)$ and a one-to-one Borel map $g : 2^{\mathbb{N}} \times C \longrightarrow A$ such that $\pi_Y(g(\alpha, y)) = y$. Let D be a Borel subset of $2^{\mathbb{N}} \times C$ such that $\pi_C(D)$ is not Borel and let $B = \pi_X(g(D))$. Since $\pi_X \circ g$ is one-to-one, B is Borel by 4.5.4. Since $f(B) = \pi_C(D)$, the result follows in this case.

The general case follows from case 1 by replacing X by graph(f) and f by $\pi_Y|\text{graph}(f)$. ■

Proof of 4.12.6.

Fix a compatible complete metric on Y and a countable base (U_n) for the topology of Y. For each $s \in 2^{<\mathbb{N}}$, we define a map $n_s(x) : \pi_X(A) \longrightarrow \mathbb{N}$ satifying the following conditions.

(i) $x \longrightarrow n_s(x)$ is $\sigma(\Sigma^1_1)$-measurable,

(ii) diameter$(U_{n_s(x)}) < \frac{1}{2^{|s|}}$,

(iii) $U_{n_s(x)} \bigcap A_x \neq \emptyset$ for all $x \in \pi_X(A)$,

(iv) $\mathrm{cl}(U_{n_{s^\frown\epsilon}(x)}) \subseteq U_{n_s(x)}$, $\epsilon = 0$ or 1, and

(v) $\mathrm{cl}(U_{n_{s^\frown 0}(x)}) \bigcap \mathrm{cl}(U_{n_{s^\frown 1}(x)}) = \emptyset$.

Such a system of functions is defined by induction on $|s|$. This is a fairly routine exercise, which we leave for the reader. Now fix a continuous probability measure P on X such that $P(\pi_X(A)) = 1$. Since every set in $\sigma(\Sigma_1^1)$ is P-measurable and since $\pi_X(A)$ is uncountable, there is a homeomorph C of the Cantor set contained in $\pi_X(A)$ such that $n_s|C$ is Borel measurable for all $s \in 2^{<\mathbf{N}}$. Take $x \in C$ and $\alpha \in 2^{\mathbf{N}}$. Note that $\bigcap_k U_{n_{\alpha|k}(x)}$ is a singleton, say $\{y\}$. Put $f(x, \alpha) = (x, y)$. The map f has the desired properties. ■

The above proof is due to R. D. Mauldin [81].

5

Selection and Uniformization Theorems

In this chapter we present some measurable selection theorems. Selection theorems are needed in several branches of mathematics such as probability theory, stochastic processes, ergodic theory, mathematical statistics ([17], [34], [89], [18], etc.), functional analysis, harmonic analysis, representation theory of groups and C^*-algebras ([4], [6], [7], [35], [36], [37], [40], [50], [54], [72], [73], [124], etc.), game theory, gambling, dynamic programming, control theory, mathematical economics ([55], [78], etc.). Care has been taken to present the results in such a way that they are readily applicable in a variety of situtations. It is impossible to present a satisfactory account of applications in a book of this size. We shall be content with giving some applications that do not require much background beyond what has been developed in this book. From time to time we give some references, where interested readers will find more applications.

The axiom of choice states that every family $\{A_i : i \in I\}$ of nonempty sets admits a choice function. For most purposes this is of no use. For instance, if X and Y are topological spaces and $f : X \longrightarrow Y$ a continuous map, one might want a continuous map $s : Y \longrightarrow X$ such that $f \circ s$ is the identity map. This is not always possible: For the map $f(t) = e^{it}$ from $\mathbb{R}$ onto S^1 no such continuous s exists. (Why?) Conditions under which a continuous selection exists are very stringent and not often met. Interested readers can consult [82] for some very useful continuous selection theorems. On the other hand, measurable selections exist under fairly mild conditions. Note that the map $f : \mathbb{R} \longrightarrow S^1$ defined above admits a Borel selection S. In what follows, we systematically present most of the major measurable selection theorems.

5.1 Preliminaries

A **multifunction** $G : X \longrightarrow Y$ is a map with domain X and whose values are nonempty subsets of Y. For $A \subset Y$, we put

$$G^{-1}(A) = \{x \in X : G(x) \bigcap A \neq \emptyset\}.$$

The set

$$\{(x, y) \in X \times Y : y \in G(x)\}$$

will be called the **graph** of the multifunction G. It will be denoted by $\text{gr}(G)$. We have

$$G^{-1}(A) = \pi_X(\text{gr}(G) \bigcap (X \times A)).$$

A **selection** of a multifunction $G : X \longrightarrow Y$ is a point map $s : X \longrightarrow Y$ such that $s(x) \in G(x)$ for every $x \in X$.

Let $\mathcal{A}$ be a class of subsets of X. *We shall consider only the cases where $\mathcal{A}$ is a σ-algebra or X a Polish space and $\mathcal{A}$ one of the additive class $\mathbf{\Sigma}^0_\alpha(X)$.* Let Y be a Polish space. A multifunction $G : X \longrightarrow Y$ is called **$\mathcal{A}$-measurable (strongly $\mathcal{A}$-measurable)** if $G^{-1}(U) \in \mathcal{A}$ for every open (closed) set U in Y. In particular, a point map $g : X \longrightarrow Y$ is $\mathcal{A}$-measurable if $g^{-1}(U) \in \mathcal{A}$ for all open U in Y. We shall drop the prefix $\mathcal{A}$ from these notions if there is no scope for confusion.

Remark 5.1.1 Suppose X is a measurable space, Y a Polish space and $F(Y)$ the space of all nonempty closed sets in Y with the Effros Borel structure. Then a closed-valued multifunction $G : X \longrightarrow Y$ is measurable if and only if $G : X \longrightarrow F(Y)$ is measurable as a point map.

A multifunction $G : X \longrightarrow Y$ is called **lower-semicontinuous (upper-semicontinuous)** if $G^{-1}(U)$ is open (closed) for every open (closed) set $U \subseteq Y$. Let X, Y be topological spaces and $g : Y \longrightarrow X$ a continuous open (closed) onto map. Then $G(x) = g^{-1}(x)$ is lower semicontinuous (upper semicontinuous).

Lemma 5.1.2 *Suppose Y is metrizable, $G : X \longrightarrow Y$ strongly $\mathcal{A}$-measurable, and $\mathcal{A}$ closed under countable unions. Then G is $\mathcal{A}$-measurable.*

Proof. Let U be open in Y. Since Y is metrizable, U is an F_σ set in Y. Let $U = \bigcup_n C_n$, C_n closed. Then

$$G^{-1}(U) = \bigcup_n G^{-1}(C_n).$$

Since G is strongly $\mathcal{A}$-measurable and $\mathcal{A}$ closed under countable unions, $G^{-1}(U) \in \mathcal{A}$. ■

Exercise 5.1.3 Let X and Y be Polish spaces and $\mathcal{A} = \mathcal{B}_X$. Give an example of a closed-valued, $\mathcal{A}$-measurable multifunction $G : X \longrightarrow Y$ that is not strongly $\mathcal{A}$-measurable.

Lemma 5.1.4 *Suppose $(X, \mathcal{A})$ is a measurable space, Y a Polish space, and $G : X \longrightarrow Y$ a closed-valued measurable multifunction. Then* $\mathrm{gr}(G) \in \mathcal{A} \bigotimes \mathcal{B}_Y$.

Proof. Let (U_n) be a countable base for Y. Note that

$$y \notin G(x) \iff \exists n[G(x) \bigcap U_n = \emptyset \ \& \ y \in U_n].$$

Therefore,

$$(X \times Y) \setminus gr(G) = \bigcup_n [(G^{-1}(U_n))^c \times U_n],$$

and the result follows. ∎

Exercise 5.1.5 Show that the converse of 5.1.4 is not true in general.

Exercise 5.1.6 Let X and Y be Polish spaces and $\mathcal{A}$ a sub σ-algebra of $\mathcal{B}_X$. Show that every compact-valued multifunction $G : X \longrightarrow Y$ whose graph is in $\mathcal{A} \bigotimes \mathcal{B}_Y$ is $\mathcal{A}$-measurable.

The problem of selection occurs in several forms. Let $B \subseteq X \times Y$. A set $C \subseteq B$ is called a **uniformization** of B if for every $x \in X$, the section C_x contains at most one point and $\pi_X(C) = \pi_X(B)$. In other words, $C \subseteq B$ is a uniformization of B if it is the graph of a function $f : \pi_X(B) \longrightarrow Y$. Such a map f will be called a **section** of B.

One of the most basic problems we shall address is the following: When does a Borel set in the product of two Polish spaces admit a Borel uniformization? Let X, Y be Polish and $B \subseteq X \times Y$ Borel. Suppose B admits a Borel uniformization C. Then $\pi_X|C$ is a one-to-one continuous map with range $\pi_X(B)$. Hence, by 4.5.1, $\pi_X(B)$ is Borel. Blackwell([16]) showed that this condition is not sufficient.

Example 5.1.7 (Blackwell[16]) Let C_1, C_2 be two disjoint coanalytic subsets of $[0, 1]$ that cannot be separated by Borel sets. The existence of such sets has been shown in (4.9.17). Let B_i be a closed subset of $[0, 1] \times \Sigma(i)$ whose projection is $[0, 1] \setminus C_i$, $i = 1$ or 2. Take $B = B_1 \bigcup B_2$. Then B is a closed subset of $[0, 1] \times \mathbb{N}^{\mathbb{N}}$ whose projection is $[0, 1]$. Suppose there exists a Borel section $f : [0, 1] \longrightarrow \mathbb{N}^{\mathbb{N}}$ of B. Then $f^{-1}(\Sigma(2))$ is a Borel set containing C_1 and disjoint from C_2. But no such Borel set exists. Thus B does not admit a Borel uniformization.

Exercise 5.1.8 Show that a Borel set $B \subseteq X \times Y$ admits a Borel uniformization if and only if $\pi_X(B)$ is Borel and B admits a Borel section.

Two questions now arise: (i) Under what conditions does a Borel set admit a Borel uniformization? (ii) Can we uniformize any Borel set by a set that is nice in some way? An answer to the second question was given by Von Neumann. We shall present Von Neumann's theorem in Section 5. In subsequent sections, we shall discuss the first problem in detail.

A **partition** $\mathbf{\Pi}$ of a set X is a family of pairwise disjoint nonempty subsets of X whose union is X. There is an obvious one-to-one correspondence between partitions of X and equivalence relations on X. We shall go back anf forth between the two notions without any explicit mention. Let $\mathbf{\Pi}$ be a partition of X and $A \subset X$. We put

$$A^* = \bigcup\{P \in \mathbf{\Pi} : A \bigcap P \neq \emptyset\}.$$

Therefore, A^* is the smallest invariant set containing A, called the **saturation** of A.

Let X be a Polish space and $\mathcal{A}$ a family of subsets of X. A partition $\mathbf{\Pi}$ is called **$\mathcal{A}$-measurable** if the saturation of every open set is in $\mathcal{A}$. Let $\mathbf{\Pi}$ be a partition of a Polish space X. We call $\mathbf{\Pi}$ closed, Borel, etc. if it is closed, Borel, etc. in $X \times X$. It is said to be **lower-semicontinuous** (**upper-semicontinuous**) if the saturation of every open (closed) set is open (closed).

A **cross section** of $\mathbf{\Pi}$ is a subset S of X such that $S \bigcap A$ is a singleton for every $A \in \mathbf{\Pi}$. A **section** of $\mathbf{\Pi}$ is a map $f : X \longrightarrow X$ such that for any x, y in X,

(a) $x \mathbf{\Pi} f(x)$, and

(b) $x \mathbf{\Pi} y \Longrightarrow f(x) = f(y)$.

To each section f we canonically associate a cross section

$$S = \{x \in X : x = f(x)\}$$

of $\mathbf{\Pi}$.

Proposition 5.1.9 *Suppose X is a Polish space and $\mathbf{\Pi}$ a Borel equivalence relation on X. Then the following statements are equivalent.*

(i) *$\mathbf{\Pi}$ has a Borel section.*

(ii) *$\mathbf{\Pi}$ admits a Borel cross section.*

Proof. If f is a Borel section of $\mathbf{\Pi}$, then the corresponding cross section is clearly Borel. On the other hand, let S be a Borel cross section of $\mathbf{\Pi}$. Let $f(x)$ be the unique point of S equivalent to x. It is clearly a section of $\mathbf{\Pi}$. Note that

$$y = f(x) \Longleftrightarrow x \mathbf{\Pi} y \ \& \ y \in S.$$

Therefore, as Π and S are Borel, the graph of f is Borel. Hence, f is Borel measurable by 4.5.2. ∎

A partition Π is called **countably separated** if there is a Polish space Y (or equivalently, a standard Borel space Y) and a Borel map $f : X \longrightarrow Y$ such that

$$x \,\Pi\, x' \iff f(x) = f(x').$$

Exercise 5.1.10 Let Π be a partition of a Polish space X. Show that the following statements are equivalent.

(i) Π is countably separated.

(ii) There is a Polish space Y and a sequence of Borel maps $f_n : X \longrightarrow Y$ such that

$$\forall x, y(x \,\Pi\, y \iff \forall n(f_n(x) = f_n(y))).$$

(iii) There is a sequence (B_n) of invariant Borel subsets of X such that

$$\forall x, y(x \,\Pi\, y \iff \forall n(x \in B_n \iff y \in B_n));$$

that is,

$$X \times Y \setminus \Pi = \bigcup_n (B_n \times B_n^c).$$

Proposition 5.1.11 *Every closed equivalence relation Π on a Polish space X is countably separated.*

Proof. Take any countable base (U_n) for the topology of X. For every x, y in X such that $(x, y) \notin \Pi$, there exist basic open sets U_n and U_m containing x and y respectively with $U_n \times U_m \subseteq (X \times Y) \setminus \Pi$. In particular, U_n^* is disjoint from U_m. Since U_n^* is the projection onto the first coordinate axis of $\pi_X(\Pi \bigcap (X \times U_n))$, which is Borel, U_n^* is analytic. Thus U_n^* is an invariant analytic set disjoint from U_m. Hence, by 4.4.5, there exists an invariant Borel set B_n containing U_n^* and disjoint from U_m. It is now fairly easy to see that

$$X \times Y \setminus \Pi = \bigcup_n (B_n \times B_n^c).$$

The result follows from 5.1.10. ∎

Proposition 5.1.12 *Every Borel measurable partition of a Polish space into G_δ sets is countably separated.*

Proof. Let X be a Polish space and Π a Borel measurable partition of X into G_δ sets. Take $Y = F(X)$, the Effros Borel space of X. Then Y is standard Borel (3.3.10). For $x \in X$, let $[x]$ be the equivalence class containing x and $p(x) = \text{cl}([x])$. For any open $U \subseteq X$,

$$\{x \in X : p(x) \bigcap U \neq \emptyset\} = U^*,$$

which is Borel, since Π is measurable. Therefore, $p : X \longrightarrow Y$ is Borel measurable (5.1.1). We now show that:

$$x \equiv y \Longleftrightarrow p(x) = p(y). \qquad (\star)$$

Clearly, $x \equiv y \Longrightarrow p(x) = p(y)$. Suppose $x \not\equiv y$ but $p(x) = p(y)$. Then $[x]$ and $[y]$ are two disjoint dense G_δ sets in $p(x)$. This contradicts the Baire category theorem, and we have proved $(\star)$. Thus, $p : X \longrightarrow Y$ witnesses the fact that Π is countably separated. ∎

Remark 5.1.13 In the above proposition, let (U_n) be a countable base for the topology of X. Let $x \not\equiv y$. By $(\star)$, there exists a basic open set U_n that intersects precisely one of $p(x)$, $p(y)$, and so it intersects precisely one of $[x]$, $[y]$. It follows that U_n^* contains exactly one of x, y. Conversely, assume that $x \equiv y$. Since the U_n^*'s are invariant, we have $\forall n(x \in U_n^* \Longleftrightarrow y \in U_n^*)$. Thus, we see that for x, y in X,

$$x \equiv y \Longleftrightarrow \forall n(x \in U_n^* \Longleftrightarrow y \in U_n^*).$$

We shall use this observation later in proving some cross section theorems.

Let Π be a partition of a Polish space X and let X/Π denote the set of all Π-equivalence classes. Suppose $q : X \longrightarrow X/\Pi$ is the canonical quotient map. X/Π equipped with the largest σ-algebra making q measurable is called the **quotient Borel space**. So the quotient σ-algebra consists of all subsets E of X/Π such that $q^{-1}(E)$ is Borel in X. The quotient of a standard Borel space by an equivalence relation need not be isomorphic to the Borel σ-algebra of a metric space. However, we have the following.

Exercise 5.1.14 Show that if Π is a countably separated partition of a Polish space X, then the quotient Borel space X/Π is Borel isomorphic to an analytic set in a Polish space.

Exercise 5.1.15 (i) Give an example of a countably separated partition of a Polish space that does not admit a Borel cross section.

(ii) Give an example of a closed equivalence relation on a Polish space X that does not admit a Borel cross section.

Lemma 5.1.16 *Let Π be a Borel partition of a Polish space X. The following statements are equivalent.*

(i) *Π is countably separated.*

(ii) *The σ-algebra $\mathcal{B}^*$ of Π-invariant Borel sets is countably generated.*

Proof. (i) implies (ii): Let Π be countably separated. Take a Polish space Y and $f : X \longrightarrow Y$ a Borel map such that

$$x \,\Pi\, x' \Longleftrightarrow f(x) = f(x').$$

We show that $\mathcal{B}^* = f^{-1}(\mathcal{B}_Y)$, which will then show that Π satisfies (ii). Clearly, $\mathcal{B}^* \supseteq f^{-1}(\mathcal{B}_Y)$. To prove the reverse inclusion, let $B \subseteq X$ be an invariant Borel set. Then $f(B)$ and $f(B^c)$ are disjoint analytic subsets of Y. By the first separation principle for analytic sets (4.4.1), there is a Borel set C such that

$$f(B) \subseteq C \text{ and } C \bigcap f(B^c) = \emptyset.$$

Therefore, $B = f^{-1}(C) \in f^{-1}(\mathcal{B}_Y)$. Hence, (i) implies (ii).

(ii) implies (i): Let $\mathcal{B}^*$ be countably generated. Take any countable generator (A_n) of $\mathcal{B}^*$. Note that the atoms of $\mathcal{B}^*$ are precisely the Π-equivalence classes. Therefore, for any x, x' in X,

$$x \,\Pi\, x' \Longleftrightarrow \forall n(x \in A_n \Longleftrightarrow x' \in A_n).$$

From this and 5.1.10, it follows that (ii) implies (i). ■

5.2 Kuratowski and Ryll-Nardzewski's Theorem

In this section we present a fairly general measurable selection theorem for closed-valued multifunctions and give some applications. The result is due to Kuratowski and Ryll-Nardzewski[63]. Because of its general nature, it can be used in a variety of situations.

In 5.2.1 - 5.2.3, Y is a Polish space, $d < 1$ a compatible complete metric on Y, X a nonempty set, and $\mathcal{L}$ an algebra of subsets of X.

Theorem 5.2.1 *(Kuratowski and Ryll-Nardzewski [63]) Every $\mathcal{L}_\sigma$-measurable, closed-valued multifunction $F : X \longrightarrow Y$ admits an $\mathcal{L}_\sigma$-measurable selection.*

To prove this, we need two lemmas. The first lemma is a straightforward generalization of the reduction principle for additive Borel classes (3.6.10). The second one generalizes the fact that the uniform limit of a sequence of class α functions is of class α (3.6.5 (ii)).

Lemma 5.2.2 *Suppose $A_n \in \mathcal{L}_\sigma$. Then there exist $B_n \subseteq A_n$ such that the B_n's are pairwise disjoint elements of $\mathcal{L}_\sigma$ and $\bigcup_n A_n = \bigcup_n B_n$.*

Proof. Write

$$A_n = \bigcup_m C_{nm},$$

$C_{nm} \in \mathcal{L}$. Enumerate $\{C_{nm} : n, m \in \mathbb{N}\}$ in a single sequence, say (D_i). Set

$$E_i = D_i \setminus \bigcup_{j<i} D_j.$$

Clearly, $E_i \in \mathcal{L}$. Take

$$B_n = \bigcup\{E_i : E_i \subseteq A_n \& (\forall m < n)(E_i \not\subseteq A_m)\}.$$

■

Lemma 5.2.3 *Suppose $f_n : X \longrightarrow Y$ is a sequence of $\mathcal{L}_\sigma$-measurable functions converging uniformly to $f : X \longrightarrow Y$. Then f is $\mathcal{L}_\sigma$-measurable.*

Proof. Replacing (f_n) by a subsequence if necessary, we assume that

$$\forall x \forall n (d(f(x), f_n(x)) < 1/(n+1)).$$

Let F be a closed set in Y and

$$F_n = \operatorname{cl}(\{y \in Y : d(y, F) < 1/(n+1)\}).$$

Then

$$f(x) \in F \iff \forall n f_n(x) \in F_n;$$

i.e., $f^{-1}(F) = \bigcap_n f_n^{-1}(F_n) \in \mathcal{L}_\delta$, and our result is proved. ■

Proof of 5.2.1. Inductively we define a sequence (s_n) of $\mathcal{L}_\sigma$-measurable maps from X to Y such that for every $x \in X$ and every $n \in \mathbf{N}$,

(i) $d(s_n(x), F(x)) < 2^{-n}$, and

(ii) $d(s_n(x), s_{n+1}(x)) < 2^{-n}$.

To define (s_n) we take a countable dense set (r_n) in Y. Define $s_0 \equiv r_0$. Let $n > 0$. Suppose that for every $m < n$, s_m satisfying conditions (i) and (ii) have been defined. Let

$$E_k = \{x \in X : d(s_{n-1}(x), r_k) < 2^{-n+1} \;\&\; d(r_k, F(x)) < 2^{-n}\}.$$

So,

$$E_k = s_{n-1}^{-1}(B(r_k, 2^{-n+1})) \bigcap F^{-1}(B(r_k, 2^{-n})),$$

where $B(y, r)$ denotes the open ball in Y with center y and radius r. It follows that $E_k \in \mathcal{L}_\sigma$.

Further, $\bigcup_k E_k = X$. To see this, take any $x \in X$. As $d(s_{n-1}(x), F(x)) < 2^{-n+1}$, there is a y in $F(X)$ such that $d(y, s_{n-1}(x)) < 2^{-n+1}$. Since (r_k) is dense, there exists an l such that $d(r_l, y) < 2^{-n}$ and $d(r_l, s_{n-1}(x)) < 2^{-n+1}$. Then $x \in E_l$.

By 5.2.2, there exist pairwise disjoint sets $D_k \subseteq E_k$ in $\mathcal{L}_\sigma$ such that $\bigcup_k D_k = \bigcup_k E_k = X$. Define

$$s_n(x) = r_k \text{ if } x \in D_k.$$

It is easy to check that the sequence (s_n) thus defined satisfies conditions (i) and (ii).

By (ii), (s_n) converges uniformly on X, say to s. By 5.2.3, s is $\mathcal{L}_\sigma$-measurable. Since, $d(s(x), F(x)) = \lim d(s_n(x), F(x)) = 0$ and $F(x)$ is closed, $s(x) \in F(x)$. ∎

Corollary 5.2.4 *Let X be a Polish space and $F(X)$ the space of nonempty closed subsets of X with Effros Borel structure. Then there is a measurable $s : F(X) \longrightarrow X$ such that $s(F) \in F$ for all $F \in F(X)$.*

Proof. Apply 5.2.1 to the multifunction $G : F(X) \longrightarrow X$, where $G(F) = F$, with $\mathcal{L}$ the Effros Borel σ-algebra on $F(X)$. ∎

Corollary 5.2.5 *Let $(T, \mathcal{T})$ be a measurable space and Y a separable metric space. Then every $\mathcal{T}$-measurable, compact-valued multifunction $F : T \longrightarrow Y$ admits a $\mathcal{T}$-measurable selection.*

Proof. Let X be the completion of Y. Then F as a multifunction from T to X is closed-valued and $\mathcal{T}$-measurable. Apply 5.2.1 now. ∎

Corollary 5.2.6 *Suppose Y is a compact metric space, X a metric space, and $f : Y \longrightarrow X$ a continuous onto map. Then there is a Borel map $s : X \longrightarrow Y$ of class 2 such that $f \circ s$ is the identity map on X.*

Proof. Let $F(x) = f^{-1}(x)$, $x \in X$, and $\mathcal{L} = \mathbf{\Delta}^0_2$. For any closed set C in Y,

$$F^{-1}(C) = \pi_X(\text{graph}(f) \bigcap (X \times C)).$$

Therefore, by 2.3.24, $F^{-1}(C)$ is closed. Hence, F is $\mathcal{L}_\sigma$-measurable. Now apply 5.2.1. ∎

Proposition 5.2.7 *(A. Maitra and B. V. Rao[77]) Let T be a nonempty set, $\mathcal{L}$ an algebra on T, and X a Polish space. Suppose $F : T \longrightarrow X$ is a closed-valued $\mathcal{L}_\sigma$-measurable multifunction. Then there is a sequence (f_n) of $\mathcal{L}_\sigma$-measurable selections of F such that*

$$F(t) = \text{cl}(\{f_n(t) : n \in \mathbb{N}\}), \quad t \in T.$$

Proof. Fix a countable base $\{U_n : n \in \mathbb{N}\}$ for the topology of X and fix also an $\mathcal{L}_\sigma$-measurable selection f for F. For each n, $T_n = F^{-1}(U_n) \in \mathcal{L}_\sigma$. Write $T_n = \bigcup_m T_{nm}$, $T_{nm} \in \mathcal{L}$. Define $F_{nm} : T_{nm} \longrightarrow X$ by

$$F_{nm}(t) = \text{cl}(F(t) \bigcap U_n), \quad t \in T_{nm}.$$

By 5.2.1, there is an $\mathcal{L}_\sigma | T_{nm}$-measurable selection h_{nm} for F_{nm}. Define

$$f_{nm}(t) = \begin{cases} h_{nm}(t) & \text{if } t \in T_{nm}, \\ f(t) & \text{otherwise.} \end{cases}$$

Then each f_{nm} is $\mathcal{L}_\sigma$-measurable. Further,

$$F(t) = \text{cl}\{f_{nm}(t) : n, m \in \mathbb{N}\}, \quad t \in T.$$ ∎

In the literature, results of the above type, showing the existence of a dense sequence of measurable selections, are called Castaing's theorems [27]. The technique of A. Maitra and B. V. Rao given above can be used to prove such results in various other situations. Finally, we have the following result.

Theorem 5.2.8 *(Srivastava[115]) Let T, $\mathcal{L}$, X, and F be as in 5.2.7. Then there is a map $f : T \times \mathbb{N}^{\mathbb{N}} \longrightarrow X$ such that*

(i) *for every $\alpha \in \mathbb{N}^{\mathbb{N}}$, $t \longrightarrow f(t, \alpha)$ is $\mathcal{L}_\sigma$-measurable, and*

(ii) *for every $t \in T$, $f(t, \cdot)$ is a continuous map from $\mathbb{N}^{\mathbb{N}}$ onto $F(t)$.*

We shall only sketch a proof of this theorem. Readers are invited to work out the details.

Exercise 5.2.9 Let T, $\mathcal{L}$, X, and F be as above. Suppose $s : T \longrightarrow X$ is an $\mathcal{L}_\sigma$-measurable selection for F and $\epsilon > 0$. Show that the multifunction $G : T \longrightarrow X$ defined by

$$G(t) = \mathrm{cl}(F(t) \bigcap B(s(t), \epsilon)), \quad t \in T,$$

is $\mathcal{L}_\sigma$-measurable.

Proof of 5.2.8 Fix a complete compatible metric d on X. Applying 5.2.9 and 5.2.7 repeatedly, for each $s \in \mathbb{N}^{<\mathbb{N}}$, we get an $\mathcal{L}_\sigma$-measurable selection $f_s : T \longrightarrow X$ for F satisfying the following condition: For every $s \in \mathbb{N}^{<\mathbb{N}}$ and every $t \in T$, $\{f_{s^\frown n}(t) : n \in \mathbb{N}\}$ is dense in $F(t) \bigcap B(f_s(t), 1/2^{-|s|}))$. Note that for every $\alpha \in \mathbb{N}^{\mathbb{N}}$ and every $t \in T$, the sequence $(f_{\alpha|n}(t))$ is Cauchy and hence convergent. Take $f : T \times \mathbb{N}^{\mathbb{N}} \longrightarrow X$ defined by

$$f(t, \alpha) = \lim_n f_{\alpha|n}(t). \qquad \blacksquare$$

In the hypothesis of the selection theorem of Kuratowski and Ryll-Nardzewski (5.2.1), further assume that F is strongly $\mathcal{L}_\sigma$-measurable. Then F is $\mathcal{L}_\sigma$-measurable (5.1.2). Therefore, F admits an $\mathcal{L}_\sigma$-measurable selection. The next theorem, due to S. Bhattacharya and S. M. Srivastava[12], shows that in this case we can say more. We shall use this finer selection theorem to prove a beautiful invariance property of Borel pointclasses.

Theorem 5.2.10 *(S. Bhattacharya and S. M. Srivastava [12]) Let $F : X \longrightarrow Y$ be closed-valued and strongly $\mathcal{L}_\sigma$-measurable. Suppose Z is a separable metric space and $g : Y \longrightarrow Z$ a Borel map of class 2. Then there is an $\mathcal{L}_\sigma$-measurable selection f of F such that $g \circ f$ is $\mathcal{L}_\sigma$-measurable.*

Proof. Let (U_n) be a countable base for the topology of Z. Write $g^{-1}(U_n) = \bigcup_m H_{nm}$, the H_{nm}'s closed. Also, take a countable base (W_n) for Y and write $W_n = \bigcup_m C_{nm}$, the C_{nm}'s closed. Let $\mathcal{B}$ be the smallest family of subsets of Y closed under finite intersections, containing each H_{nm} and each C_{nm}. Let $\mathcal{T}'$ be the topology on Y with $\mathcal{B}$ a base. Note that $\mathcal{T}'$ is finer than $\mathcal{T}$, the original topology of Y. By Observations 1 and 2 of Section 2, Chapter 3, we see that $\mathcal{T}'$ is a Polish topology on Y. Note that with Y equipped with the topology $\mathcal{T}'$, g is continuous and F $\mathcal{L}_\sigma$-measurable. By 5.2.1, there is an $\mathcal{L}_\sigma$-measurable selection f of F. This f works. ∎

Theorem 5.2.11 *Let X, Y be compact metric spaces, $f : X \longrightarrow Y$ a continuous onto map. Suppose $A \subseteq Y$ and $1 \leq \alpha < \omega_1$. Then*

$$f^{-1}(A) \in \mathbf{\Pi}^0_\alpha(X) \Longleftrightarrow A \in \mathbf{\Pi}^0_\alpha(Y).$$

To prove this we need a lemma.

Lemma 5.2.12 *Let X, Y, and f be as in 5.2.11. Suppose $1 \leq \alpha < \omega_1$, Z is a separable metric space, and $g : X \longrightarrow Z$ is a Borel map of class α. Then there is a class 2 map $s : Y \longrightarrow X$ such that $g \circ s$ is of class α and $f(s(y)) = y$ for all y.*

Proof. Let $F(y) = f^{-1}(y)$, $y \in Y$. Then $F : Y \longrightarrow X$ is an upper-semicontinuous closed-valued multifunction. By 5.2.1 there is a selection s of F that is Borel of class 2. This s works if either $\alpha = 1$ (i.e., if g is continuous) or if $\alpha \geq \omega_0$ (in this case $g \circ s$ is of class $1 + \alpha = \alpha$). So, we need to prove the result for $2 \leq \alpha < \omega_0$ only. We prove this by induction on α.

For $\alpha = 2$ we get this by 5.2.10. Let $n \geq 2$, and the result is true for $\alpha = n$. Let $g : X \longrightarrow Z$ be of class $n+1$. By 3.6.15, there is a sequence (g_n) of Borel measurable functions from X to Z of class n converging pointwise to g. By 3.6.5, $h = (g_n) : X \longrightarrow Z^{\mathbb{N}}$ is of class n. So, by the induction hypothesis, there is a selection s of F of class 2 such that $h \circ s$ is of class n. In particular, each $g_n \circ s$ is of class n. As $g_n \circ s \to g \circ s$ pointwise, $g \circ s$ is of class $(n + 1)$ by 3.6.5. ∎

Proof of 5.2.11 We need to prove the "only if" part of the result only. Let $f^{-1}(A) \in \mathbf{\Pi}^0_\alpha(X)$. There is a sequence (A_n) of ambiguous class α sets such that $f^{-1}(A) = \bigcap_n A_n$. Define $g : X \longrightarrow 2^{\mathbb{N}}$ by

$$g(x) = (\chi_{A_0}(x), \chi_{A_1}(x), \chi_{A_2}(x), \ldots).$$

The map g is of class α. By 5.2.12, there is a class 2 map $s : Y \longrightarrow X$ such that $g \circ s$ is of class α and $f(s(y)) = y$. As

$$A = (g \circ s)^{-1}(\bar{1}),$$

where $\bar{1}$ is the constant sequence 1, it is of multiplicative class α. ∎

For a more general version of this theorem see [12].

5.3 Dubins – Savage Selection Theorems

In this section we present a selection theorem due to Schäl[103], [104] that is very useful in dynamic programming, gambling, discrete-time stochastic control, etc.

Theorem 5.3.1 *(Schäl) Suppose $(T, \mathcal{T})$ is a measurable space and let Y be a separable metric space. Suppose $G : T \longrightarrow Y$ is a $\mathcal{T}$-measurable compact-valued multifunction. Let v be a real-valued function on $\text{gr}(G)$, the graph of G, that is the pointwise limit of a nonincreasing sequence (v_n) of $\mathcal{T} \bigotimes \mathcal{B}_Y|\text{gr}(G)$-measurable functions on $\text{gr}(G)$ such that for each n and each $t \in T$, $v_n(t, .)$ is continuous on $G(t)$. Let*

$$v^*(t) = \sup\{v(t, y) : y \in G(t)\}, \quad t \in T.$$

Then there is a $\mathcal{T}$-measurable selection $g : T \longrightarrow Y$ for G such that

$$v^*(t) = v(t, g(t))$$

for every $t \in T$.

In the dynamic programming literature, theorems of the above type are called Dubins – Savage selection theorems. Theorem 5.3.1 is the culmination of many attempts to improve on the original result of Dubins and Savage[39]. For applications and discussions on this selection theorem see [74], [104].

Proof of 5.3.1. (Burgess and Maitra[24]) Without any loss of generality we assume that Y is Polish. Fix a complete metric d on Y compatible with its topology. By 5.2.7, we get $\mathcal{T}$-measurable selections $g_n : T \longrightarrow Y$ of G such that

$$G(t) = \text{cl}(\{g_n(t) : n \in \mathbb{N}\}), \quad t \in T.$$

Then $v^*(t) = \sup\{v(t, g_n(t)) : n \in \mathbb{N}\}$. Hence, v^* is $\mathcal{T}$-measurable.

We first prove the result when v is $\mathcal{T} \bigotimes \mathcal{B}_Y|\text{gr}(G)$-measurable with $v(t, .)$ continuous for all t. Set

$$H(t) = \{y \in G(t) : v(t, y) = v^*(t)\}, \quad t \in T.$$

Clearly, H is a compact-valued multifunction. Let C be any closed set in Y and let

$$C_n = \{y \in Y : d(y, C) < 1/n\}, \quad n \geq 1.$$

We easily check that

$$\{t : H(t) \bigcap C \neq \emptyset\} = \bigcap_n \bigcup_i \{t : v(t, g_i(t)) > v^*(t) - 1/n \text{ and } g_i(t) \in C_n\}.$$

It follows that H is $\mathcal{T}$-measurable. To complete the proof in the special case, apply the Kuratowski and Ryll-Nardzewski selection theorem (5.2.1) and take any $\mathcal{T}$-measurable selection g for H.

We now turn to the general case. By the above case, for each n there is a $\mathcal{T}$-measurable selection $g_n : T \longrightarrow Y$ of G such that

$$v_n(t, g_n(t)) = \sup\{v_n(t, y) : y \in G(t)\}$$

for every $t \in T$. For $t \in T$, set

$$H(t) = \{y \in G(t) : \text{ there is a subsequence } (g_{n_i}(t)) \text{ such that } g_{n_i}(t) \to y\}.$$

Since $G(t)$ is nonempty and compact, so is $H(t)$. We now show that H is $\mathcal{T}$-measurable. Let C be closed in Y. Then

$$\{t \in T : H(t) \bigcap C \neq \emptyset\} = \bigcap_{k \geq 1} \bigcup_{m > k} \{t \in T : d(g_m(t), C) < 1/k\}.$$

It follows that H is $\mathcal{T}$-measurable. By the Kuratowski and Ryll-Nardzewski selection theorem, there is a $\mathcal{T}$-measurable selection g of H.

To complete the proof, fix $t \in T$. Then, there is a subsequence $g_{n_i}(t)$ such that $g_{n_i}(t) \to g(t)$. By our hypothesis and 2.3.28, we have

$$\lim_i v_{n_i}(t, g_{n_i}(t)) \leq v(t, g(t)).$$

It follows that

$$v(t) = v(t, g(t)).$$

■

Example 5.3.2 It is not unreasonable to conjecture that 5.3.1 remains true even for v that are $\mathcal{T} \bigotimes \mathcal{B}_Y | gr(G)$-measurable such that $v(t, .)$ is upper-semicontinuous for every t. However, this is not true. Recall that in the last chapter, using Solovay's coding of Borel sets, we showed that there is a coanalytic set T and a function $g : T \longrightarrow 2^{\mathbb{N}}$ whose graph is relatively Borel in $T \times 2^{\mathbb{N}}$ but that is not Borel measurable. Take $\mathcal{T} = \mathcal{B}_T$, $G(t) = 2^{\mathbb{N}}$ $(t \in T)$, and $v : T \times 2^{\mathbb{N}} \longrightarrow \mathbb{R}$ the characteristic function of graph(g).

5.4 Partitions into Closed Sets

In this section we prove several cross section theorems for partitions of Polish spaces into closed sets and give some applications of these results.

Theorem 5.4.1 *(Effros [40]) Every lower-semicontinuous or upper-semicontinuous partition* Π *of a Polish space* X *into closed sets admits a Borel measurable section* $f : X \longrightarrow X$ *of class 2. In particular, they admit a* G_δ *cross section.*

Proof. In 5.2.1, take $Y = X$, $\mathcal{L}$ the family of invariant sets that are simultaneously F_σ and G_δ, and $F(x) = [x]$, the equivalence class containing x. So, there is an $\mathcal{L}_\sigma$-measurable selection $f : X \longrightarrow X$ of F. This means that f is a Borel measurable section of Π of class 2. The corresponding cross section $S = \{x \in X : x = f(x)\}$ is a G_δ cross section of Π. ■

Theorem 5.4.2 *(Effros - Mackey cross section theorem) Suppose H is a closed subgroup of a Polish group G and Π the partition of G consisting of all the right cosets of H. Then Π admits a Borel measurable section of class 2. In particular, it admits a G_δ cross section.*

Proof. Note that for any open set U in G,

$$U^* = \bigcup\{g \cdot U : g \in H\}.$$

So, U^* is open. Thus Π is lower semicontinuous. The result follows from Effros's cross section theorem (5.4.1). ■

Theorem 5.4.3 *Every Borel measurable partition Π of a Polish space X into closed sets admits a Borel measurable section $f : X \longrightarrow X$. In particular, it admits a Borel cross section.*

Proof. Let $\mathcal{A}$ be the σ-algebra of all invariant Borel subsets of X and $F : X \longrightarrow X$ the multifunction that assigns to each $x \in X$ the member of Π containing x. By our assumptions, F is $\mathcal{A}$-measurable. By 5.2.1, we get a measurable selection f for F. Note that f is a section of Π. The corresponding cross section $S = \{x \in X : x = f(x)\}$ of Π is clearly a Borel cross section of Π. ■

This is one of the most frequently used cross section theorems. As an application we consider the classical problem of classifying complex irreducible $n \times n$ matrices with respect to the unitary equivalence. We refer the reader to [4] for the terminology used here.

Define an equivalence relation $\sim$ on $irr(n)$ by $A \sim B$ if and only if A and B are unitarily equivalent; i.e., there is an unitary $n \times n$ matrix U such that $A = UBU^*$. The quotient Borel space $irr(n)/\sim$ is called the **classification space** for irreducible $n \times n$ matrices. The **classification problem** is the problem of finding a countable and complete set of unitary invariants. This amounts to finding suitable real- or complex-valued Borel functions f_i, $i \in \mathbb{N}$, on $irr(n)$ such that for every A, B in $irr(n)$,

$$A \sim B \Longleftrightarrow \forall i(f_i(A) = f_i(B)).$$

Several countable and complete sets of unitary invariants have been found. (See [4], p. 74.) In particular, $\sim$ is countably separated. Therefore, by 2.4.5 and 5.1.14, *the classification space $irr(n)/\sim$ is Borel isomorphic to an analytic subset of a Polish space.* Using our results, we can say more.

Theorem 5.4.4 *The classification space $irr(n)/\sim$ is standard Borel.*

Proof. Fix any irreducible A. Then the $\sim$-equivalence class $[A]$ containing A equals

$$\pi_1\{(B,U) \in irr(n) \times U(n) : A = UBU^*\},$$

where $\pi_1 : irr(n) \times U(n) \longrightarrow irr(n)$ is the projection map to the first coordinate space. (Recall that $U(n)$ denotes the set of all $n \times n$ unitary matrices.) As the set

$$\{(B,U) \in irr(n) \times U(n) : A = UBU^*\}$$

is closed and $U(n)$ compact, $[A]$ is closed by 2.3.24.

Now let $\mathcal{O}$ be any open set in $irr(n)$. Its saturation is

$$\bigcup_{U\in U(n)} \{A \in irr(n) : UAU^* \in \mathcal{O}\},$$

which is open. Thus $\sim$ is a lower-semicontinuous partition of $irr(n)$ into closed sets. By 5.4.3, let C be a Borel cross section of $\sim$. Then $q|C$ is a one-to-one Borel map from C onto $irr(n)/\sim$, where $q : irr(n) \longrightarrow irr(n)/\sim$ is the canonical quotient map. By the Borel isomorphism theorem (3.3.13), q is a Borel isomorphism, and our result is proved. ■

We give some more applications of 5.4.2. Recall that if G is a Polish group, H a closed subgroup, and E the equivalence relation induced by the right cosets, then the σ-algebra of invariant Borel sets is countably generated. Elsewhere (4.8.1) we used the theory of analytic sets to prove this result. As an application of 5.4.2 we give an alternative proof of this fact without using the theory of analytic sets. As a second application we show that the orbit of any point under a Borel action is Borel.

An alternative proof of 4.8.1. Let G, H, and Π be as in 5.4.2. Let $\mathcal{B}$ be the σ-algebra of invariant Borel sets. As proved in 5.4.2, there is a Borel section $s : G \longrightarrow G$ of Π. Then,

$$\mathcal{B} = \{s^{-1}(B) : B \in \mathcal{B}_G\}.$$

Hence, $\mathcal{B}$ is countably generated. ■

Theorem 5.4.5 *(Miller[84]) Let $(G,\cdot)$ be a Polish group, X a Polish space, and $a(g,x) = g \cdot x$ an action of G on X. Suppose for a given $x \in X$ that $g \longrightarrow g \cdot x$ is Borel. Then the orbit*

$$\{g \cdot x : g \in G\}$$

of x is Borel.

Proof. Let $H = G_x$ be the stabilizer of x. By 4.8.4, H is closed in G. Let S be a Borel cross section of the partition Π consisting of the left cosets of H. The map $g \longrightarrow g \cdot x$ restricted to S is one-to-one, Borel, and onto the orbit of x. The result follows from 4.5.4. ■

5.5 Von Neumann's Theorem

In Section 1, we showed that a Borel set need not admit a Borel uniformization. So, what is the best we can do? Von Neumann answered this question, and his theorem has found wide application in various areas of mathematics. He showed that every Borel set admits a coanalytic uniformization and something more: It admits a section that is measurable with respect to all continuous probability measures (such functions are called **universally measurable**) and that is Baire measurable.

The following reasonably simple argument shows that an analytic uniformization of a Borel set must be Borel. Hence, a Borel set need not have an analytic uniformization.

Proposition 5.5.1 *Let X, Y be Polish spaces, $B \subseteq X \times Y$ Borel, and C an analytic uniformization of B. Then C is Borel.*

Proof. We show that C is also coanalytic. The result will then follow from Souslin's theorem. That C is coanalytic follows from the following relation:

$$(x, y) \in C \Longleftrightarrow (x, y) \in B \ \& \ \forall z((x, z) \in C \Longrightarrow y = z).$$

■

Before we prove Von Neumann's theorem, we make a simple observation. Let C be a nonempty closed set in $\mathbf{N}^{\mathbf{N}}$. Then there exists a unique point α in C such that for all $\beta \neq \alpha$ in C, there exists an $n \in \mathbf{N}$ such that $\alpha(n) < \beta(n)$ and for all $m < n$, $\alpha(m) = \beta(m)$; i.e., α is the lexicographic minimum of the elements of C. To show the existence of such an α, we define a sequence (α_n) in C by induction as follows. Let α_0 be any point of C such that

$$\alpha_0(0) = \min(\{\beta(0) : \beta \in C\}).$$

Having defined α_i for $i < n$, let

$$\alpha_n \in C \bigcap \Sigma(\alpha_0(0), \alpha_1(1), \ldots, \alpha_{n-1}(n-1))$$

be such that

$$\alpha_n(n) = \min\{\beta(n) : \beta \in C \bigcap \Sigma(\alpha_0(0), \alpha_1(1), \ldots, \alpha_{n-1}(n-1))\}.$$

(α_n) converges to some point α. Since C is closed, $\alpha \in C$. Clearly, α is the lexicographic minimum of C.

Theorem 5.5.2 *(Von Neumann[124]) Let X and Y be Polish spaces, $A \subseteq X \times Y$ analytic, and $\mathcal{A} = \sigma(\Sigma^1_1(X))$, the σ-algebra generated by the analytic subsets of X. Then there is an $\mathcal{A}$-measurable section $u : \pi_X(A) \longrightarrow Y$ of A.*

Proof. Let $f : \mathbb{N}^{\mathbb{N}} \longrightarrow A$ be a continuous surjection. Consider

$$B = \{(x,\alpha) \in X \times \mathbb{N}^{\mathbb{N}} : \pi_X(f(\alpha)) = x\}.$$

Then B is a closed set with $\pi_X(B) = \pi_X(A)$. For $x \in \pi_X(A)$, define $g(x)$ to be the lexicographic minimum of B_x; i.e.,

$$\begin{aligned} g(x) = \alpha \iff & (x,\alpha) \in B \\ & \& \forall \beta\{(x,\beta) \in B \Longrightarrow \\ & \quad \exists n[\alpha(n) < \beta(n) \text{ and } \forall m < n(\alpha(m) = \beta(m))]\}. \end{aligned}$$

By induction on $|s|$, we prove that $g^{-1}(\Sigma(s)) \in \mathcal{A}$ for every $s \in \mathbb{N}^{<\mathbb{N}}$. Since $\{\Sigma(s) : s \in \mathbb{N}^{<\mathbb{N}}\}$ is a base for $\mathbb{N}^{\mathbb{N}}$, it follows that g is $\mathcal{A}$-measurable. Suppose $g^{-1}(\Sigma(t)) \in \mathcal{A}$ and $s = t\hat{\ }k$, $k \in \mathbb{N}$. Then for any x,

$$\begin{aligned} x \in g^{-1}(\Sigma(s)) \iff & x \in g^{-1}(\Sigma(t)) \\ & \& \ \exists \alpha((x,\alpha) \in B \ \& \ s \prec \alpha) \\ & \& \ \forall l < k \neg \exists \beta((x,\beta) \in B \ \& \ t\hat{\ }l \prec \beta). \end{aligned}$$

Hence, $g^{-1}(\Sigma(s)) \in \mathcal{A}$. Now, define $u(x) = \pi_Y(f(g(x)))$, $x \in \pi_X(A)$. Then u is an $\mathcal{A}$-measurable section of A. ∎

Theorem 5.5.3 *Every analytic subset A of the product of Polish spaces X, Y admits a section u that is universally measurable as well as Baire measurable.*

Proof. The result follows from 5.5.2, 4.3.1, and 4.3.2. ∎

Proposition 5.5.4 *In 5.5.3, further assume that A is Borel. Then the graph of the section u is coanalytic.*

Proof. Note that

$$\begin{aligned} u(x) = y \iff & (x,y) \in A \ \& \ (\forall \alpha \in \mathbb{N}^{\mathbb{N}})(\forall \beta \in \mathbb{N}^{\mathbb{N}})([(x,\alpha) \in B \\ & \&(x,\beta) \in B \ \& \ f(\alpha) = (x,y)] \Longrightarrow \alpha \leq_{\text{lex}} \beta), \end{aligned}$$

where $\leq_{\text{lex}}$ is the lexicographic ordering on B. ∎

In a significant contribution to the theory, M. Kondô showed that every coanalytic set can be uniformized by a coanalytic graph [56]. We present this remarkable result in the last section of this chapter.

Example 5.5.5 Let $A \subseteq X \times Y$ be a Borel set whose projection is X and that cannot be uniformized by a Borel graph. By 5.5.4, there is a coanalytic uniformization C of A. By 5.5.1, C is not analytic. Now, the one-to-one continuous map $f = \pi_X|C$ is not a Borel isomorphism. Thus *a one-to-one Borel map defined on a coanalytic set need not be a Borel isomorphism, although those with domain analytic are* (4.5.1).

Further, let $\mathcal{B} = \{f^{-1}(B) : B \in \mathcal{B}_X\}$. Then $\mathcal{B}$ is a countably generated sub σ-algebra of $\mathcal{B}_C$ containing all the singletons and yet different from $\mathcal{B}_C$. This shows that in general, the Blackwell – Mackey theorem (4.5.10) does not hold for a coanalytic set.

Exercise 5.5.6 Let X, Y be Polish spaces and $f : X \longrightarrow Y$ Borel. Show that there is a coanalytic set $C \subseteq X$ such that $f|C$ is one-to-one and $f(C) = f(X)$.

The next theorem is a generalization of Von Neumann's theorem. Its corollary is essentially the form in which Von Neumann proved his theorem originally.

Theorem 5.5.7 *Let $(X, \mathcal{E})$ be a measurable space with $\mathcal{E}$ closed under the Souslin operation, Y a Polish space, and $A \in \mathcal{E} \bigotimes \mathcal{B}_Y$. Then $\pi_X(A) \in \mathcal{E}$, and there is an $\mathcal{E}$-measurable section of A.*

Proof. By 3.1.7, there exists a countable sub σ-algebra $\mathcal{D}$ of $\mathcal{E}$ such that $A \in \mathcal{D} \bigotimes \mathcal{B}_Y$. Let (B_n) be a countable generator of $\mathcal{D}$ and $\chi : X \longrightarrow \mathcal{C}$ the map defined by

$$\chi(x) = (\chi_{B_0}(x), \chi_{B_1}(x), \chi_{B_2}(x), \ldots), \quad x \in X.$$

Let $Z = \chi(X)$. Then χ is a bimeasurable map from $(X, \mathcal{D})$ onto $(Z, \mathcal{B}_Z)$. Let

$$B = \{(\chi(x), y) \in Z \times Y : (x, y) \in A\}.$$

B is Borel in $Z \times Y$. Take a Borel set C in $\mathcal{C} \times Y$ such that $B = C \bigcap (Z \times Y)$.

Let $E = \pi_{\mathcal{C}}(C)$. Then E is analytic, and therefore it is the result of the Souslin operation on a system $\{E_s : s \in \mathbb{N}^{<\mathbb{N}}\}$ of Borel subsets of $\mathcal{C}$. Note that

$$\pi_X(A) = \chi^{-1}(E) = \mathcal{A}(\chi^{-1}(\{E_s\})).$$

Since $\mathcal{E}$ is closed under the Souslin operation, $\pi_X(A) \in \mathcal{E}$.

By 5.5.2, there is a $\sigma(\Sigma_1^1(\mathcal{C}))$-measurable section $v : E \longrightarrow Y$ of C. Take $f = v \circ \chi$. Then f is an $\mathcal{E}$-measurable section of A. ■

Corollary 5.5.8 *Let $(X, \mathcal{A}, P)$ be a complete probability space, Y a Polish space, and $B \in \mathcal{A} \bigotimes \mathcal{B}_Y$. Then $\pi_X(B) \in \mathcal{A}$, and B admits an $\mathcal{A}$-measurable section.*

Proof. Since $\mathcal{A}$ is closed under the Souslin operation, the result follows from 5.5.7. ■

The reader is referred to [28] for some applications of Von Neumann's selection theorem.

5.6 A Selection Theorem for Group Actions

Many interesting partitions encountered in the representation theory of groups and C^*-algebras are induced by group actions. In this section, we show the existence of a Borel cross section for such partitions under a fairly mild restriction. This remarkable result is due to J. P. Burgess.

Theorem 5.6.1 *(Burgess[23]) Let a Polish group G act continuously on a Polish space X, inducing an equivalence relation E_G. Suppose E_G is countably separated. Then it admits a Borel cross section.*

Proof. Fix a sequence of invariant Borel sets $Z_0, Z_1, Z_2, \ldots$, closed under complementation, such that for all $x, y \in X$,

$$xE_G y \iff \forall m(x \in Z_m \iff y \in Z_m). \tag{0}$$

Also, fix a complete metric d compatible with the topology of X.

The construction of the required cross section proceeds in four steps.

Step 1. For each $s \in \mathbb{N}^{<\mathbb{N}}$ of even length, we define a Borel set $A(s)$.

Case 1. $s = e$, the empty sequence. Set $A(e) = X$.

Case 2. Let $s = (m, n)$ be a sequence of length 2. Set

$$A((m,n)) = \begin{cases} Z_m & \text{if } n = 0, \\ X \setminus Z_m & \text{otherwise.} \end{cases}$$

Case 3. $s = t\hat{}m\hat{}n$, where t has length ≥ 2 and $A(t)$ is a closed set. For such t we define $A(t\hat{}m\hat{}n)$ for all m and n at once. For each m we let $\{A(t\hat{}m\hat{}n) : n \in \mathbb{N}\}$ be a family of closed sets of d-diameter $< 1/(m+1)$ whose union is $A(t)$. Note that in every case so far we have

$$A(t) = \bigcap_m \bigcup_n A(t\hat{}m\hat{}n). \tag{1}$$

Case 4. $s = t\hat{}m\hat{}n$, where t has length ≥ 2 and $A(t)$ is not a closed set. Again, for such t we define all $A(t\hat{}m\hat{}n)$ at once.

First we introduce by induction on countable ordinals α a slight modification of the usual hierarchies of Borel sets. Let $\mathcal{M}_0$ be the family of all closed subsets of X. For a countable ordinal $\alpha > 0$, let $\mathcal{M}_\alpha$ be the family of all sets of the form $\bigcap_m \bigcup_n W_{mn}$ with $W_{mn} \in \bigcup_{\beta<\alpha} \mathcal{M}_\beta$. Thus $\mathcal{M}_1 = \mathbf{\Pi}^0_3$, $\mathcal{M}_2 = \mathbf{\Pi}^0_5$. For present purposes, the rank of a Borel set W will mean the least α with $W \in \mathcal{M}_\alpha$. Now, let $A(t)$ be of rank $\alpha > 0$. Choose Borel sets $A(t\hat{}m\hat{}n)$ of rank $< \alpha$ satisfying (1) above. This completes the first step of the construction.

Step 2. Let us fix an enumeration $s_0, s_1, s_2, \ldots$ of nonempty members of $\mathbb{N}^{<\mathbb{N}}$ such that $s_m \preceq s_n \implies m \leq n$. Let $\mathcal{F}_n$ be the set of all functions from $\{s_0, s_1, \ldots, s_{n-1}\}$ to $\mathbb{N}$. (So $\mathcal{F}_0$ contains only the empty function $\emptyset$.) Let $\mathcal{F} = \bigcup_n \mathcal{F}_n$ and let $\mathcal{F}_\infty$ be the set of all functions from $\{s_i : i \in \mathbb{N}\}$ to $\mathbb{N}$. Throughout this proof, the letters σ, τ with or without suffix will range over $\mathcal{F}$. We say that τ is an **immediate proper extension** of σ, and write $\sigma \ll \tau$, if for some n, $\sigma \in \mathcal{F}_n$, we have $\tau \in \mathcal{F}_{n+1}$ and τ extends σ.

For $\psi \in \mathcal{F} \bigcup \mathcal{F}_\infty$ and $s = (m_0, m_1, \ldots, m_{k-1}) \in \text{domain}(\psi)$ we define

$$\psi^+(s) = (m_0, n_0, m_1, n_1, \ldots, m_{k-1}, n_{k-1}),$$

where $n_0 = \psi((m_0))$, $n_1 = \psi((m_0, m_1))$, ..., $n_{k-1} = \psi(s)$.

To complete the second step of the construction, we define $B(\sigma)$ to be the intersection of all $A(\sigma^+(s))$ for $s \in \text{domain}(\sigma)$. Using all these definitions, we see that

$$B(\sigma) = \bigcup_{\sigma \ll \tau} B(\tau). \tag{2}$$

Further, by Step 1, Case 2,

$$x \in B(\sigma)\&(m) \in \text{domain}(\sigma) \Longrightarrow (a \in Z_m \Longleftrightarrow \sigma((m)) = 0). \tag{3}$$

The following fact is one of the two important observations that give a clue to defining the required cross section.

(A) Suppose $\emptyset = \sigma_0 \ll \sigma_1 \ll \sigma_2 \ll \cdots$ is a sequence in $\mathcal{F}$ such that each $B(\sigma_n) \neq \emptyset$. Then $\bigcap_n B(\sigma_n)$ is a singleton.

To see this, recall that

$$B(\sigma_n) = \bigcap\{A(\sigma_n^+(s_i)) : i < n\} = \bigcap\{A(\psi^+(s_i)) : i < n\},$$

where $\psi \in \mathcal{F}_\infty$ is the union of the σ_n's.

Set

$$L_n = \bigcap\{A(\psi^+(s_i)) : i < n \text{ and } A(\psi^+(s_i)) \text{ is closed}\}.$$

Then the L_n are closed, $L_{n+1} \subseteq L_n$, and L_n contains $B(\sigma_n)$ and hence is nonempty. Further, the d – diameter$(L_n) \to 0$. To see this, consider for any given m the sets $A(\psi^+(m))$, $A(\psi^+(m, m))$, $A(\psi^+(m, m, m))$, ... By Step 1, Case 4 of our construction, the ranks of these sets decrease until at some step we reach a closed set; then by Step 1, Case 3, at the very next step we get a closed set of diameter ¡ $1/(m+1)$. By the Cantor intersection theorem there is an $x \in X$ such that $\bigcap_n L_n = \{x\}$.

To prove our claim (A) we show that $x \in A(\psi^+(s))$ for all s. This is established by induction on the rank of the set involved.

We know already that the claim holds for sets of rank 0; i.e., for closed sets. Suppose then that $A(\psi^+(s))$ has rank $\alpha > 0$, and assume as induction hypothesis that the claim holds for sets of rank $< \alpha$, e.g., for the various $A(\psi^+(s) \hat{} m \hat{} n)$. For any m, letting $n = \psi(s \hat{} m)$, we have $\psi^+(s\hat{}m) = \psi^+(s) \hat{} m \hat{} n$. Hence $A(\psi^+(s\hat{}m))$ is of rank less than α, and so by the induction hypothesis,

$$x \in A(\psi^+(s\hat{}m)) = A(\psi^+(s)\hat{}m\hat{}n).$$

This shows that

$$x \in \bigcap_m \bigcup_n A(\psi^+(s)\hat{}m\hat{}n) = A(\psi^+(s)),$$

as required to prove the claim.

Step 3. Let us define

$$C(\sigma) = B(\sigma)^{\Delta},$$

the Vaught transform of $B(\sigma)$. By 3.5.19, $C(\emptyset) = X$, and each $C(\sigma)$ is invariant and Borel. Further,

$$C(\sigma) = \bigcup_{\sigma \ll \tau} C(\tau). \tag{4}$$

Now, if $x \in C(\sigma)$, then by the Baire category theorem, some $g \cdot x \in B(\sigma)$; so applying (3) above and recalling that the Z_m are invariant, we conclude that

$$x \in C(\sigma) \& (m) \in \text{domain}(\sigma) \Longrightarrow (a \in Z_m \Longleftrightarrow \sigma((m)) = 0). \tag{5}$$

Step 4. We say that σ **lexicographically precedes** τ, and write $\sigma \rhd \tau$, if for some n and $i < n$ we have $\sigma \in \mathcal{F}_n$, $\tau \in \mathcal{F}_n$, $\sigma(s_j) = \tau(s_j)$ for all $j < i$ and $\sigma(s_i) < \tau(s_i)$. The relation $\rhd$ well-orders each $\mathcal{F}_n$. Let

$$D(\sigma) = C(\sigma) \setminus \bigcup \{C(\tau) : \tau \rhd \sigma\}.$$

Thus $D(\sigma)$ is an invariant Borel set with $D(\emptyset) = X$, and by (4) and (5) we have

$$D(\sigma) = \Sigma_{\sigma \ll \tau} D(\tau) \tag{6}$$

and

$$x \in D(\sigma) \ \& \ (m) \in \text{domain}(\sigma) \Longrightarrow (x \in Z_m \Longleftrightarrow \sigma((m)) = 0). \tag{7}$$

In (6), Σ denotes *disjoint* union.

Now we make the second crucial observation. Though we do not need it in its full strngth, this together with (A) gives a good clue to defining the required cross section.

(B) Let K be an E_G-equivalence class. From (6) it is evident that there exists a sequence $\emptyset = \sigma_0 \ll \sigma_1 \ll \sigma_2 \ll \cdots$ of elements of $\mathcal{F}$ such that $K \subseteq D(\sigma_n)$ for each n, but $K \bigcap D(\sigma) = \emptyset$ for any other $\sigma \in \mathcal{F}$. Then $K = \bigcap_n D(\sigma_n)$.

Since $K \subseteq \bigcap_n D(\sigma_n)$, we take any $x \in X \setminus K$ and show that $x \notin D(\sigma_n)$ for some n. Since $\{Z_n : n \in \mathbb{N}\}$ is closed under complementation, by (0), there is an m such that $K \subseteq Z_m$ but $x \notin Z_m$. Take a large enough n such that $(m) \in \text{domain}(\sigma_n)$. Suppose $x \in D(\sigma_n)$. We shall arive at a contradiction. Since $x \notin Z_m$, $\sigma_n((m)) > 0$ by (7). On the other hand, take any $y \in K \subseteq Z_m$. Then $y \in D(\sigma_n)$. So $\sigma_n((m)) = 0$, and we have arrived at a contradiction.

Finally, we are in a position to introduce the Borel set

$$S = \bigcap_n \bigcup_{\sigma \in \mathcal{F}_n} (B(\sigma) \bigcap D(\sigma)).$$

We aim to show that S is a cross section of E_G. To this end we consider an arbitrary E_G-equivalence class K and verify that $S\bigcap K$ is a singleton.

Take a sequence $\emptyset = \sigma_0 \ll \sigma_1 \ll \sigma_2 \ll \cdots$ of elements of $\mathcal{F}$ such that $K \subseteq D(\sigma_n)$ for each n, but $K\bigcap D(\sigma) = \emptyset$ for any other $\sigma \in \mathcal{F}$. As before, let $\psi \in \mathcal{F}_\infty$ be the union of these σ_n.

Since $K \subseteq D(\sigma_n) \subseteq C(\sigma_n)$, $K\bigcap B(\sigma_n) \neq \emptyset$. In particular, $B(\sigma_n) \neq \emptyset$. Therefore, $\bigcap_n B(\sigma_n) = \{x\}$ for some $x \in X$ by (A). By (3), for any m, $x \in Z_m \iff \psi((m)) = 0$. On the other hand, by (7), for any m, $K \subseteq Z_m \iff \psi((m)) = 0$. But then by (0), $x \in K$. This implies that $x \in \bigcap_n D(\sigma_n)$. Now it is easily seen that $S\bigcap K = \{x\}$ as required. ∎

5.7 Borel Sets with Small Sections

We have seen that a Borel set with projection a Borel set need not admit a Borel uniformization. However, under suitable conditions on the sections of the Borel set, there does exist a Borel uniformization. Such results are among the most basic results on Borel sets, and in the next few sections we present several such results.

Generally, the conditions on sections under which a Borel uniformization exists can be divided into two kinds: *large-section conditions* and *small-section conditions.* A large-section condition is one where sections do not belong to a σ-ideal having an appropriate computability property, e.g., the σ-ideal of meager sets or the σ-ideal of null sets. A small section-condition is one where sections do belong to a σ-ideal having an appropriate computability property, e.g., the σ-ideal of countable sets or the σ-ideal of K_σ sets. In this section we prove two very famous uniformization theorems for Borel sets with small sections.

Theorem 5.7.1 *(Novikov [90]) Let X, Y be Polish spaces and $\mathcal{A}$ a countably generated sub σ-algebra of $\mathcal{B}_X$. Suppose $B \in \mathcal{A}\bigotimes\mathcal{B}_Y$ is such that the sections B_x are compact. Then $\pi_X(B) \in \mathcal{A}$, and B admits an $\mathcal{A}$-measurable section.*

Proof. Since the projection of a Borel set with compact sections is Borel (4.7.11), $\pi_X(B)$ is Borel. Since $\pi_X(B)$ is a union of atoms of $\mathcal{A}$, by the Blackwell – Mackey theorem (4.5.7), it is in $\mathcal{A}$.

Let U be an open set in Y. Write $U = \bigcup_n F_n$, the F_n's closed. Then

$$\pi_X(B\bigcap(X \times U)) = \bigcup_n \pi_X(B\bigcap(X \times F_n)).$$

Hence, by 4.7.11 and 4.5.7, $\pi_X(B\bigcap(X \times U)) \in \mathcal{A}$. It follows that the multifunction $x \longrightarrow B_x$ defined on $\pi_X(B)$ is $\mathcal{A}$-measurable. The result follows from the selection theorem of Kuratowski and Ryll-Nardzewski (5.2.1). ∎

Theorem 5.7.2 *(Lusin) Let X, Y be Polish spaces and $B \subseteq X \times Y$ Borel with sections B_x countable. Then B admits a Borel uniformization.*

Proof. By 3.3.17, there is a closed set E in $\mathbb{N}^{\mathbb{N}}$ and a one-to-one continuous map $f : E \longrightarrow X \times Y$ with range B. Set

$$H = \{(x, \alpha) \in X \times E : \pi_X(f(\alpha)) = x\}.$$

Then H is a closed set in $X \times \mathbb{N}^{\mathbb{N}}$ with sections H_x countable. Further, $\pi_X(B) = \pi_X(H)$. Fix a countable base (V_n) for $\mathbb{N}^{\mathbb{N}}$. Let

$$Z_n = \{x \in X : H_x \bigcap V_n \text{ is a singleton}\}.$$

By 4.12.2, Z_n is coanalytic. Each H_x is countable and closed, and so if nonempty must have an isolated point. Therefore,

$$\bigcup_n Z_n = \pi_X(H) = \pi_X(B).$$

Hence, $\pi_X(B)$ is both coanalytic and analytic, and so by Souslin's theorem, Borel. By the weak reduction principle for coanalytic sets (4.6.5), there exist pairwise disjoint Borel sets $B_n \subseteq Z_n$ such that $\bigcup_n B_n = \bigcup_n Z_n$. Let

$$D = \bigcup_n [(B_n \times V_n) \bigcap H].$$

Then D is a Borel uniformization of H. Let $g : D \longrightarrow X \times X$ be the map defined by $g(x, \alpha) = f(\alpha)$. Since g is one-to-one, the set

$$C = \{f(\alpha) : (x, \alpha) \in D\}$$

is Borel (4.5.4). It clearly uniformizes B. ■

Proposition 5.7.3 *Let X be a Polish space and Π a countably separated partition of X with all equivalence classes countable. Then Π admits a Borel cross section.*

Proof. Let Y be a Polish space and $f : X \longrightarrow Y$ a Borel map such that

$$x \, \Pi \, x' \Longleftrightarrow f(x) = f(x').$$

Define

$$B = \{(y, x) \in Y \times X : f(x) = y\}.$$

Then B is a Borel set with sections B_y countable. By 5.7.2, $\pi_Y(B)$ is Borel and there is a Borel section $g : \pi_Y(B) \longrightarrow X$ of B. Note that g is one-to-one. Take S to be the range of g. Then S is Borel by 4.5.4. Evidently, it is a cross section of Π. ■

In Section 6 of this chapter we shall generalize this result to partitions of Polish spaces into K_σ sets.

5.8 Borel Sets with Large Sections

Let X and Y be Polish spaces. A map $\mathcal{I} : X \longrightarrow \mathcal{P}(\mathcal{P}(Y))$ is called **Borel on Borel** if for every Borel $B \subseteq X \times Y$, the set

$$\{x \in X : B_x \in \mathcal{I}(x)\}$$

is Borel. The following are some important Borel on Borel maps.

Example 5.8.1 Let P be a transition probability on $X \times Y$; X, Y Polish. By 3.4.24, the map $\mathcal{I} : X \longrightarrow \mathcal{P}(\mathcal{P}(Y))$ defined by

$$\mathcal{I}(x) = \{N \subseteq Y : P(x, N) = 0\}$$

is Borel on Borel.

Example 5.8.2 Let X, Y be Polish spaces and $\mathcal{I}(x)$ the σ-ideal of all meager sets in Y. By 3.5.18, $\mathcal{I}$ is Borel on Borel.

Example 5.8.3 Let X, Y be Polish spaces and $G : X \longrightarrow Y$ a closed-valued Borel measurable multifunction. Define $\mathcal{I} : X \longrightarrow \mathcal{P}(\mathcal{P}(Y))$ by

$$\mathcal{I}(x) = \{I \subseteq Y : I \text{ is meager in } G(x)\}.$$

By imitating the proof of 3.5.18 we can show the following:

For every open set U in Y and every Borel set B in $X \times Y$, the sets

$$\begin{aligned} B^{*U} &= \{x \in X : G(x) \bigcap U \neq \emptyset \\ &\quad \;\&\; B_x \bigcap G(x) \bigcap U \textit{ is comeager in } G(x) \bigcap U\} \end{aligned}$$

and

$$\begin{aligned} B^{\Delta U} &= \{x \in X : G(x) \bigcap U \neq \emptyset \\ &\quad \;\&\; B_x \bigcap G(x) \bigcap U \textit{ is nonmeager in } G(x) \bigcap U\} \end{aligned}$$

are Borel.

It follows that $\mathcal{I}$ is Borel on Borel.

Theorem 5.8.4 *(Kechris [52]) Let X, Y be Polish spaces. Assume that $x \longrightarrow \mathcal{I}_x$ is a Borel on Borel map assigning to each $x \in X$ a σ-ideal $\mathcal{I}_x$ of subsets of Y. Suppose $B \subseteq X \times Y$ is a Borel set such that for every $x \in \pi_X(B)$, $B_x \notin \mathcal{I}_x$. Then $\pi_X(B)$ is Borel, and B admits a Borel section.*

Proof. Since $x \longrightarrow \mathcal{I}_x$ is Borel on Borel,

$$\pi_X(B) = \{x : B_x \in \mathcal{I}_x\}^c$$

is Borel.

It remains to prove that B admits a Borel section. Fix a closed subset F of $\mathbb{N}^{\mathbb{N}}$ and a continuous bijection $f : F \longrightarrow B$. For each $s \in \mathbb{N}^{<\mathbb{N}}$ we define a Borel subset B_s of X such that for every $s, t \in \mathbb{N}^{<\mathbb{N}}$,

(i) $B_e = \pi_X(B)$;

(ii) $|s| = |t|$ & $s \neq t \Longrightarrow B_s \bigcap B_t = \emptyset$;

(iii) $B_s = \bigcup_n B_{s\hat{}n}$; and

(iv) $B_s \subseteq \{x \in X : (f(\Sigma(s) \bigcap F))_x \notin \mathcal{I}_x\}$.

We define such a system of sets by induction on $|s|$. Suppose B_t have been defined for every $t \in \mathbb{N}^{<\mathbb{N}}$ of length $< n$, and $s \in \mathbb{N}^{<\mathbb{N}}$ is of length $n-1$. For any $k \in \mathbb{N}$, let

$$D_k = \{x \in B_s : (f(\Sigma(s\hat{}k) \bigcap F))_x \notin \mathcal{I}_x\}.$$

Since f is one-to-one and continuous, $f(\Sigma(s\hat{}k) \bigcap F)$ is Borel (4.5.4). Hence, as $x \longrightarrow \mathcal{I}_x$ is Borel on Borel, each D_k is Borel. By (iv), $B_s = \bigcup_k D_k$. Take

$$B_{s\hat{}k} = D_k \setminus \bigcup_{l<k} D_l.$$

We define $u : \pi_X(B) \longrightarrow Y$ as follows. Given any $x \in \pi_X(B)$ there is a unique $\alpha \in F$ (call it $p(x)$) such that $x \in B_{\alpha|k}$ for every k. Define u by

$$u(x) = \pi_Y(f(p(x))).$$

We wish to check that u is a Borel section of B.

We first check that u is a section of B. Let $x \in \pi_X(B)$. It is sufficient to show that $\pi_X(f(p(x))) = x$. Let $p(x) = \alpha$. Then $x \in B_{\alpha|k}$ for all k. So, $(f(\Sigma(\alpha|k) \bigcap F))_x \notin \mathcal{I}_x$. In particular, $(f(\Sigma(\alpha|k) \bigcap F))_x \neq \emptyset$. Choose $\alpha_k \in \Sigma(\alpha|k) \bigcap F$ such that $\pi_X(f(\alpha_k)) = x$. Since $\alpha_k \to \alpha$, $\pi_X(f(\alpha)) = x$.

It remains to show that u is Borel. It is sufficient to prove that p is Borel. For evey $s \in \mathbb{N}^{<\mathbb{N}}$, we shall prove that $p^{-1}(\Sigma(s))$ is Borel. This will complete the proof. We proceed by induction on $|s|$. Suppose $p^{-1}(\Sigma(s))$ is Borel and $k \in \mathbb{N}$. Then

$$\begin{aligned} x \in p^{-1}(\Sigma(s\hat{}k)) \iff\ & x \in p^{-1}(\Sigma(s)) \ \&\ (f(\Sigma(s\hat{}k) \bigcap F))_x \notin \mathcal{I}_x \\ & \&\ \forall (l < k)((f(\Sigma(s\hat{}l) \bigcap F))_x \in \mathcal{I}_x). \end{aligned}$$

Since $x \longrightarrow \mathcal{I}_x$ is Borel on Borel and f is bimeasurable, $p^{-1}(\Sigma(s\hat{}k))$ is Borel, and our result is proved. ∎

(See also [75].)

Theorem 5.8.5 *(Kechris [52] and Sarbadhikari [100]) If B is a Borel subset of the product of two Polish spaces X and Y such that B_x is nonmeager in Y for every $x \in \pi_X(B)$, then B admits a Borel uniformization.*

Proof. Apply 5.8.4 with $\mathcal{I}_x$ as in example 5.8.2. ∎

Example 5.8.6 As a special case of 5.8.5 we see that every Borel set $B \subseteq X \times Y$ with B_x a dense G_δ set admits a Borel uniformization. However, there is an F_σ subset E of $[0,1] \times \mathbb{N}^{\mathbb{N}}$ with sections E_x dense and that does not admit a Borel uniformization. Here is an example.

Let $C \subseteq [0,1] \times \mathbb{N}^{\mathbb{N}}$ be a closed set with projection to the first coordinate space $[0,1]$, that does not admit a Borel uniformization. Such a set exists by 5.1.7. For each $s \in \mathbb{N}^{<\mathbb{N}}$, fix a homeomorphism $f_s : \Sigma \longrightarrow \Sigma(s)$. Take

$$E = \bigcup_{s \in \mathbb{N}^{<\mathbb{N}}} \{(x, f_s(\alpha)) : (x, \alpha) \in B\}.$$

This E works.

Theorem 5.8.7 *(Blackwell and Ryll-Nardzewski [17]) Let X, Y be Polish spaces, P a transition probability on $X \times Y$, and B a Borel subset of $X \times Y$ such that $P(x, B_x) > 0$ for all $x \in \pi_X(B)$. Then $\pi_X(B)$ is Borel, and B admits a Borel uniformization.*

Proof. Apply 5.8.4 with $\mathcal{I}_x$ as in Example 5.8.1. ■

The selection theorem of Blackwell and Ryll-Nardzewski holds in a more general situation.

Theorem 5.8.8 *(Blackwell and Ryll-Nardzewski) Let X, Y be Polish spaces, $\mathcal{A}$ a countably generated sub σalgebra of $\mathcal{B}_X$, and P a transition probability on $X \times Y$ such that for every $B \in \mathcal{B}_Y$, $x \longrightarrow P(x, B)$ is $\mathcal{A}$-measurable. Suppose $B \in \mathcal{A} \bigotimes \mathcal{B}_Y$ is such that $P(x, B_x) > 0$ for all $x \in \pi_X(B)$. Then $\pi_X(B) \in \mathcal{A}$, and B admits an $\mathcal{A}$-measurable section.*

We prove a lemma first.

Lemma 5.8.9 *Let X, Y, $\mathcal{A}$, and P be as above. For every $E \in \mathcal{A} \bigotimes \mathcal{B}_Y$ and every $\epsilon > 0$, there is an $F \in \mathcal{A} \bigotimes \mathcal{B}_Y$ contained in E such that F_x is compact and $P(x, F_x) \geq \epsilon \cdot P(x, E_x)$.*

Proof. Let $\mathcal{M}$ be the class of all sets in $\mathcal{A} \bigotimes \mathcal{B}_Y$ such that the conclusion of the lemma holds for every P and every $\epsilon > 0$. By 3.4.20, $\mathcal{M}$ contains all rectangles $A \times B$, where $A \in \mathcal{A}$ and B Borel in Y. So, $\mathcal{M}$ contains all finite disjoint unions of such rectangles. It is fairly routine to check that $\mathcal{M}$ is a monotone class. Therefore, the result follows from the monotone class theorem. ■

Proof of 5.8.8. By a slight modification of the argument contained in the proof of 3.4.24 we see that for every $E \in \mathcal{A} \bigotimes \mathcal{B}_Y$, $x \longrightarrow P(x, E_x)$ is $\mathcal{A}$-measurable. As $\pi_X(B) = \{x \in X : P(x, B_x) > 0\}$, it follows that $\pi_X(B) \in \mathcal{A}$.

By 5.8.9, there is a $C \subseteq B$ in $\mathcal{A} \bigotimes \mathcal{B}_Y$ with compact x-sections such that $P(x, C_x) > 0$ for every $x \in \pi_X(B)$. In particular, $\pi_X(B) = \pi_X(C)$. The result follows from Novikov's uniformization theorem (5.7.1). ■.

Here is an application of 5.8.8 to probability theory. Let X be a Polish space. For any probability P on $\mathcal{B}_X$ and $f : X \longrightarrow \mathbb{R}$ any Borel map, a **conditional distribution** given f is a transition probability Q on $X \times X$ such that

(i) for every $B \in \mathcal{B}_X$, $x \longrightarrow Q(x, B)$ is $\mathcal{A}$-measurable, where $\mathcal{A} = \{f^{-1}(C) : C \text{ Borel in } \mathbb{R}\}$; and

(ii) for every $A \in \mathcal{A}$ and every $B \in \mathcal{B}_X$,

$$\int_A Q(x, B) dP(x) = P(A \bigcap B).$$

A conditional distribution Q is called **proper** at x_0 if

$$Q(x_0, A) = 1 \text{ for } x_0 \in A \in \mathcal{A};$$

i.e., we assign conditional probability 1 to $\{x \in X : f(x) = f(x_0)\}$. It is known that conditional distributions always exist that are proper at all points of X except at a P-null set N. Using 5.8.7 we show that, in general, the exceptional set N cannot be removed.

Proposition 5.8.10 *Let X, f, and $\mathcal{A}$ be as above. An everywhere proper conditional distribution given f exists if and only if there is an $\mathcal{A}$-measurable $g : X \longrightarrow X$ such that $f(g(x)) = f(x)$ for all x.*

Proof. Suppose an $\mathcal{A}$-measurable $g : X \longrightarrow X$ such that $f \circ g$ is the identity exists. Define

$$Q(x, B) = \begin{cases} 1 & \text{if } g(x) \in B, \\ 0 & \text{otherwise.} \end{cases}$$

It is easy to verify that Q has the desired properties.

Conversely, let an everywhere proper conditional distribution Q given f exist. Let

$$S = \{(x, y) \in X \times X : f(x) = f(y)\}.$$

Then $S \in \mathcal{A} \bigotimes \mathcal{B}_Y$ and $Q(x, S_x) = 1$. By 5.8.8, there is an $\mathcal{A}$-measurable section g of S, which is what we are looking for. ■

Since g is $\mathcal{A}$-measurable, $g(x) = g(y)$ whenever $f(x) = f(y)$. It follows that there is a Borel function $h : \mathbb{R} \longrightarrow X$ such that $g(x) = h(f(x))$ for all x. Then the range of f equals $\{y \in \mathbb{R} : f(h(y)) = y\}$, which is a Borel set. It follows from the above proposition that whenever the range of f is not a Borel set, everywhere proper conditional distributions given f cannot exist.

As another application of 5.8.5, we present a proof of Lusin's famous theorem on Borel sets with countable sections.

Theorem 5.8.11 *(Lusin) Let X, Y be Polish spaces and B a Borel set with B_x countable. Then B is a countable union of Borel graphs.*

Proof. (Kechris) Without loss of generality we assume that for each $x \in X$, B_x is countably infinite. Using 5.8.4, we shall show that there is a Borel map $f : X \longrightarrow Y^{\mathbb{N}}$ such that $B_x = \{f_n(x) : n \in \mathbb{N}\}$.

Granting this, we complete the proof by taking

$$B_n = \{(x, f_n(x)) : n \in \mathbb{N}\}.$$

We now show the existence of the map $f : X \longrightarrow Y^{\mathbb{N}}$ satisfying the above conditions.

(i) Let

$$E = \{(x, (e_n)) \in X \times Y^{\mathbb{N}} : \{e_n : n \in \mathbb{N}\} = B_x\}.$$

The set E is Borel. This follows from the following observation.

$$(x, (e_n)) \in E \Longleftrightarrow \forall n((x, e_n) \in B) \;\&\; \neg\exists y((x, (e_n), y) \in S),$$

where

$$S = \{(x, (e_n), y) \in X \times Y^{\mathbb{N}} \times Y \;:\; (x, y) \in B \;\&\; \forall n(y \neq e_n)\}.$$

Since $S_{(x,(e_n))}$ is countable, by 4.12.3 E is Borel.

(ii) Let $x \in X$. Give B_x the discrete toplogy and $B_x^{\mathbb{N}}$ the product topology. So, $B_x^{\mathbb{N}}$ is homeomorphic to $\mathbb{N}^{\mathbb{N}}$. We show that E_x is a dense G_δ set in $B_x^{\mathbb{N}}$. Note that $E_x \subseteq B_x^{\mathbb{N}}$. Let $(e_n) \in B_x^{\mathbb{N}}$. Then

$$\{e_n : n \in \mathbb{N}\} = B_x \Longleftrightarrow \forall y \in B_x \exists n(y = e_n).$$

So E_x is a G_δ set in $B_x^{\mathbb{N}}$. It remains to show that E_x is dense in $B_x^{\mathbb{N}}$. Take a finite sequence $(y_0, y_1, \ldots, y_{n-1})$, each y_i in B_x. Since B_x is countable, there exists a sequence (e_k) in Y enumerating B_x such that $e_i = y_i$ for all $i < n$. It follows that E_x is dense in $B_x^{\mathbb{N}}$.

(iii) For $x \in X$, let

$$\mathcal{I}_x = \{I \subseteq Y^{\mathbb{N}} : I \bigcap E_x \text{ is meager in } B_x^{\mathbb{N}}\}.$$

Clearly, each $\mathcal{I}_x$ is a σ-ideal and $E_x \notin \mathcal{I}_x$. Further, $x \longrightarrow \mathcal{I}_x$ is Borel on Borel. To see this, take a Borel set A in $X \times Y^{\mathbb{N}}$. We need to show that

$$\{x : A_x \in \mathcal{I}_x\} = \{x : A_x \bigcap E_x \text{ is meager in } B_x^{\mathbb{N}}\}$$

is Borel. Without loss of generality we assume that $A \subseteq E$.

For the rest of the proof, $e = (e_n) : \mathbb{N} \longrightarrow B_x$ will stand for a bijection and $\pi_e : \mathbb{N}^{\mathbb{N}} \longrightarrow B_x^{\mathbb{N}}$ will denote the homeomorphism defined by

$$\pi_e(\alpha) = e \circ \alpha, \; \alpha \in \mathbb{N}^{\mathbb{N}}.$$

Consider the set $Q \subseteq X \times Y^{\mathbb{N}}$ defined by

$$(x,e) \in Q \iff (x,e) \in E \;\&\; (\forall n \neq m)(e_n \neq e_m) \\ \& \; \{\alpha \in \mathbb{N}^{\mathbb{N}} : (x, e \circ \alpha) \in A\} \text{ is meager in } \mathbb{N}^{\mathbb{N}}.$$

By 3.5.18, Q is Borel. Now note the following:

$$\begin{aligned} A_x \in \mathcal{I}_x &\iff A_x \text{ is meager in } B_x^{\mathbb{N}} \\ &\iff \pi_e^{-1}(A_x) \text{ is meager in } \mathbb{N}^{\mathbb{N}} \text{ for some } e \\ &\iff \{\alpha \in \mathbb{N}^{\mathbb{N}} : e \circ \alpha \in A_x\} \text{ is meager in } \mathbb{N}^{\mathbb{N}} \text{ for some } e \\ &\iff \exists e (x,e) \in Q. \end{aligned}$$

Hence, $\{x : A_x \in \mathcal{I}_x\}$ is analytic. We also have

$$\begin{aligned} A_x \in \mathcal{I}_x &\iff A_x \text{ is meager in } B_x^{\mathbb{N}} \\ &\iff \pi_e^{-1}(A_x) \text{ is meager in } \mathbb{N}^{\mathbb{N}} \text{ for all } e \\ &\iff \{\alpha \in \mathbb{N}^{\mathbb{N}} : e \circ \alpha \in A_x\} \text{ is meager in } \mathbb{N}^{\mathbb{N}} \text{ for all } e \\ &\iff \forall f \in B_x^{\mathbb{N}} \{[(x,f) \in E \;\&\; \forall m \neq n (f_n \neq f_m)] \\ &\qquad\qquad \implies (x,f) \in Q\}. \end{aligned}$$

So, $\{x : A_x \in \mathcal{I}_x\}$ is also coanalytic. Hence, $\{x : A_x \in \mathcal{I}_x\}$ is Borel by Souslin's theorem (4.4.3).

The existence of $f : X \longrightarrow Y^{\mathbb{N}}$ with the desired properties now follows from 5.8.4. ■

Exercise 5.8.12 Let Π be a countably separated partition of a Polish space into countable sets. Show that there is a sequence (G_n) of partial Borel cross sections of Π such that $\bigcup_n G_n = X$ and if G_n and G_m are distinct, then $G_n \bigcup G_m$ is not a partial cross section. (A subset A of X is a partial cross section if $A \bigcap C$ is at most a singleton for every member C of Π.)

Lusin, in fact, proved a much stronger result: *Every analytic set in the product with countable sections can be covered by countably many Borel graphs.* We shall give a proof of this later.

We close this section by giving another refinement of Lusin's theorem. For an application of this result see [41].

Let X be a Polish space and G a group of Borel automorphisms on X; i.e., each member of G is a Borel isomorphism of X onto itself and G is a group under composition. Define

$$x \; E_G \; y \iff (\exists g \in G)(y = g(x)).$$

E_G is clearly an equivalence relation on X. E_G is called the **equivalence relation induced by** G. It is clearly analytic; it is Borel if G is countable. We show next that the converse of this result is also true.

Proposition 5.8.13 *(Feldman and Moore [41]) Every Borel equivalence relation on a Polish space X with equivalence classes countable is induced by a countable group of Borel automorphisms.*

Proof. Let Π be a Borel equivalence relation on X with equivalence classes countable. By 5.8.11, write

$$\Pi = \bigcup_n G_n,$$

where $\pi_1 | G_n$ is one-to-one, $\pi_1(x, y) = x$; i.e., the G_n's are graphs of Borel functions. Let

$$H_n = \varphi(G_n),$$

where $\varphi(x, y) = (y, x)$. Then $\pi_2 | H_n$ is one-to-one, where $\pi_2(x, y) = y$. Let

$$X \times X \setminus \Delta = \bigcup_k (U_k \times V_k),$$

U_k, V_k open, where $\Delta = \{(x, x) : x \in X\}$. Note that $U_k \bigcap V_k = \emptyset$. Put

$$D_{nmk} = (G_n \bigcap H_m) \bigcap (U_k \times V_k).$$

Note that $\pi_1 | D_{nmk}$ and $\pi_2 | D_{nmk}$ are one-to-one, and

$$\pi_1(D_{nmk}) \bigcap \pi_2(D_{nmk}) = \emptyset.$$

So, there is a Borel automorphism g_{nmk} of X given by

$$g_{nmk}(x) = \begin{cases} y & \text{if } (x, y) \in D_{nmk} \text{ or } (y, x) \in D_{nmk}, \\ x & \text{otherwise.} \end{cases}$$

Clearly,

$$\Pi = \Delta \bigcup \bigcup_{nmk} \text{graph}(g_{nmk}).$$

Now take G to be the group of automorphisms generated by $\{g_{nmk} : n, m, k \in \mathbb{N}\}$. ■

5.9 Partitions into G_δ Sets

We return to the problem of existence of nice cross sections for partitions of Polish spaces. In an earlier section we dealt with this problem when the equivalence classes are closed. How important is the condition that the members of Π be closed? Does every Borel partition Π of a Polish space into Borel sets admit a Borel cross section? We consider this problem now.

The next result generalizes 5.4.1 for partitions into G_δ sets.

Theorem 5.9.1 *(Miller [85]) Every partition* Π *of a Polish space X into* G_δ *sets such that the saturation of every basic open set is simultaneously* F_σ *and* G_δ *admits a section* $s : X \longrightarrow X$ *that is Borel measurable of class 2. In particular, such partitions admit a* G_δ *cross section.*

Proof. Let (U_n) be a countable base for the topology of X. Let (V_n) be an enumeration of $\{U_n^* : n \in \mathbb{N}\} \bigcup \{(U_n^*)^c : n \in \mathbb{N}\}$. Let $\mathcal{T}'$ be the topology on X generated by $\{U_n : n \in \mathbb{N}\} \bigcup \{V_n : n \in \mathbb{N}\}$. Note that every $\mathcal{T}'$ open set is an F_σ set in X relative to the original topology of X. Consider the map $f : X \longrightarrow X \times 2^{\mathbb{N}}$ defined by

$$f(x) = (x, \chi_{V_0}(x), \chi_{V_1}(x), \chi_{V_2}(x), \ldots), \quad x \in X.$$

The map f is one-to-one and of class 2. Let G be the range of f. It is quite easy to see that

$$\mathcal{T}' = \{f^{-1}(W) : W \text{ open in } G\}.$$

Arguing as in the proof of 3.2.5, it is easily seen that G is a G_δ set in $X \times 2^{\mathbb{N}}$. Therefore, by 2.2.1, $(X, \mathcal{T}')$ is Polish. As argued in 5.1.13,

$$[x] = \bigcap \{U_n^* : U_n^* \supseteq [x]\}.$$

So, each Π-equivalence class is closed relative to $\mathcal{T}'$.

Let $\mathcal{L}$ be the set of all invariant subsets of X that are clopen relative to $\mathcal{T}'$. We claim that the multifunction $x \longrightarrow [x]$ is $\mathcal{L}_\sigma$-measurable. Let $\mathcal{S} = \{V_n : n \in \mathbb{N}\}_d$, the set of all finite intersections of sets in $\{V_n : n \in \mathbb{N}\}$. Any $\mathcal{T}'$-open W is of the form $U \bigcup V$, U open relative to the original toplogy of X and V a union of sets in $\mathcal{S}$. Then $W^* = U^* \bigcup V$, which proves our claim.

By the selection theorem of Kuratowski and Ryll-Nardzewski, there exists an $\mathcal{L}_\sigma$-measurable selection s for $x \longrightarrow [x]$. In particular, s is continuous with respect to $\mathcal{T}'$. The associated cross section $S = \{x \in s(x) = x\}$ is $\mathcal{T}'$-closed and so is a G_δ set relative to the original toplogy of X. ■

Here is a generalization of 5.4.3.

Theorem 5.9.2 *(Srivastava [114]) Every Borel measurable partition* Π *of a Polish space X into* G_δ *sets admits a Borel cross section.*

Proof. (Kechris) For $x \in X$ let $[x]$ denote the member of Π containing x. Consider the multifunction $p : X \longrightarrow X$ defined by

$$p(x) = \text{cl}([x]).$$

Then $p : X \longrightarrow X$ is a closed-valued measurable multifunction. Further, for every $x, y \in X$, $x \equiv y \Longleftrightarrow p(x) = p(y)$ (5.9.1).

Now consider $F(X)$, the set of nonempty closed subsets of X with Effros Borel structure. By 3.3.10, it is standard Borel. Note that p considered as a map from X to $F(X)$ is measurable. Let

$$P = \{(F, x) \in F(X) \times X : p(x) = F\}.$$

The set P is Borel. For $F \in F(X)$, let $\mathcal{I}_F$ be the σ-ideal of subsets of X that are meager in F. As the multifunction $F \longrightarrow F$ from $F(x)$ to X is measurable, by 5.8.3, $F \longrightarrow \mathcal{I}_F$ is Borel on Borel. By the Baire category theorem, $P_F \notin \mathcal{I}_F$ for each F. Therefore, by 5.8.4, $D = \pi_{F(X)}(P)$ is Borel, and there is a Borel section $q : D \longrightarrow X$ of P. Let

$$S = \{x \in X : x = q(p(x))\}.$$

Clearly S is a Borel cross section of Π. ∎

Remark 5.9.3 Recall the Vitali partition of $\mathbb{R}$ discussed in 3.4.18. Each of its members is countable and hence an F_σ. If U is an open set of real numbers, then

$$U^* = \bigcup_{r \in \mathbb{Q}} (r + U),$$

which is open. Hence, the Vitali partition is a lower-semicontinuous partition of $\mathbb{R}$ into F_σ sets. In 3.4.18, we showed that the Vitali partition does not admit even a Lebesgue measurable cross section. Members of the Vitali partition are homeomorphic to the set of rationals. So, they are not G_δ sets. It follows that 5.9.2 is the best possible result on the existence of Borel cross sections.

For more on selections for G_δ-valued multifunctions see [114], [101], [116].

Now we outline an important application of our selection theorem in the representation theory of C^*-algebras. We consider only separable C^*-algebras A here. An important class of such C^*-algebras is known as **GCR C^*-algebras** which by well-known theorems due to Kaplanski and Glimm[43], [51], are precisely the **type I C^*-algebras** (meaning these are the C^*-algebras having tractable representation theory). (We refer the reader to [4] for the terminology.) The class of all irreducible $*$-representations of a C^*-algebra by operators on a Hilbert space H_n of dimension n is denoted by $irr(A, H_n)$, $n = 1, 2, \ldots, \infty$. $irr(A, H_n)$ is given the so-called weak topology, and $irr(A)$ stands for the topological sum $\bigoplus_n irr(A, H_n)$. Following the ideas contained in the proof of 2.4.6, we have the following result.

Proposition 5.9.4 *$irr(A)$ is Polish.*

For each $n = 1, 2, \ldots, \infty$, we have a natural equivalence relation $\sim$ on $irr(A, H_n)$, namely $\pi \sim \sigma$ if π and σ are unitarily equivalent. We denote the topological quotient of $irr(A)$ under unitary equivalence of representations by $irr(A)/\sim$ and the canonical quotient map by $q : irr(A) \longrightarrow irr(A)/\sim$. We have the following celebrated result of the theory.

Theorem 5.9.5 *$irr(A)/\sim$ is standard Borel if and only if A is GCR.*

Its proof makes crucial uses of 5.4.3 and 4.5.4. We refer the interested reader to [4] and [43] for a proof.

A third important object in this circle of ideas is the space $Prim(A)$ of $*$-ideals of A that are kernels of irreducible $*$-representations of A, given the **hull – kernel topology**. Let $\kappa : irr(A) \longrightarrow Prim(A)$ be the map

$$\kappa(\pi) = \text{kernel}(\pi), \quad \pi \in irr(A).$$

The map κ is continuous and open and induces a map

$$\hat{\kappa} : irr(A)/\sim \longrightarrow Prim(A).$$

A pleasant property of GCR algebras is that *$\hat{\kappa}$ is one-to-one on $irr(A)/\sim$* (the class of a $*$-representation is determined by its kernel), but in general, $\hat{\kappa}$ is not a one-to-one map.

The following concept of "locally type I" was introduced by Moore [87]: A C^*-algebra A is of **locally type I** on a Borel subset B of $irr(A)/\sim$ if

(i) $\hat{\kappa}|B$ is one-to-one, and

(ii) there exists a Borel selection $s : B \longrightarrow irr(A)$ for $q^{-1}|B$.

It may be mentioned that Auslander and Konstant[6] make essential use of this concept (and a theorem due to Moore) in giving a criterion for a solvable group (equivalently, the group C^*-algebra) to be of type I.

The cross section theorem Srivastava 5.9.2 was conjectured in [50] and it was pointed out that 5.9.2 would make condition (ii) in the definition of locally type I redundant. Both [85] and [50] replaced condition (ii) by some additional hypothesis. For instance, Kallman and Mauldin showed that condition (i) can be dropped from the definition of locally type I, provided that the relative Borel structure of B separates points. Below, we explain the implication of 5.9.2 on condition (ii) of Moore's definition.

Let B be a Borel subset of $irr(A)/\sim$ such that $\hat{\kappa}$ is one-to-one on B. A standard argument will show that

$$\{C \in \mathcal{B}_{irr(A)/\sim} : \hat{\kappa}^{-1}(\hat{\kappa}(C)) = C\}$$

is a σ-algebra containing all open sets of $irr(A)/\sim$. Hence,

$$\hat{\kappa}^{-1}(\hat{\kappa}(B)) = B.$$

This means that

$$\kappa^{-1}(\hat{\kappa}(\hat{\sigma})) = q^{-1}(\hat{\sigma})$$

for each $\hat{\sigma} \in B$. We now look at the equivalence relation Π induced by q on the Borel subset $q^{-1}(B)$ of $irr(A)$. From what we have just shown, this equivalence relation coincides with the equivalence relation Φ induced by κ on $q^{-1}(B)$. Now, each equivalence class of the equivalence relation Φ is a G_δ set in $irr(A)$. This is because $Prim(A)$ is a second countable T_0 space. Again, as κ is an open continuous map, the saturation under Φ of a relatively open set in $q^{-1}(B)$ is relatively open. Now consider the partition Ψ of $irr(A)$ whose equivalence classes are the Φ-equivalence classes and $\{A\}$ for $A \notin q^{-1}(B)$. Theorem 5.9.2 now gives a Borel selection of $q^{-1}|B$.

5.10 Reflection Phenomenon

In this section we show a rather interesting reflection phenomenon discovered by Burgess[21]. We give several applications of this, including Lusin'stheorem on analytic sets with countable sections.

Let X be a Polish space and $\Phi \subseteq \mathcal{P}(X)$. We say that Φ is $\mathbf{\Pi}^1_1$ on $\mathbf{\Pi}^1_1$ if for every Polish space Y and every $\mathbf{\Pi}^1_1$ subset D of $Y \times X$,

$$\{y \in Y : D_y \in \Phi\} \in \mathbf{\Pi}^1_1.$$

Theorem 5.10.1 *(The reflection theorem) Let X be a Polish space and $\Phi \subseteq \mathcal{P}(X)$ $\mathbf{\Pi}^1_1$ on $\mathbf{\Pi}^1_1$. For every $\mathbf{\Pi}^1_1$ set $A \in \Phi$ there is a Borel $B \subseteq A$ in Φ.*

Proof. Suppose there is a $\mathbf{\Pi}^1_1$ set $A \subseteq X$ in Φ that does not contain a Borel set belonging to Φ. We shall get a contradiction. Let φ be a $\mathbf{\Pi}^1_1$-norm on A and

$$C = \{(x, y) : y <^*_\varphi x\}.$$

We claim that

$$x \notin A \Longleftrightarrow C_x \in \Phi. \quad (*)$$

Suppose $x \notin A$. Then $C_x = A \in \Phi$. Conversely, if $x \in A$, then C_x is a Borel subset of A. So by our assumptions, $C_x \notin \Phi$.

Since Φ is $\mathbf{\Pi}^1_1$ on $\mathbf{\Pi}^1_1$, A^c is $\mathbf{\Pi}^1_1$ by $(*)$. Hence, by Souslin's theorem, it is Borel, contradicting our assumption again. ■

See [21] for more on reflections.

Theorem 5.10.2 *Let X, Y be Polish spaces and $A \subseteq X \times Y$ analytic with sections A_x countable. Then every coanalytic set B containing A contains a Borel set $E \supseteq A$ with all sections countable.*

Proof. Let $C = B^c$. Define $\Phi \subseteq \mathcal{P}(X \times Y)$ by

$$D \in \Phi \Longleftrightarrow D^c \subseteq B \;\&\; \forall x((D^c)_x \text{ is countable}).$$

Using 4.3.7 we can easily check that Φ is $\mathbf{\Pi}^1_1$ on $\mathbf{\Pi}^1_1$. Since $A^c \in \Phi$, by 5.10.1 there is a Borel set D in Φ contained in A^c. Take $E = D^c$. ■

Theorem 5.10.3 *(Lusin) Every analytic set with countable sections, in the product of two Polish spaces, can be covered by countably many Borel graphs.*

Proof. The result immediately follows from 5.10.2 and 5.8.11. ■

Proposition 5.10.4 *(Burgess) Let X be Polish, E an analytic equivalence relation on X, and $C \subseteq X \times X$ a coanalytic set containing E. Then there is a Borel equivalence relation B such that $E \subseteq B \subseteq C$.*

We need a lemma to prove this proposition. For any $P \subseteq X \times X$, define $\mathcal{E}(P) \subseteq X \times X$ by

$$(x, y) \in \mathcal{E}(P) \iff x = y \text{ or } ((x, y) \text{ or } (y, x) \in P) \text{ or } \exists z((x, z), (z, y) \in P).$$

Note that $P \subseteq \mathcal{E}(P)$, and if P is analytic, so is $\mathcal{E}(P)$.

Lemma 5.10.5 *Let X be a Polish space, P analytic, C coanalytic, and $\mathcal{E}(P) \subseteq C$. Then there is a Borel set B containing P such that*

$$\mathcal{E}(B) \subseteq C.$$

Proof. Define $\Phi \subseteq \mathcal{P}(X \times X)$ by

$$D \in \Phi \iff \mathcal{E}(D^c) \subseteq C.$$

Φ is $\mathbf{\Pi}_1^1$ on $\mathbf{\Pi}_1^1$. Further, $P^c \in \Phi$. By the reflection theorem (5.10.1), there is a Borel set D in Φ that is contained in P^c. Take $B = D^c$. ■

Proof of 5.10.4. Applying 5.10.5 repeatedly, by induction on n we can define a sequence of Borel sets (B_n) such that

$$E \subseteq B_n \subseteq \mathcal{E}(B_n) \subseteq B_{n+1} \subseteq C$$

for all n. Take $B = \bigcup_n B_n$. ■

Corollary 5.10.6 *For every analytic equivalence relation E on a Polish space X there exist Borel equivalence relations B_α, $\alpha < \omega_1$, such that $E = \bigcap_{\alpha<\omega_1} B_\alpha$.*

Proof. By 4.3.17, write $E = \bigcap_{\alpha<\omega_1} C_\alpha$, C_α coanalytic. By 5.10.4, for each α there exists a Borel equivalence relation B_α such that $E \subseteq B_\alpha \subseteq C_\alpha$. ■

Exercise 5.10.7 Let X be a Polish space, Y a separable Banach space, $A \subseteq X \times Y$ an analytic set with sections A_x convex, and $C \supseteq A$ coanalytic. Using the reflection theorem, show that there is a Borel set B in $X \times Y$ with convex sections such that $A \subseteq B \subseteq C$.

The above result was first proved by Saint Pierre, albeit by a different method.

5.11 Complementation in Borel Structures

Let X be a Polish space and $\mathcal{C}$ a sub σ-algebra of the Borel σ-algebra $\mathcal{B}_X$. A **weak complement** of $\mathcal{C}$ is a sub σ-algebra $\mathcal{D}$ of $\mathcal{B}_X$ such that

$$\mathcal{C}\bigvee\mathcal{D} = \mathcal{B}_X,$$

where $\mathcal{C}\bigvee\mathcal{D} = \sigma(\mathcal{C}\bigcup\mathcal{D})$. A weak complement $\mathcal{D}$ is **minimal** if no proper sub σ-algebra is a weak complement. A **complement** of $\mathcal{C}$ is a sub σ-algebra $\mathcal{D}$ such that

$$\mathcal{C}\bigvee\mathcal{D} = \mathcal{B}_X \text{ and } \mathcal{C}\bigcap\mathcal{D} = \{\emptyset, X\}.$$

The following exercises are reasonably simple.

Exercise 5.11.1 Let X be Polish and $\mathcal{C}$ a countably generated sub σ-algebra of $\mathcal{B}_X$. Show that every weak complement of $\mathcal{C}$ contains a countably generated weak complement.

Exercise 5.11.2 Let X be Polish, $\mathcal{C} \subseteq \mathcal{B}_X$. If $\mathcal{D}$ is a minimal weak complement, then show that $\mathcal{C}\bigcap\mathcal{D} = \{\emptyset, X\}$; i.e., $\mathcal{D}$ is also a complement.

Exercise 5.11.3 Let X be an uncountable Polish space. Show that the countable – cocountable σ-algebra does not have a complement.

A question arises: When does a sub σ-algebra of the Borel σ-algebra $\mathcal{B}_X$ admit a complement? This question was posed by D. Basu [10] in his study of maximal and minimal elements of families of statistics. We answer this question now.

Theorem 5.11.4 *Every countably generated sub σ-algebra of the Borel σ-algebra of a Polish space has a minimal complement.*

This beautiful result is due to E. Grzegorek, K. P. S. B. Rao, and H. Sarbadhikari[46].

Lemma 5.11.5 *Let X be Polish and $\mathcal{C}$ a countably generated sub σ-algebra of $\mathcal{B}_X$. Suppose $\mathcal{D}$ is a countably generated sub σ-algebra of $\mathcal{B}_X$ such that every atom A of $\mathcal{D}$ is a partial cross section of the atoms of $\mathcal{C}$. Further, assume that for any two distinct atoms C_1, C_2 of $\mathcal{D}$, $C_1 \bigcup C_2$ is not a partial cross section of the set of atoms of $\mathcal{C}$. Then $\mathcal{D}$ is a minimal complement of $\mathcal{C}$.*

Proof. Under the hypothesis, $\mathcal{C}\bigvee\mathcal{D}$ is a countably generated sub σ-algebra of $\mathcal{B}_X$ with atoms singletons. Hence, by 4.5.7, $\mathcal{C}\bigvee\mathcal{D} = \mathcal{B}_X$.

Let $\mathcal{D}^*$ be a proper countably generated sub σ-algebra of $\mathcal{D}$. By the corollary to 4.5.7, there is an atom A of $\mathcal{D}^*$ that is not an atom of $\mathcal{D}$. Hence, it is a union of more than one atom of $\mathcal{D}$. Hence, there exist two distinct points x, y of A that belong to the same atom of $\mathcal{C}$. This implies that there is no $E \in \mathcal{C} \bigvee \mathcal{D}^*$ containing exactly one of x, y. So, $\mathcal{C} \bigvee \mathcal{D}^* \neq \mathcal{B}_X$. The result now follows from 5.11.1 and 5.11.2. ■

Proof of 5.11.4. Let X be Polish and $\mathcal{C}$ a countably generated sub σ-algebra of $\mathcal{B}_X$.

Case 1. There is a cocountable atom A of $\mathcal{C}$.

Let $f : X \setminus A \longrightarrow A$ be a one-to-one map. Take

$$\mathcal{D} = \sigma(\{\{x, f(x)\} : x \in X \setminus A\} \bigcup \mathcal{B}_{A \setminus f(A^c)}).$$

By 5.11.5, $\mathcal{D}$ is a minimal complement of $\mathcal{C}$.

Case 2. There is an uncountable atom A of $\mathcal{C}$ such that $X \setminus A$ is also uncountable.

Let $f : A \longrightarrow A^c$ be a Borel isomorphism and $g : X \longrightarrow X$ the map that equals f on A and the identity on A^c. Take

$$\mathcal{D} = g^{-1}(\mathcal{B}_X).$$

By 5.11.5, $\mathcal{D}$ is a minimal complement of $\mathcal{C}$.

Case 3. All atoms of $\mathcal{C}$ are countable. Since $\mathcal{C}$ is countably generated with all atoms countable, by 5.8.12 there exists a countable partition G_n of X such that each G_n is a partial cross section of the set of atoms of $\mathcal{C}$. It is easy to choose the G_n's in such a way that for distinct G_n and G_m, $G_n \bigcup G_m$ is not a partial cross section of the set of atoms of $\mathcal{C}$. The result follows by 5.11.5 by taking

$$\mathcal{D} = \sigma(\{G_n : n \in \mathbb{N}\}).$$

■

5.12 Borel Sets with σ-Compact Sections

Our main goal in this section is to give a proof of the following uniformization theorm.

Theorem 5.12.1 *(Arsenin, Kunugui [60]) Let $B \subseteq X \times Y$ be a Borel set, X, Y Polish, such that B_x is σ-compact for every x. Then $\pi_X(B)$ is Borel, and B admits a Borel uniformization.*

Our proof of 5.12.1 is based on the following result.

Theorem 5.12.2 *(Saint Raymond [97]) Let X, Y be Polish spaces and $A, B \subseteq X \times Y$ analytic sets. Assume that for every x, there is a σ-compact set K such that $A_x \subseteq K \subseteq B_x^c$. Then there exists a sequence of Borel sets (B_n) such that the sections $(B_n)_x$ are compact,*

$$A \subseteq \bigcup_n B_n, \text{ and } B \bigcap \bigcup_n B_n = \emptyset.$$

This result of Saint Raymond is not only powerful, but the technique employed in its proof is very useful. The main idea is taken from Lusin's original proof of the following: *Every analytic set in the product of two Polish spaces with vertical sections countable can be covered by countably many Borel graphs* (5.10.3).

We assume 5.12.2 and give several consequences first.

Theorem 5.12.3 *Let X, Y be Polish spaces and $A \subseteq X \times Y$ a Borel set with sections A_x σ-compact. Then $A = \bigcup_n B_n$, where each B_n is Borel with $(B_n)_x$ compact for all x and all n.*

Proof. The result trivially follows from 5.12.2 by taking $B = A^c$. ■

Proof of 5.12.1. Write $B = \bigcup_n B_n$, the B_n's Borel with compact sections. That this can be done follows from 5.12.3. Then

$$\pi_X(B) = \bigcup_n \pi_X(B_n).$$

Since the projection of a Borel set with compact sections is Borel (4.7.11), each $\pi_X(B_n)$, and hence $\pi_X(B)$, is Borel. Let

$$D_n = \pi_X(B_n) \setminus \bigcup_{m<n} \pi_X(B_m).$$

Then the D_n's are Borel and pairwise disjoint. Further, the set

$$C = \bigcup_n (B_n \bigcap (D_n \times Y))$$

is a Borel subset of B with compact sections such that $\pi_X(C) = \pi_X(B)$. By Novikov's uniformization theorem (5.7.1), C admits a Borel uniformization, and our result follows. ■

Proposition 5.12.4 *Let $B \subseteq X \times Y$ be a Borel set with sections B_x that are G_δ sets in Y. Then there exist Borel sets B_n with open sections such that $B = \bigcap_n B_n$.*

Proof. Let Z be a compact metric space containing (a homeomorph of) Y. Then B is Borel in $X \times Z$ with sections G_δ sets (2.2.7). By 5.12.3, there exist Borel sets C_n in $X \times Z$ with sections compact such that $(X \times Z) \setminus B = \bigcup_n C_n$. Take $B_n = (X \times Y) \setminus C_n$. ■

Corollary 5.12.5 *Let $B \subseteq X \times Y$ be a Borel set with sections B_x that are F_σ sets in Y. Then there exist Borel sets B_n with closed sections such that $B = \bigcup_n B_n$.*

Before we present a proof of 5.12.2, we make a series of important observations.

(I) Recall the following from 4.9.6:
For any $x \in 2^{\mathbf{N}\times\mathbf{N}}$,

$$\begin{array}{rcl} D(x) & = & \{m \in \mathbf{N} : x(m,m) = 1\}, \\ m \leq_x^* n & \Longleftrightarrow & x(m,n) = 1, \end{array}$$

and

$$m <_x^* n \Longleftrightarrow m \leq_x^* n \;\&\; \neg(n \leq_x^* m).$$

Further,

$$LO^* = \{x \in LO : x(0,m) = 1 \text{ for every } m \in D(x)\},$$

and

$$WO^* = \{x \in WO : x(0,m) = 1 \text{ for every } m \in D(x)\}.$$

Thus, LO^* is the set of all x that encode linear orders on subsets of $\mathbf{N}$ with 0 the first element. It is Borel. Similarly, WO^* is the set of all x that encode well-orders on subsets of $\mathbf{N}$ with 0 the first element. We know that WO is a coanalytic non-Borel set (4.2.2), which easily implies that WO^* is a coanalytic non-Borel set.

(II) Let X be a Polish space. Recall that $F(X)$, the set of all closed subsets of X with the Effros Borel structure, is a standard Borel space. A family $\mathcal{B} \subseteq F(X)$ is called **hereditary** if whenever $A \in \mathcal{B}$ and B is a closed subset of A, then $B \in \mathcal{B}$. A **derivative** on X is a map $D : F(X) \longrightarrow F(X)$ such that for $A, B \in F(X)$,

(i) $D(A) \subseteq A$, and

(ii) $A \subseteq B \Longrightarrow D(A) \subseteq D(B)$.

Here are some interesting examples of derivatives.
Let $\mathcal{B} \subseteq F(X)$ be hereditary. Define

$$D_{\mathcal{B}}(A) = \{x \in X : (\forall \text{ open } U \ni x)(\mathrm{cl}(A \bigcap U) \notin \mathcal{B})\}.$$

Since $\mathcal{B}$ is hereditary, $D_{\mathcal{B}}$ is a derivative on X. If $\mathcal{B}$ consists of sets with at most one point, $D_{\mathcal{B}}(A)$ is the usual derived set of A. Another important example is obtained by taking $\mathcal{B}$ to be the family of all compact subsets of X.

We shall use the following property of $D_{\mathcal{B}}$, $\mathcal{B}$ hereditary $\mathbf{\Pi}^1_1$, without explicit mention. The set

$$\{(A,B) \in F(X) \times F(X) : A \subseteq D_{\mathcal{B}}(B)\}$$

is analytic. To see this, fix a countable base (U_n) for X. We have

$$A \subseteq D_{\mathcal{B}}(B) \Longleftrightarrow \forall n(U_n \bigcap A \neq \emptyset \Longrightarrow \mathrm{cl}(U_n \bigcap B) \notin \mathcal{B}).$$

Since $B \longrightarrow \mathrm{cl}(U_n \bigcap B)$ is a Borel map from $F(X)$ to $F(X)$, our assertion follows.

(III) Let X be Polish, $D : F(X) \longrightarrow F(X)$ a derivative on X, $A \subseteq X$ closed, and α any countable ordinal. We define $D^\alpha(A)$ by induction on α as follows:

$$D^0(A) = A,$$

$$D^\alpha(A) = D(D^\beta(A)), \text{ if } \alpha = \beta + 1, \text{ and}$$

$$D^\alpha(A) = \bigcap_{\beta<\alpha} D^\beta(A), \text{ if } \alpha \text{ is limit.}$$

So, $\{D^\alpha(A) : \alpha < \omega_1\}$ is a nonincreasing transfinite sequence of closed sets. Hence, by 2.1.13, there is an $\alpha < \omega_1$ such that $D^\alpha(A) = D^{\alpha+1}(A)$. The least such α will be denoted by $|A|_D$. We set

$$D^\infty(A) = D^{|A|_D}(A)$$

and

$$\Omega_D = \{A \in F(X) : D^\infty(A) = \emptyset\}.$$

Proposition 5.12.6 *Let X be a Polish space and $\mathcal{B} \subseteq F(X)$ hereditary. Then $\Omega_{D_\mathcal{B}} = \mathcal{B}_\sigma \bigcap F(X)$.*

Proof. Fix a closed set $A \subseteq X$ and a countable base (U_n) for X. Let $D^\infty(A) = \emptyset$. Then

$$\begin{aligned} A &= \bigcup_{\alpha<|A|_D}(D^\alpha(A) \setminus D^{\alpha+1}(A)) \\ &= \bigcup_{\alpha<|A|_D} \bigcup_n \{U_n \bigcap D^\alpha(A) : \ \mathrm{cl}(U_n \bigcap D^\alpha(A)) \in \mathcal{B}\} \\ &= \bigcup_{\alpha<|A|_D} \bigcup_n \{\mathrm{cl}(U_n \bigcap D^\alpha(A)) : \ \mathrm{cl}(U_n \bigcap D^\alpha(A)) \in \mathcal{B}\}. \end{aligned}$$

The last equality holds because A is closed. Thus, $A \in \mathcal{B}_\sigma$.

To prove the converse, take an $A \in \mathcal{B}_\sigma \bigcap F(X)$. Suppose $D^\infty(A) \neq \emptyset$. We shall get a contradiction. Write $A = \bigcup_m B_m$, $B_m \in \mathcal{B}$. By the Baire category theorem, there exist n and m such that

$$\emptyset \neq D^\infty(A) \bigcap U_n \subseteq D^\infty(A) \bigcap B_m.$$

This implies that

$$D^{|A|_D+1}(A) \neq D^{|A|_D}(A).$$

We have arrived at a contradiction. ■

Proposition 5.12.7 *Let X be Polish and D a derivative on X such that*

$$\{(A, B) \in F(X) \times F(X) : A \subseteq D(B)\}$$

is analytic. Then

(i) *Ω_D is coanalytic, and*

(ii) *for every analytic $\mathcal{A} \subseteq \Omega_D$,*

$$\sup\{|A|_D : A \in \mathcal{A}\} < \omega_1.$$

Proof. Assertion (i) follows from the following equivalence:

$$A \notin \Omega_D \iff \exists B(B \neq \emptyset \ \& \ B \subseteq A \ \& \ B \subseteq D(B)).$$

(The sets A and B are closed in X.)

Suppose (ii) is false for some analytic $\mathcal{A} \subseteq \Omega_D$. Then,

$$\sup\{|A|_D : A \in \mathcal{A}\} = \omega_1.$$

Define $R \subseteq 2^{\mathbb{N}\times\mathbb{N}} \times F(X)$ as follows:

$$\begin{aligned}(x, A) \in R \iff\ & x \in LO^* \ \& \\ & \exists f \in F(X)^{\mathbb{N}}[f(0) = A \ \& \\ & \forall m \in D(x)\{f(m) \neq \emptyset \ \& \\ & (m \neq 0 \Longrightarrow \forall n <_x^* m(f(m) \subseteq D(f(n))))\}].\end{aligned}$$

It is fairly easy to check that R is analytic and that for $\emptyset \neq A \in \Omega_D$,

$$R(x, A) \iff x \in WO^* \ \& \ |x| \leq |A|_D.$$

By our assumptions,

$$x \in WO^* \iff \exists A \in \mathcal{A}(R(x, A)).$$

This implies that WO^* is analytic, which is not the case, and our result is proved. ■

Lemma 5.12.8 *Let $\mathcal{F} \subseteq F(\mathbb{N}^{\mathbb{N}})$ be a hereditary $\mathbf{\Pi}_1^1$ family. Suppose X is a Polish space and $H \subseteq X \times \mathbb{N}^{\mathbb{N}}$ a closed set such that $H_x \in \mathcal{F}_\sigma$. Then there exists a sequence (H_n) of Borel sets such that $H = \bigcup_n H_n$ and $(H_n)_x \in \mathcal{F}$ for all x.*

Proof. Since H is closed and $\mathcal{F}$ hereditary, it is sufficient to show that there exist Borel sets H_n with sections in $\mathcal{F}$ covering H.

Let $D : F(\mathbb{N}^{\mathbb{N}}) \longrightarrow F(\mathbb{N}^{\mathbb{N}})$ be the derivative $D_{\mathcal{F}}$. For $\alpha < \omega_1$, define

$$H^\alpha = \{(x, y) \in X \times \mathbb{N}^{\mathbb{N}} : y \in D^\alpha(H_x)\}.$$

For each $\alpha < \omega_1$, we show that H^α is analytic. Towards showing this, let E be an analytic subset of $X \times \mathbb{N}^{\mathbb{N}}$ with closed sections, and observe that

$$\begin{aligned}y \in D(E_x) \iff\ & (x, y) \in E \ \& \\ & \forall s \in \mathbb{N}^{<\mathbb{N}}[y \in \Sigma(s) \Longrightarrow \\ & \exists F \in F(\mathbb{N}^{\mathbb{N}})(F \subseteq \Sigma(s) \textstyle\bigcap E_x \ \& \ F \notin \mathcal{F})].\end{aligned}$$

Thus $\{(x,y) \in X \times \mathbb{N}^{\mathbb{N}} : y \in D(E_x)\}$ is analytic. Using this observation, by induction on α it is quite easy to see that H^α is analytic.

Since $H_x \in \mathcal{F}_\sigma$, by 5.12.6, $D^\infty(H_x) = \emptyset$. Let

$$\mathcal{A} = \{F \in F(\mathbb{N}^{\mathbb{N}}) : \exists x(F \subseteq H_x)\}.$$

$\mathcal{A}$ is an analytic subset of Ω_D. Hence, by 5.12.7, there is an $\alpha_0 < \omega_1$ such that $H^{\alpha_0} = \emptyset$.

We claim the following.

Claim 1. For every $\alpha < \omega_1$ and every Borel set $B \supseteq H^\alpha$ with closed sections, there exist Borel sets H_n with closed sections such that

$$D((H_n)_x) = \emptyset$$

and

$$H \setminus B \subseteq \bigcup_n H_n.$$

Claim 2. If $B \subseteq X \times \mathbb{N}^{\mathbb{N}}$ is a Borel set with closed sections such that $D(B_x) = \emptyset$, then there is a sequence (H_n) of Borel sets such that $B = \bigcup_n H_n$ and $(H_n)_x \in \mathcal{F}$ for all n and all x.

Assuming these two claims, we obtain our result by taking $B = \emptyset$ and $\alpha = \alpha_0$.

Proof of claim 1. The proof is by induction on α. Let $\alpha < \omega_1$ and suppose that Claim 1 is true for all $\beta < \alpha$.

Case 1: $\alpha = \beta + 1$ for some β.

We first prove the following: Let $A \subseteq X \times \mathbb{N}^{\mathbb{N}}$ be an analytic set with sections closed, $B \supseteq A^1$ a Borel set with closed sections, where

$$(x,y) \in A^1 \iff y \in D(A_x).$$

Then $A \setminus B$ can be covered by a sequence of Borel sets (C_n) with closed sections such that $D((C_n)_x) = \emptyset$ for all n.

Since B^c is a Borel set with open sections and $\{\Sigma(s) : s \in \mathbb{N}^{<\mathbb{N}}\}$ a base for $\mathbb{N}^{\mathbb{N}}$, by 4.7.2, for each $s \in \mathbb{N}^{<\mathbb{N}}$ there is a Borel set B_s such that

$$B^c = \bigcup_s (B_s \times \Sigma(s)).$$

So,

$$A \setminus B = \bigcup_s ((B_s \times \Sigma(s)) \bigcap A).$$

From 4.7.1 it follows that $A \setminus B \subseteq \bigcup_n C_n$, where the C_n's are Borel sets with closed sections disjoint from A^1. As $(C_n)_x \subseteq A_x \setminus D((A)_x)$,

$$D((C_n)_x) \subseteq (C_n)_x \bigcap D((A)_x) = \emptyset.$$

Now, let $B \supseteq H^{\alpha}$ be a Borel set with closed sections. By the above observation, there exist Borel sets C_n with closed sections such that $D((C_n)_x) = \emptyset$ and $H^{\beta} \setminus B \subseteq \bigcup_n C_n = C$, say. So, $H^{\beta} \subseteq B \bigcup C$. By 4.7.1, there is a Borel set B' with closed sections such that $H^{\beta} \subseteq B' \subseteq B \bigcup C$. By the induction hypothesis, there exists a sequence (D_n) of Borel sets with closed sections such that $D((D_n)_x) = \emptyset$ and whose union contains $H \setminus B'$. As $H \setminus B \subseteq \bigcup_n D_n \bigcup \bigcup_n C_n$, our claim is proved in this case.

Case 2: α is a limit ordinal.

Let $H^{\alpha} = \bigcap_{\beta<\alpha} H^{\beta} \subseteq B$, B Borel. By the generalized first separation theorem (4.6.1), there exist Borel sets C^{β}, $\beta < \alpha$, such that $H^{\beta} \subseteq C^{\beta}$ and $\bigcap_{\beta<\alpha} C^{\beta} \subseteq B$. By 4.7.1, there exists a Borel set B_{β} with closed sections such that $H^{\beta} \subseteq B^{\beta} \subseteq C^{\beta}$. Then $\bigcap_{\beta<\alpha} B_{\beta} \subseteq B$. By the induction hypothesis, each $H \setminus B_{\beta}$ can be covered by a sequence (C_n) of Borel sets with closed sections such that $D((C_n)_x) = \emptyset$. As $H \setminus B \subseteq \bigcup_{\beta<\alpha}(H \setminus B_{\beta})$, it also can be so covered. ■

Proof of claim 2. Let $B \subseteq X \times \mathbb{N}^{\mathbb{N}}$ be a Borel set with closed sections such that $D(B_x) = \emptyset$. Then, for every $x \in X$ and every $y \in B_x$, there exists an $s \in \mathbb{N}^{<\mathbb{N}}$ such that $y \in \Sigma(s)$ and $\Sigma(s) \bigcap B_x \in \mathcal{F}$. Let

$$C(s) = \{x \in X : \Sigma(s) \bigcap B_x \in \mathcal{F}\}.$$

Since $\mathcal{F} \in \mathbf{\Pi}_1^1$, $C(s)$ is coanalytic and $B \subseteq \bigcup_s (C(s) \times \Sigma(s))$. Consider the Polish space $Z = \mathbb{N}^{<\mathbb{N}} \times X$ ($\mathbb{N}^{<\mathbb{N}}$ has the discrete topology) and $\Phi \subseteq \mathcal{P}(Z)$ defined by

$$E \in \Phi \iff B \subseteq \bigcup_s (E_s \times \Sigma(s)).$$

Then Φ is $\mathbf{\Pi}_1^1$ on $\mathbf{\Pi}_1^1$, and $\bigcup_s(\{s\} \times C(s)) \in \Phi$. Therefore, by the reflection theorem (5.10.1), there is a Borel set $D \subseteq \bigcup_s(\{s\} \times C(s))$ in Φ. Clearly,

$$B \subseteq \bigcup_s (D_s \times \Sigma(s)).$$

Let

$$B(s) = (D_s \times \Sigma(s)) \bigcap B, \quad s \in \mathbb{N}^{<\mathbb{N}}.$$

Then the $B(s)$'s are Borel sets with closed sections, and $\bigcup_s B(s) = B$. Further,

$$(B(s))_x = \begin{cases} B_x \bigcap \Sigma(s) & \text{if } x \in D_s, \\ \emptyset & \text{otherwise.} \end{cases}$$

In either case, $(B(s))_x \in \mathcal{F}$. This completes the proof. ■

Proof of 5.12.2. Let $f : \mathbb{N}^{\mathbb{N}} \longrightarrow A$ be a continuous onto map. Define

$$H = \{(x, \alpha) \in X \times \mathbb{N}^{\mathbb{N}} : x = \pi_X(f(\alpha))\}.$$

Clearly, H is closed. Take

$$\mathcal{F} = \{F \in F(\mathbb{N}^{\mathbb{N}}) : \text{cl}(f(F)) \subseteq B^c \ \& \ \text{cl}(\pi_Y(f(F))) \text{ is compact}\}.$$

The family $\mathcal{F}$ is clearly hereditary. By 3.3.11, $\{K \in F(Y) : K \text{ is compact}\}$ is Borel. Similarly, for every continuous function $g : \mathbb{N}^{\mathbb{N}} \longrightarrow Y$, the map $F \longrightarrow \text{cl}(g(F))$ from $F(\mathbb{N}^{\mathbb{N}})$ to $F(Y)$ is Borel measurable. Hence, $\mathcal{F}$ is Π^1_1. Suppose $x \in X$ and the K_n's are compact sets such that $A_x \subseteq \bigcup_n K_n \subseteq B^c_x$. Then

$$H_x = \bigcup_n f^{-1}(\{x\} \times K_n),$$

and each $f^{-1}(\{x\} \times K_n) \in \mathcal{F}$. Therefore, by 5.12.8, there exist Borel sets H_n with $(H_n)_x \in \mathcal{F}$ and $H = \bigcup H_n$. Now consider

$$A_n = \{f(\alpha) \in X \times Y : (x, \alpha) \in H_n\}.$$

Then A_n is analytic and $\bigcup_n A_n = A$. Let

$$\hat{A}_n = \{(x, y) \in X \times Y : y \in \text{ cl}((A_n)_x)\}.$$

So,

$$(x, y) \in \hat{A}_n \iff \forall m(y \in V_m \Longrightarrow (A_n)_x \bigcap V_m \neq \emptyset),$$

where (V_m) is a countable base for Y. It follows that $\hat{A}_n$ is analytic. Since $(H_n)_x \in \mathcal{F}$, sections of $\hat{A}_n$ are compact and $\hat{A}_n \bigcap B = \emptyset$. Hence, there is a Borel set B_n with compact sections such that $\hat{A}_n \subseteq B_n \subseteq B^c$ by 4.7.5. The Borel sets B_n serve our purpose. ■

Exercise 5.12.9 Show that every countably separated partition of a Polish space into σ-compact sets admits a Borel cross section.

Using the same technique, we can prove the following results.

Proposition 5.12.10 *Let X and Y be Polish spaces and A, B two disjoint analytic subsets of $X \times Y$ such that A_x is closed and nowhere dense for all x. Then there is a Borel $C \subseteq X \times Y$ such that the sections C_x are closed and nowhere dense, and such that*

$$A \subseteq C \text{ and } C \bigcap B = \emptyset.$$

Proposition 5.12.11 *(i) (Hillard [48]) Let X and Y be Polish spaces and A, B disjoint analytic subsets of $X \times Y$. Assume that the sections A_x are meager in Y. Then there is a sequence (C_n) of Borel sets with sections nowhere dense such that*

$$A \subseteq \bigcup_n C_n \text{ and } (\bigcup_n C_n) \bigcap B = \emptyset.$$

(ii) (H. Sarbadhikari [100]) For every Borel set $B \subseteq X \times Y$ with sections B_x comeager in Y, there is a sequence (B_n) of Borel sets such that $(B_n)_x$ is dense and open for every x and $\bigcap B_n \subseteq B$.

For proofs of the above two results see also [53].

We return to 5.12.4 and 5.12.5. We have seen that every Borel set with G_δ sections is a countable intersection of Borel sets with open sections, or equivalently, every Borel set with F_σ sections is a countable union of Borel sets with closed sections. Is a similar result true for all Borel pointclasses? In a significant contribution to the theory of Borel sets, Alain Louveau[66] showed that this is indeed the case. Unfortunately, no classical proof of this beautiful result is known. Known proofs use effective methods or forcing which are beyond the scope of our notes. Here we simply state Louveau's theorem. For a proof see [66] or [83].

Let X, Y be Polish spaces. For $1 \leq \alpha < \omega_1$, let $\mathcal{F}_\alpha$ denote the family of all Borel subsets of $X \times Y$ with x-sections of multiplicative class α and let $\mathcal{G}_\alpha = \neg\mathcal{F}_\alpha$. Again, by induction on α, $1 \leq \alpha < \omega_1$, we define families $\mathbf{\Sigma}^*_\alpha$, $\mathbf{\Pi}^*_\alpha$ of subsets of $X \times Y$ as follows. Take $\mathbf{\Pi}^*_0$ to be the subsets of $X \times Y$ of the form $B \times V$, B Borel and V open. For $\alpha > 0$, set

$$\mathbf{\Sigma}^*_\alpha = (\bigcup_{\beta<\alpha} \mathbf{\Pi}^*_\beta)_\sigma$$

and

$$\mathbf{\Pi}^*_\alpha = \neg\mathbf{\Sigma}^*_\alpha.$$

Clearly, $\mathbf{\Sigma}^*_\alpha \subseteq \mathcal{G}_\alpha$ and $\mathbf{\Pi}^*_\alpha \subseteq \mathcal{F}_\alpha$. We have already shown that $\mathbf{\Pi}^*_2 = \mathcal{G}_2$ (5.12.4) and $\mathbf{\Sigma}^*_2 = \mathcal{F}_2$ (5.12.5). We have also seen that $\mathbf{\Sigma}^*_1$ is precisely the family of all Borel sets with sections open (4.7.1). In a remarkable contribution to the theory of Borel sets, Louveau showed that this identity holds at all levels.

Theorem 5.12.12 *(A. Louveau [66]) For every* $1 \leq \alpha < \omega_1$, $\mathbf{\Sigma}^*_\alpha = \mathcal{F}_\alpha$.

5.13 Topological Vaught Conjecture

In this section we shall discuss one of the outstanding open problems in descriptive set theory. The study of this problem led to a rich subbranch of descriptive set theory now known as **invariant descriptive set theory.**

The Weak Topological Vaught Conjecture (WTVC) *Suppose a Polish group G acts continuously on a Polish space X. Then the number of orbits is* $\leq \aleph_0$ *or equals* $2^{\aleph_0}$. ■

WTVC is, of course, true under **CH**. The problem is to prove it without using **CH**. A statement equivalent to **WTVC** for $G = S_\infty$, the group of permutations of $\mathbb{N}$, first appeared as an open problem in [122]. We shall assume a little familiarity with first order logic to state this. Let L be a countable first order language. Assume first that the only non-logical symbols of L are relation symbols, say R_0, R_1, R_2, ... Suppose that R_i is

n_i-ary. Set

$$X_L = \prod_i 2^{\mathbb{N}^{n_i}}.$$

We equip X_L with the product of discrete topologies on $2 = \{0,1\}$. It is homeomorphic to the Cantor set. Elements of X_L can be identified with the structures of L with universe $\mathbb{N}$ as follows: To each $x \in X_L$ associate a countable structure $\mathcal{A}_x$ of L whose universe is $\mathbb{N}$, and in which R_i is interpretated by the set $\{s \in \mathbb{N}^{n_i} : x_i(s) = 1\}$. Define an action of S_∞ on X_L by

$$(g \cdot x)_i(n_0, n_1, \ldots, n_{i-1}) = 1 \iff x_i(g(n_0), g(n_1), \ldots, g(n_{i-1})) = 1.$$

This action is called the **logic action** on X_L. Clearly, the logic action is continuous. Further, $x, y \in X_L$ are in the same orbit if and only if $\mathcal{A}_x$ and $\mathcal{A}_y$ are isomorphc.

In the general situation (when L also has function symbols), we modify the definition of X_L and the logic action as follows: Corresponding to each k-ary function symbol, we add a coordinate axis consisting of all maps from $\mathbb{N}^k$ to $\mathbb{N}$ to X_L. Finally, modify the action of S_∞ to X_L in an obvious way so that each orbit represents an isomorphism class of countable structures of L. In what follows, for simplicity, we shall restrict our discussion to languages whose non-logical symbols are relation symbols only.

Let $L_{\omega_1\omega}$ be the set of formulas built up from symbols of L using countable conjunctions and disjunctions as well as the usual first order logical operations. Thus, in the inductive definition of formulae of $L_{\omega_1\omega}$, whenever (ϕ_n) is a sequence of formulae such that no variable other than $v_0, v_1, \ldots, v_{k-1}$ are free in any of ϕ_n, $\bigvee_n \phi_n$ is also a formula of $L_{\omega_1\omega}$. For any sentence σ of $L_{\omega_1\omega}$, put

$$A_\sigma = \{x \in X_L : \mathcal{A}_x \models \sigma\},$$

where "$\mathcal{A}_x \models \sigma$" means that σ is valid in $\mathcal{A}_x$. A basic result in this circle of ideas is the following. We shall give only the essential idea of the proof of this result. Readers are invited to complete the proof themselves.

Theorem 5.13.1 *(Lopez-Escobar) A subset A of X_L is invariant (with respect to the logic action) and Borel, if and only if there is a sentence σ of $L_{\omega_1\omega}$ such that $A = A_\sigma$.*

Proof. The sufficient part of this result is proved by induction on formulae of $L_{\omega_1\omega}$ as follows:

For every formula $\phi[v_0, v_1, \ldots, v_{k-1}]$, the set

$$A_{\phi,k} = \{(x, n_0, n_1, \ldots, n_{k-1}) : \mathcal{A}_x \models \phi[n_0, n_1, \ldots, n_{k-1}]\}$$

is Borel.

The necessary part is also proved by induction, but the induction in this case is a bit subtle. We proceed as follows. Let $(\mathbb{N})^k$ denote the set of all one-to-one finite sequences in $\mathbb{N}$ of length k and for any $s \in (\mathbb{N})^k$,

$$[s] = \{g \in S_\infty : s \prec g^{-1}\}.$$

Clearly, $\{[s] : s \in (\mathbb{N})^k\}$ form a base for the topology of S_∞.

Suppose A is a Borel set in X_L. Then, for every k there is a formula $\phi[v_0, v_1, \ldots, v_{k-1}]$ of $L_{\omega_1\omega}$ such that

$$A_{\phi,k} = \{(x, s) : s \in (\mathbb{N})^k \ \&\ x \in A^{*[s]}\}.$$

This is proved by induction on A using basic identities on Vaught transforms. We invite readers to complete the proof themselves. (Otherwise consult [[53], p.97].)

Now, if $A \subseteq X_L$ is an invariant Borel set, then $A^* = A$ and the result follows from the above assertion by taking $k = 0$. ■

The original conjecture of Vaught was the following.

Vaught Conjecture (VC) *Suppose L is a countable first order language. Then the number of countable nonisomorphic models of any sentence σ of $L_{\omega_1\omega}$ is $\leq \aleph_0$ or equals $2^{\aleph_0}$.* ■

In other words, **VC** states that A_σ is a union of countably many or $2^{\aleph_0}$ many orbits with respect to the logic action on X_L.

We now show how **VC** follows from **WTVC**. By the theorem of Lopez-Escobar, A_σ is an invariant Borel set. However, A_σ need not be Polish. Now we use the following remarkable result of Becker and Kechris[11] to immediately conclude **VC** from **WTVC**.

Theorem 5.13.2 *(Becker – Kechris) Suppose a Polish group G acts continuously on a Polish space X and A is an invariant Borel subset of X. Then there is a finer Polish topology on X making A clopen such that the action still remains continuous.*

We may also use the following similar result of Becker and Kechris to prove Vaught conjecture from **WTVC**.

Theorem 5.13.3 *(Becker – Kechris) Suppose a Polish group G acts on a Polish space X and the action is Borel. Then there is a finer Polish topology on X making the action continuous.*

The proofs of the above theorems are somewhat elaborate and make use of Vaught transforms and Borel generated topologies. The reader is referred to [11] for proofs of these results.

There are certain metamathematical difficulties with $2^{\aleph_0}$ (namely, it is not "absolute"). Consequently, **VC** may be independent of **ZFC**. To avoid independence proofs, one considers a stronger version of the conjecture. For brevity we introduce the following terminology. Let E be an equivalence relation on a Polish space X. We say that E **has perfectly many equivalence classes** if there is a nonempty, perfect subset of X consisting of pairwise E-inequivalent elements.

The Topological Vaught Conjecture (TVC) *Suppose a Polish group G acts continuously on a Polish space X. Then the number of equivalence classes is countable or perfectly many.*

TVC clearly implies **WTVC**. Further, under $\neg$**CH**, **WTVC** implies **TVC**. This follows immediately from the following result of Burgess [22].

Theorem 5.13.4 *(Burgess) Suppose E is an analytic equivalence relation on a Polish space X. Then the number of equivalence classes is $\leq \aleph_1$ or perfectly many.*

We shall give a prrof of this result later in the section.

Remark 5.13.5 It is easy to see that Burgess's theorem can be extended to analytic equivalence relations on analytic sets X.

Exercise 5.13.6 Show that **TVC** is equivalent to the following statement: *Suppose a Polish group G acts on a standard Borel space X and the action is Borel. Then the number of orbits is $\leq \aleph_0$ or perfectly many.*

Remark 5.13.7 Kunen([112]) has shown that **TVC** does not hold for analytic sets X. His example is from logic which we omit.

There are strong indications that **TVC** is decidable in **ZFC**. For these reasons, in the rest of this section we shall consider **TVC** only.

We now give some sufficient conditions under which **TVC** holds.

Theorem 5.13.8 *Topological Vaught conjecture holds if G is a locally compact Polish group.*

We shall need the following result of Stern([118]) to prove 5.13.8.

Theorem 5.13.9 *Let E be an analytic equivalence relation on a Polish space X with all equivalence classes F_σ. Then the number of equivalence classes is $\leq \aleph_0$ or perfectly many.*

Assuming 5.13.9, we prove 5.13.8 as follows: Let G be a locally compact Polish group acting continuously on a Polish space X. Write $G = \bigcup_n K_n$, K_n compact. Then, for $x, y \in X$,

$$\exists g \in G(y = g \cdot x) \iff \exists n \exists g \in K_n(y = g \cdot x).$$

Since K_n is compact and the set $\{(x, y, g) \in X \times X \times K_n : y = g \cdot x\}$ is closed, the equivalence relation induced by the group action is an F_σ set. Our result now follows from 5.13.9. ■

To prove 5.13.9, we shall need the following result which is interesting on its own right.

Proposition 5.13.10 *Suppose X is a Polish space and E an equivalence relation on X which is meager in X^2. Then E has perfectly many equivalence classes.*

Proof. Let $E \subseteq \bigcup_n F_n$, F_n closed and nowhere dense in X^2. Without any loss of generality, we further assume that the diagonal $\{(x, y) \in X^2 : x = y\}$ is contained in each of F_n.

For each $s \in 2^{<\mathbb{N}}$, we define a nonempty open set $U(s)$ in X satisfying the following properties.

(i) diameter$(U(s)) \leq 2^{-|s|}$.

(ii) $s \prec t \Longrightarrow \mathrm{cl}(U(t)) \subseteq U(s)$.

(iii) If $s \neq s'$ and $|s| = |s'|$, then $(U(s) \times U(s')) \bigcap F_{|s|} = \emptyset$. In particular, U_s and $U_{s'}$'s are disjoint.

We define $\{U(s) : s \in 2^{<\mathbb{N}}\}$ by induction on $|s|$. Take $U(e)$ to be any nonempty open set of diameter less than 1 disjoint from F_0. Since F_0 is closed nowhere dense, such a set exists. Suppose n is a positive integer and $U(s)$ has been defined for every sequence s of length less than n. Consider the set $F_n^{2^n}$. It is closed and nowhere dense in $X^{2^{n+1}}$. Hence, there is an open set of the form $\prod_{s\in 2^{n-1}}(U(s\hat{\ }0) \times U(s\hat{\ }1))$ contained in $\prod_{s\in 2^{n-1}}(U(s) \times U(s))$ disjoint from $F_n^{2^n}$. We can further assume that the diameter of $U(s\hat{\ }\epsilon)$, $|s| = n - 1$ and $\epsilon = 0$ or 1, is less than 2^{-n}, and that its closure is contained in $U(s)$.

For $\alpha \in 2^{\omega}$, let $f(\alpha)$ be the unique element of X that belongs to each of $U(\alpha|n)$. It is easy to see that the range of f is a perfect set of pairwise E-inequivalent elements. ■

Proof of 5.13.9. Let X be a Polish space and E an analytic equivalence relation on X with all its equivalence classes F_σ sets. Further assume that there are uncountably many E-equivalence classes. Fix a countable base (V_n) for the topology of X. Let P be the union of all basic open sets which is contained in countably many equivalence classes and Q its saturation; i.e., $Q = \mathrm{proj}(E\bigcap(P \times P))$. Thus Q is analytic. Set $Y = X \setminus Q$ and $E' = E\bigcap(Y \times Y)$. Note that E' has the Baire property. Also note that every section of E' is meager. So, by Kuratowski – Ulam theorem, E' is meager. Our result now follows from 5.13.10. ■

In the rest of this section, the following result of Silver[106] will play a very important role.

Theorem 5.13.11 *(Silver's theorem) Suppose E is a coanalytic equivalence relation on a Polish space X. Then the number of equivalence classes is countable or perfectly many.*

By 5.13.9, the above result holds for F_σ equivalence relations. Known proofs of Silver's result, even for Borel equivalence relations, use either effective methods or forcing. This is beyond the scope of this book.

Recently Sami([98]) showed that **TVC** is true if G is abelian. We give the proof below.

Theorem 5.13.12 *(Sami) Topological Vaught conjecture holds if G is abelian.*

Proof. Assume that the number of orbits is uncountable. We shall show that there is a perfect set of inequivalent elements.

Let E be the equivalence relation on X defined by

$$xEy \Longleftrightarrow G_x = G_y,$$

where G_x is the stabilizer of x. Let $y = g \cdot x$ for some $g \in G$. Then $G_x = g^{-1} \cdot G_y \cdot g = G_y$, as G is abelian. Thus,

$$xE_ay \Longrightarrow xEy,$$

where E_a is the equivalence relation induced by the action. Now note that

$$xEy \Longleftrightarrow \forall g(g \cdot x = x \Longleftrightarrow g \cdot y = y).$$

Hence, E is coanalytic.

Suppose there are uncountably many E-equivalence classes. Then by Silver's theorem, there is a perfect set of E-inequivalent elements. In particular, there is a perfect set of E_a-inequivalent elements.

Now assume that the set of E-equivalence classes is countable. We shall show that E_a is Borel. Our proof will then follow from Silver's theorem.

Let $Y \subseteq X$ be an E-equivalence class. It is sufficient to show that $E_a \bigcap (Y \times Y)$ is Borel. Let $x \in Y$ and $H = G_x$. The partition of G by the cosets of H is lower-semicontinuous. Hence, there is a Borel cross-section S for this partition. For $x, y \in Y$, we have the following:

$$xE_ay \Longleftrightarrow (\exists \text{ a unique } g \in S)(y = g \cdot x);$$

i.e., $E_a \bigcap (Y \times Y)$ is a one-to-one projection of the Borel set

$$\{(x, y, g) : g \in S \text{ and } y = g \cdot x\}.$$

Hence, E_a is Borel. ∎

Remark 5.13.13 Recently Solecki [108] showed that the equivalence relation induced by a continuous action of an abelian Polish group on a Polish space need not be Borel.

Proof of Burgess's theorem.

The proof of this theorem is based on Silver's theorem, reflection principle and the following combinatorial lemma.

Lemma 5.13.14 *Suppose $\{A_\alpha : \alpha < \omega_1\}$ is a family of Borel subsets of a Polish space X and E the equivalence relation on X defined by*

$$xEy \iff \forall\alpha(x \in A_\alpha \iff y \in A_\alpha),\ x, y \in X. \qquad (*)$$

Then the number of E-equivalence classes is $\leq \aleph_1$ or perfectly many.

Assuming the lemma, Burgess's theorem is proved as follows. Let E be an analytic equivalence relation. By 5.10.6, there exist Borel equivalence relations B_α, $\alpha < \omega_1$, such that

$$E = \bigcap_{\alpha<\omega_1} B_\alpha.$$

If for some $\alpha < \omega_1$, B_α has uncountably many equivalence classes, then by Silver's theorem there is a perfect set P of pairwise B_α-inequivalent elements. In particular, elements of P are pairwise E-inequivalent.

Now assume that the set of B_α-equivalence classes is countable for all $\alpha < \omega_1$. Let $\{A_\beta : \beta < \omega_1\}$ be the set of all B_α-equivalence classes, $\alpha < \omega_1$. Clearly, for any two x, y in X

$$xEy \iff \forall\beta < \omega_1(x \in A_\beta \iff y \in A_\beta).$$

Thus, the result follows from the above lemma in this case also. ■

Proof of 5.13.14. Although the proof of the lemma is messy looking, ideawise it is quite simple. Assume that the number of E-equivalence classes is $> \aleph_1$. We shall then show that there are perfectly many E-equivalence classes. The following fact will be used repeatedly in the proof of the lemma.

Fact. *Suppose Z is a subset of X of cardinality $> \aleph_1$ such that no two distinct elements Z are E-equivalent. Then there is an $\alpha < \omega_1$ such that both $Z \bigcap A_\alpha$ and $Z \bigcap A_\alpha^c$ are of cardinality $> \aleph_1$.*

We prove this fact by contradiction. If possible, let for every $\alpha < \omega_1$ at least one of $Z \bigcap A_\alpha$ and $Z \bigcap A_\alpha^c$ be of cardinality $\leq \aleph_1$. Denote one such set by M_α. We claim that $Z \setminus \bigcup_\alpha M_\alpha$ is a singleton. Suppose not. Let x, y be two distinct elements of $Z \setminus \bigcup_\alpha M_\alpha$. Since x, y are E-inequivalent, by ($\star$) there exists an $\alpha < \omega_1$ such that exactly one of x and y belong to A_α. It follows that at least one of x, y belong to M_α. But this is not the case. Hence, $Z \setminus \bigcup_\alpha M_\alpha$ contains at most one point. It follows that the cardinality of Z is at most $\aleph_1$, and we have arrived at a contradiction.

Fix a compatible complete metric on X. Following our usual notation, for $\epsilon = 0$ or 1, we set

$$A_\alpha^\epsilon = \begin{cases} A_\alpha & \text{if } \epsilon = 0, \\ A_\alpha^c & \text{if } \epsilon = 1. \end{cases}$$

Since A_α^ϵ analytic, there is a continuous map $f_\alpha^\epsilon : \mathbb{N}^\mathbb{N} \longrightarrow X$ whose range is A_α^ϵ. We can arrange matters so that for every $s \in \mathbb{N}^{<\mathbb{N}}$, the diameter of $f_\alpha^\epsilon(\Sigma(s))$ is at most $2^{-|s|}$.

Fix any subset Z of X of cardinality $> \aleph_1$ consisting of pairwise E-inequivalent elements. By the above fact, there exists an ordinal $\alpha(e) < \omega_1$ such that both $Z \bigcap A_{\alpha(e)}$ and $Z \bigcap A^c_{\alpha(e)}$ are of cardinality $> \aleph_1$. Let $\epsilon_0 =$ 0 or 1. As

$$Z \bigcap A^{\epsilon_0}_{\alpha(e)} = \bigcup_m (Z \bigcap f^{\epsilon_0}_{\alpha(e)}(\Sigma(m))),$$

there is an $m_{\epsilon_0} \in \mathbb{N}$ such that $Z \bigcap f^{\epsilon_0}_{\alpha(e)}(\Sigma(m_{\epsilon_0}))$ is of cardinality $> \aleph_1$. Set $s(\epsilon_0, 0) = (m_{\epsilon_0})$.

Now fix any finite sequence $(\epsilon_0\epsilon_1)$ of 0's and 1's of length 2. Applying the fact again, there is an ordinal $\alpha(\epsilon_0) < \omega_1$ such that both $Z \bigcap f^{\epsilon_0}_{\alpha(e)}(\Sigma(s(\epsilon_0, 0))) \bigcap A_{\alpha(\epsilon_0)}$ and $Z \bigcap f^{\epsilon_0}_{\alpha(e)}(\Sigma(s(\epsilon_0, 0))) \bigcap A^c_{\alpha(\epsilon_0)}$ are of cardinality $> \aleph_1$. Note that

$$\begin{aligned} & Z \bigcap f^{\epsilon_0}_{\alpha(e)}(\Sigma(s(\epsilon_0, 0))) \bigcap A^{\epsilon_1}_{\alpha(\epsilon_0)} \\ = & \bigcup_m \bigcup_{s \in \mathbb{N}^2} (Z \bigcap f^{\epsilon_0}_{\alpha(e)}(\Sigma(s(\epsilon_0, 0)\hat{\ } m)) \bigcap f^{\epsilon_1}_{\alpha(\epsilon_0)}(\Sigma(s))). \end{aligned}$$

Hence there exists an $m_{\epsilon_0\epsilon_1} \in \mathbb{N}$ and an $s(\epsilon_0\epsilon_1, 1) \in \mathbb{N}^2$ such that the set

$$Z \bigcap f^{\epsilon_0}_{\alpha(e)}(\Sigma(s(\epsilon_0, 0)\hat{\ } m_{\epsilon_0\epsilon_1})) \bigcap f^{\epsilon_1}_{\alpha(\epsilon_0)}(\Sigma(s(\epsilon_0\epsilon_1, 1)))$$

is of cardinality $> \aleph_1$. Set $s(\epsilon_0\epsilon_1, 0) = s(\epsilon_0, 0)\hat{\ } m_{\epsilon_0\epsilon_1}$.

Proceeding similarly, we can show the following: For every $l \in \mathbb{N}$, for every $\sigma \in \mathbb{N}^l$ and for every $k < l$, there exists an ordinal $\alpha(\sigma) < \omega_1$, and there exists an $s(\sigma, k) \in \mathbb{N}^l$ such that setting

$$T_\sigma = \bigcap_{k<l} f^{\sigma(k)}_{\alpha(\sigma|k)}(\Sigma(s(\sigma, k))),$$

the cardinality of the set $Z \bigcap T_\sigma$ is $> \aleph_1$. Further, if $\sigma \prec \tau$, $s(\sigma, k) \prec s(\tau, k)$ for all $k < |\sigma|$.

Now take any $g \in 2^{\mathbb{N}}$. Then $(\mathrm{cl}(T_{g|k}))$ is a nested sequence of nonempty closed sets of diameters converging to 0. Let $u(g)$ be the unique point of $\bigcap \mathrm{cl}(T_{g|k})$. It is easily seen that the map $u : 2^{\mathbb{N}} \longrightarrow X$ is continuous.

Let g and h be two distinct elements of $2^{\mathbb{N}}$. Let m be the first positive integersuch that $g(m) \neq h(m)$. Without any loss of generality, we can assume that $g(m) = 0$ and $h(m) = 1$. Then $u(g) \in A_{\alpha(g|m)}$ and $u(h) \notin A_{\alpha(g|m)}$. Thus $u(2^{\mathbb{N}})$ is a perfect set of E-inequivalent elements. ■

By Silver's theorem **TVC** holds if the equivalence relation induced by a continuous group action is always Borel. Below we give an example showing that the equivalence relation induced by a continuous action need not be Borel.

Example 5.13.15 Let L be a first order language whose non-logical symbols consists of exactly one binary relation symbol. So, $X_L = 2^{\omega \times \omega}$. We claim that in this case the equivalence relation E_a induced by the logic action is not Borel. Suppose not. Then $E_a \in \mathbf{\Sigma}^0_\beta$ for some $\beta < \omega_1$. It follows that $WO^\alpha = \{x \in WO : |x| \leq \alpha\} \in \mathbf{\Sigma}^0_\beta$ for every $\alpha < \omega_1$. Now take any Borel set A in $\mathbf{N}^\mathbf{N}$ which is not of additive class β. Since WO is $\mathbf{\Pi}^1_1$-complete, there is a continuous function $f : \mathbf{N}^\mathbf{N} \longrightarrow LO$ such that $A = f^{-1}(WO)$. But by the boundedness theorem, $A = f^{-1}(WO^\alpha)$ for some α. It follows that $A \in \mathbf{\Sigma}^0_\beta$, and we have arrived at a contradiction.

The following example shows that Silver's theorem (or TVC type result) is not true for analytic equivalence relations.

Exercise 5.13.16 For $\alpha, \beta \in 2^{\mathbf{N}\times\mathbf{N}}$, define

$$\alpha \sim \beta \Longleftrightarrow \text{ either } \alpha, \beta \notin WO \text{ or } |\alpha| = |\beta|.$$

Show that $\sim$ is an analytic equivalence relation with the number of equivalence classes $\aleph_1$ but not perfectly many.

Remark 5.13.17 Recall that the orbit of every point of a Polish space X under a continuous action of a Polish group is Borel (5.4.5). So, the equivalence relation E_a on X induced by the action is analytic with all equivalence classes Borel. A natural question arises: *Suppose E is an analytic equivalence relation on a Polish space X with all equivalence classes Borel. Is it true that the number of equivalence classes is $\leq \aleph_0$ or perfectly many?* The answer to this question is no. However, known examples use effective methods or logic. Therefore, we omit them.

In all the known examples of analytic equivalence relations such that

(i) all equivalence classes are Borel, and

(ii) there are uncountably many equivalence classes but not perfectly many,

the equivalence classes are of unbounded Borel rank. So, the following question arises: *Suppose E is an analytic equivalence relation on a Polish space such that all its equivalence classes are Borel of additive class α for some $\alpha < \omega_1$. Is it true that the number of equivalence classes is $\leq \aleph_0$ or perfectly many?* In [118] and [119], Stern considered this problem. He proved the following results.

Theorem 5.13.18 *(Stern) Let E be an analytic equivalence relation on a Polish space X such that all but countably many equivalence classes are F_σ or G_δ. The the number of equivalence classes is $\leq \aleph_0$ or perfectly many.*

Note that, earlier in this section we proved this result in the special case when all equivalence classes are F_σ sets. As the proof of this result is long, we omit it.

Theorem 5.13.19 *(Stern) Assume* **analytic determinacy**. *Let E be an analytic equivalence relation on a Polish space X such that all but countably many equivalence classes are of bounded Borel rank. Then the number of equivalence classes is $\leq \aleph_0$ or perfectly many.*

The proof this result is beyond the scope of this book.

5.14 Uniformizing Coanalytic Sets

In this section we prove the famous uniformization theorem of Kondô.

Theorem 5.14.1 *(Kondô's theorem) Let X, Y be Polish spaces. Every coanalytic set $C \subseteq X \times Y$ admits a coanalytic uniformization.*

We shall show that there is a sequence of coanalytic norms on a given coanalytic set with certain "semicontinuity" properties. The existence of such a sequence of norms gives a procedure for selecting a point from a given nonempty coanalytic set. The procedure is then applied to each nonempty section of a conanalytic set, thus yielding a uniformization. The semicontinuity properties guarantee that the uniformizing set is coanalytic. We now describe this procedure in detail.

Let A be a subset of a Polish space X. A **scale** on A is a sequence of norms φ_n on A such that $x_i \in A$, $x_i \to x$, and $\forall n(\varphi_n(x_i) \to \mu_n)$ (i.e., $\varphi_n(x_i)$ is eventually constant and equals μ_n after a certain stage) imply that $x \in A$ and $\forall n(\varphi_n(x) \leq \mu_n)$.

If for each $n, \varphi_n : A \longrightarrow \kappa$, then we say that (φ_n) is a **κ-scale**.

Given an ordinal κ, define the lexicographical ordering $<_{\text{lex}}$ on κ^n as follows.

$$(\mu(0), \mu(1), \ldots, \mu(n-1)) <_{\text{lex}} (\lambda(0), \lambda(1), \ldots, \lambda(n-1))$$
$$\iff \exists i < n[\forall j < i(\mu(j) = \lambda(j)) \ \& \ (\mu(i) < \lambda(i))].$$

This is a well-ordering with order type κ^n. Denote by

$$\langle \mu(0), \mu(1), \ldots, \mu(n-1) \rangle$$

the ordinal $< \kappa^n$ corresponding to $(\mu(0), \mu(1), \ldots, \mu(n-1))$ under the isomorphism of $(\kappa^n, <_{\text{lex}})$ with κ^n.

Remark 5.14.2 Given a scale (φ_n) on $A \subseteq \mathbb{N}^{\mathbb{N}}$ we can define a new scale (ψ_n) as follows.

$$\psi_n(\alpha) = \langle \varphi_0(\alpha), \alpha(0), \varphi_1(\alpha), \alpha(1), \ldots, \varphi_n(\alpha), \alpha(n) \rangle. \qquad (1)$$

The scale (ψ_n) has additionally the following properties.

1. $\psi_n(\alpha) \leq \psi_n(\beta) \Longrightarrow \forall m \leq n(\psi_m(\alpha) \leq \psi_m(\beta))$.

2. If $\alpha_i \in A$ and $\psi_n(\alpha_i) \to \mu_n$ for all n, then $\alpha_i \to \alpha$ for some $\alpha \in A$.

Let A be a subset of a Polish space X. A scale (φ_n) on A is called a **very good scale** if

1. $\varphi_n(x) \leq \varphi_n(y) \Longrightarrow \forall m \leq n(\varphi_m(x) \leq \varphi_m(y))$.

2. If $x_i \in A$ and $\varphi_n(x_i) \to \mu_n$ for all n, then $x_i \to x$ for some $x \in A$.

Given a very good scale (φ_n) on A, we can select a point from A as follows. Let

$$A_0 = \{x \in A : \varphi_0(x) \text{ is least, say } \mu_0\},$$

$$A_1 = \{x \in A_0 : \varphi_1(x) \text{ is least, say } \mu_1\},$$

$$A_2 = \{x \in A_1 : \varphi_2(x) \text{ is least, say} \mu_2\},$$

and so on. We have

$$A_0 \supseteq A_1 \supseteq A_2 \supseteq \cdots$$

and if $x_i \in A_i$, then $\varphi_n(x_i) = \mu_n$ for all $i > n$. Since (φ_n) is a very good scale, there is an $x \in A$ such that $x_i \to x$. Moreover, it is quite easy to see that $x \in A_n$ for all n. Let y be any other point in $\bigcap_n A_n$. Consider the sequence $x, y, x, y, \ldots$ Since (φ_n) is a very good scale, the sequence $x, y, x, y, \ldots$ is convergent. Hence, $x = y$. Thus $\bigcap_n A_n$ is a singleton. The above procedure thus selects a unique point from A, called the **canonical element of A determined by (φ_n).**

A scale (φ_n) on a coanalytic subset A of a Polish space X is called a **$\mathbf{\Pi}^1_1$-scale** if each φ_n is a $\mathbf{\Pi}^1_1$-norm.

Exercise 5.14.3 If (ϕ_n) is a $\mathbf{\Pi}^1_1$-scale on a coanalytic $A \subseteq \mathbb{N}^{\mathbb{N}}$, then show that (ψ_n) defined by (1) is also a $\mathbf{\Pi}^1_1$-scale.

We are now in a position to state the main result needed to prove Kondō's theorem.

Theorem 5.14.4 *Every coanalytic subset of $\mathbb{N}^{\mathbb{N}}$ admits a very good $\mathbf{\Pi}^1_1$-scale.*

Corollary 5.14.5 *Let X be a Polish space and $A \subseteq X$ coanalytic. Then A admits a very good $\mathbf{\Pi}^1_1$-scale.*

Proof. By 2.6.9 there is a closed set $D \subseteq \mathbb{N}^{\mathbb{N}}$ and a continuous bijection $f : D \longrightarrow X$. Now, $f^{-1}(A) \bigcap D$ is a $\mathbf{\Pi}^1_1$ subset of $\mathbb{N}^{\mathbb{N}}$ and hence admits a very good $\mathbf{\Pi}^1_1$-scale by 5.14.4. The scale on A is now obtained by transfer via the function f. ■

Assuming 5.14.5, we prove Kondô's theorem.

Proof of Kondô's theorem (5.14.1). By 5.14.5 there is a very good $\mathbf{\Pi}^1_1$-scale (φ_n) on C. Then (φ^x_n), where $\varphi^x_n(y) = \varphi_n(x, y)$, is a very good scale on the section C_x, if $C_x \neq \emptyset$. Let $y(x)$ be the canonical element of C_x determined by (φ_n^x). Set

$$(x, y) \in C^* \iff y = y(x).$$

Clearly, C^* uniformizes C. To see that C^* is coanalytic, observe that

$$(x, y) \in C^* \iff \forall n \forall z((x, y) \leq^*_{\varphi_n} (x, z)).$$

■

Before proving 5.14.4 we make some general observations.

For $\alpha \in 2^{\mathbf{N}}$, let $\leq_\alpha$ be the binary relation on $\mathbf{N}$ defined by

$$n \leq_\alpha m \iff \alpha(\langle n, m\rangle) = 1$$

and

$$D(\alpha) = \{m \in \mathbf{N} : \alpha(\langle m, m\rangle) = 1\},$$

the field of the relation $\leq_\alpha$. In what follows we shall consider only those α for which $\leq_\alpha$ is a linear order on $D(\alpha)$. For $n \in \mathbf{N}$, set

$$W_n = \{p \in \mathbf{N} : p <_\alpha n\},$$

and let $\leq_\alpha |n$ denote the restriction of $\leq_\alpha$ to W_n. So,

$$\leq_\alpha |n = \{(p, q) : p \leq_\alpha q \ \&\ q <_\alpha n\}.$$

Clearly, $W_n = \emptyset$ if $n \notin D(\alpha)$.

If $\leq_\alpha$ is a well-ordering with rank function ρ, then for each n, $\leq_\alpha |n$ is a well-ordering, and

$$\rho(n) = |\leq_\alpha |n|,$$

where $|\leq_\alpha |n|$ is the ordinal corresponding to the well-ordering $\leq_\alpha |n$. Thus

$$n \leq_\alpha m \iff |\leq_\alpha |n| \leq |\leq_\alpha |m|.$$

We make one more general observation. Let (α_i) be a sequence in $2^{\mathbf{N}}$ such that each $\leq_{\alpha_i}$ is a well-ordering and for each n, $|\leq_{\alpha_i} |n|$ is eventually constant, say λ_n. Suppose (α_i) converges to some $\alpha \in 2^{\mathbf{N}}$. Then $\leq_\alpha$ is a well-ordering and $|\leq_\alpha |n| \leq \lambda_n$ for all n.

This fact will follow if we show that the map $n \longrightarrow \lambda_n$ from $(D(\alpha), \leq_\alpha)$ into ordinals is order-preserving. We prove this now. Let $n, m \in \mathbf{N}$. We have

$$\begin{aligned} n <_\alpha m &\implies \alpha(\langle n, m\rangle) = 1 \ \&\ \alpha(\langle m, n\rangle) \neq 1 \\ &\implies \text{for all large } i, \alpha_i(\langle n, m\rangle) = 1 \\ &\qquad \text{and } \alpha_i(\langle m, n\rangle) \neq 1, \text{ since } \alpha_i \to \alpha \\ &\implies \text{for all large } i, n <_{\alpha_i} m \\ &\implies \text{for all large } i, |\leq_{\alpha_i} |n| < |\leq_{\alpha_i} |m| \\ &\implies \lambda_n < \lambda_m. \end{aligned}$$

Proof of 5.14.4. Take any coanalytic $A \subseteq \mathbf{N}^{\mathbf{N}}$. We need to show that A admits a very good $\mathbf{\Pi}_1^1$-scale. By 5.14.2 and 5.14.3, it is sufficient to show that A admits a $\mathbf{\Pi}_1^1$-scale.

By 4.2.2, there exists a continuous function $f : \mathbf{N}^{\mathbf{N}} \longrightarrow 2^{\mathbf{N}}$ such that for all x, $\leq_{f(x)}$ is a linear ordering and

$$x \in A \Longleftrightarrow f(x) \in WO.$$

Let $(\mu, \lambda) \rightarrow \langle \mu, \lambda \rangle$ be an order-preserving map of $\omega_1 \times \omega_1$, ordered lexicographically, into the ordinals. For $x \in A$, set

$$\varphi_n(x) = \langle | \leq_{f(x)} |, | \leq_{f(x)} |n| \rangle.$$

Claim: (φ_n) is a $\mathbf{\Pi}_1^1$-scale on A.

To prove this, first assume that x_i is a sequence in A such that $x_i \rightarrow x$, and suppose that for all n and all large i,

$$\varphi_n(x_i) = \langle \lambda, \lambda_n \rangle.$$

This implies that for each n and all large i,

$$| \leq_{f(x_i)} |n| = \lambda_n.$$

Since f is continuous, $f(x_i) \rightarrow f(x)$. Thus by the observations made above, $f(x) \in WO$, and hence $x \in A$. Furthermore, for every n,

$$| \leq_{f(x)} |n| \leq \lambda_n.$$

Hence,

$$\sup\{| \leq_{f(x)} |n| : n \in \mathbf{N}\} \leq \sup\{\lambda_n : n \in \mathbf{N}\}.$$

This means that

$$| \leq_{f(x)} | \leq \lambda,$$

since for all large i,

$$\lambda_n = | \leq_{f(x_i)} |n| \leq | \leq_{f(x_i)} | = \lambda.$$

Hence

$$\varphi_n(x) = \langle | \leq_{f(x)} |, | \leq_{f(x)} |n| \rangle \leq \langle \lambda, \lambda_n \rangle,$$

and so (φ_n) is a scale on A.

To show that it is a $\mathbf{\Pi}_1^1$-scale, for each n define a function $g_n : 2^{\mathbf{N}} \longrightarrow 2^{\mathbf{N}}$ as follows:

$$g_n(\alpha)(\langle p, q \rangle) = 1 \Longleftrightarrow$$
$$\alpha(\langle p, q \rangle) = 1 \;\&\; \alpha(\langle q, n \rangle) = 1 \;\&\; \alpha(\langle n, q \rangle) = 0.$$

Note that g_n is continuous and that whenever $\leq_\alpha$ is a linear ordering, $g_n(\alpha)$ is a code of the ordering $\leq_\alpha |n$.

Now define

$$x \leq_{\varphi_n}^{\Pi_1^1} y \iff f(x) \leq_{|.|}^{\Pi} f(y)$$
$$\&[\neg(f(y) \leq_{|.|}^{\Sigma_1^1} f(x)) \text{ or } g_n(f(x)) \leq_{|.|}^{\Pi_1^1} g_n(f(y))],$$

$$x \leq_{\varphi_n}^{\Sigma_1^1} y \iff f(x) \leq_{|.|}^{\Sigma} f(y)$$
$$\&[\neg(f(y) \leq_{|.|}^{\Pi_1^1} f(x)) \text{ or } g_n(f(x)) \leq_{|.|}^{\Sigma_1^1} g_n(f(y))].$$

(Recall that for any $\alpha \in WO$, $|\alpha|$ denotes the order type of $\leq_\alpha$.) The relations $\leq_{\varphi_n}^{\Pi_1^1}$ and $\leq_{\varphi_n}^{\Sigma_1^1}$ are respectively $\mathbf{\Pi}_1^1$ and $\mathbf{\Sigma}_1^1$ by definition. It is easily seen that they witness that φ_n is a $\mathbf{\Pi}_1^1$-norm. ∎

Exercise 5.14.6 Show that every $\mathbf{\Pi}_2^1$ set in the product of two Polish spaces can be uniformized by a $\mathbf{\Pi}_2^1$ set.

References

[1] J. W. Addison. Separation principles in the hierarchies of classical and effective descriptive set theory. Fund. Math., 46 (1959), 123 – 135.

[2] P. Alexandrov. Sur la puissance des ensembles mesurables *B*. C. R. Acad. des Sci., Paris, 162 (1916), 323 – 325.

[3] W. J. Arsenin. Sur les projections de certain ensembles mesurable *B*. C. R. (Doklady) Acad. Sci. USSR, 27 (1940), 107 – 109.

[4] W. Arveson. *An Invitation to C^*-Algebra*. Graduate Texts in Mathematics 39, Springer-Verlag, New York, Heidelberg, Berlin, 1976.

[5] R. J. Aumann. Measurable utility and the measurable choice theorem. Coll. Int. CNRS, 171 (1969), 15 – 28.

[6] L. Auslander and B. Konstant. Polarization and unitary representations of solvable Lie groups. Invent. Math., 14 (1971), 255 – 354.

[7] L. Auslander and C. C. Moore. Unitary representations of solvable Lie groups. Mem. Amer. Math. Soc., 62 (1966).

[8] R. Baire. Sur les fonctions de variables réelles. Annali Matematica Pura ed Applicata, Serie III^a, 3 (1899), 1 – 122.

[9] T. Bartoszyński and H. Judah. *Set Theory—On the structure of the Real Line*. A. K. Peters, Wellesley, Massachusetts, 1995.

[10] D. Basu. Problems relating to the existence of maximal elements in some fields of statistics (subfields). Proc. V Berkeley Symp. on Mathematical Statistics and Probability, 1 (1965), 41 – 50.

[11] H. Becker and A. S. Kechris. The descriptive set theory and Polish group actions. Preprint.

[12] S. Bhattacharya and S. M. Srivastava. Selection theorems and invariance of Borel pointclasses. Proc. Amer. Math. Soc., 97 (1986), 707 – 711.

[13] D. Blackwell. On a class of probability spaces. Proc. 3rd. Berkeley symp. on Mathematical Statistics and Probability, 2 (1956), 1 – 6.

[14] D. Blackwell. Discounted dynamic programming. Ann. Math. Statis., 36 (1965), 226 – 235.

[15] D. Blackwell. Infinite games and analytic sets. Proc. Nat. Acad. Sci. U.S.A., 58 (1967), 1836 – 1837.

[16] D. Blackwell. A Borel set not containing a graph. Ann. Math. Statis., 39 (1968), 1345 – 1347.

[17] D. Blackwell and C. Ryll-Nardzewski. Non-existence of everywhere proper conditional distributions. Ann. Math. Statis., 34 (1963), 223 – 225.

[18] J. V. Bondar. Borel cross-sections and maximal invariants. The Annals of Statistics, 4 (1976), 866 – 877.

[19] E. Borel. *Leçonssur les fonctions de réelles.* Gauthiers – Villars, Paris, 1905.

[20] E. Borel. Sur la classification des ensembles de mesure nulle. Bull. Soc. Math. France, 47 (1919), 97 – 125.

[21] J. P. Burgess. A reflection phenomenon in descriptive set theory. Fund. Math., 104 (1979), 127 – 139.

[22] J. P. Burgess. Equivalences generated by families of Borel sets. Proc. Amer. Math. Soc., 69 (1978), 323 – 326.

[23] J. P. Burgess. A selection theorem for group actions. Pacific Journal of Mathematics, 80 (1979), 333 – 336.

[24] J. P. Burgess and A. Maitra. Non-existence of measurable optimal selections. Proc. Amer. Math. Soc., 116 (1992), 1101 – 1106.

[25] J. P. Burgess and D. E. Miller. Remarks on invariant descriptive set theory. Fund. Math., 90 (1975), 53 – 75.

[26] Georg Cantor. ***Contributions to the Founding of the Theory of Transfinite Numbers.*** Dover Publications, Inc., New York, 1955.

[27] C. Castaing. Sur les multi-applications mesurables, Revue Française d'Informatique et de Recherche Opérationnelle, 1 (1967), 91 – 126.

[28] C. Castaing and M. Valadier. ***Convex analysis and measurable multifunctions.*** Lecture Notes in Mathematics 580, Springer-Verlag, Berlin, Heidelberg, New York, 1977.

[29] C. C. Chang and H. J. Kiesler. ***Model Theory.*** North-Holland, Amsterdam, 1973.

[30] G. Choquet. Theory of capacities. Ann. Inst. Fourier. Grenoble, 5 (1955), 131 – 295.

[31] P. J. Cohen. The independence of the continuum hypothesis, I, II. Proc. Nat. Acad. Sci. USA, 50 (1963), 1143 – 1148, 51 (1964), 105 – 110.

[32] J. K. Cole. A selector theorem in Banach spaces. J. Optimization theory, 7 (1971), 170 – 172.

[33] C. Dellacherie. Capacities and analytic sets. Cabal Seminar 77 – 79(Ed. Kechris, Martin, and Moschovakis), 1 – 31.

[34] C. Dellacherie and P. A. Meyer. ***Probabilities and Potential.*** Math. Studies 29, 72, 151, North-Holland, Amsterdam, 1978.

[35] J. Diestel. ***Sequences and series in Banach spaces.*** Springer-Verlag, New york, 1984.

[36] J. Dixmier. ***Les C^*-algèbras et leurs représentations.*** Gauthier – Villars, Paris, 1969.

[37] J. Dixmier. Dual et quasi-dual d'une algèbre de Banach involutive. Trans. Amer. Math. Soc., 104 (1962), 278 – 283.

[38] J. L. Doob. ***Classical potential theory and its probabilistic counterpart.*** Springer-Verlag, New York, Berlin, Heidelberg, Tokyo, 1984.

[39] L. E. Dubins and L. J. Savage. ***Inequalities for stochastic processes.*** Dover, New York, 1976.

[40] E. G. Effros. Transformation groups and C^*-algebras. Annals of Mathematics, 81 (1965), 38 – 55.

[41] J. Feldman and C. C. Moore. Ergodic equivalence relations and Von Neumann algebras I. Trans. Amer. Math. Soc., 234 (1977), 289 – 324.

[42] A. F. Filipov. On certain questions in the theory of optimal control. SIAM J. Control, Ser A, (1959), 76 - 84.

[43] J. Glimm. Type I C^*-algebras. Annals of Mathematics, 73 (1961), 572 - 612.

[44] K. Gödel. Über formal unentscheidbare Sätze der Principia Mathematica und verwandter Systeme I. Monatshefte für Mathematik und Physik, 38 (1931), 173 - 198.

[45] K. Gödel. The consistency of the axiom of choice and the generalized continuum hypothesis. Proc. Nat. Acad. Sci. USA, 24 (1938), 556 - 557.

[46] E. Grzegorek, K. P. S. B. Rao and H. Sarbadhikari. Complementation in the lattice of Borel structures. Coll. Math., 31 (1974), 29 - 32.

[47] L. Harrington, A. S. Kechris, and A. Louveau. A Glimm-Effros dichotomy for Borel equivalence relations. J. American Math. Soc., 3 (1990), 903 - 928.

[48] G. Hillard. Une généralisation du théorème de Saint-Raymond sur les boréliens à coupes K_σ. C. R. Acad. Sci., Paris, 288 (1979), 749 - 751.

[49] T. Jech. *Set Theory.* Academic Press, New York, San Francisco, London, 1978.

[50] R. R. Kallman and R. D. Mauldin. A cross section theorem and an application to C^*-algebras. Proc. Amer. Math. Soc., 69 (1978), 57 - 61.

[51] I. Kaplanski. The structure of certain operator algebras. Trans. Amer. Math. Soc., 70 (1951), 219 - 255.

[52] A. S. Kechris. Measure and category in effective descriptive set theory. Ann. Math. Logic, 5 (1973), 337 - 384.

[53] A. S. Kechris. *Classical Descriptive Set Theory.* Graduate Texts in Mathematics 156, Springer-Verlag, New York, 1994.

[54] A. S. Kechris and A. Louveau. *Descriptive set theory and the structure of sets of uniqueness.* London Math. Soc. Lecture Note Series, 128, Cambridge University Press, Cambridge, 1989.

[55] E. Klewin and A. C. Thompson. *Theory of correspondences including applications to Mathematical Economics.* Wiley-Interscience, New York, 1984.

[56] M. Kondô. Sur l'uniformisation des complémentaires analytiques et les ensembles projectifs de la seconde classe. Japan J. Math., 15 (1939), 197 - 230.

[57] D. König. Über eine Schlussweise aus dem Endlichen ins Unendliche. Acta Litt. Ac. Sci. Hung. Fran. Joseph, 3 (1927), 121 - 130.

[58] J. König. Zum Kontinuumproblem. Math. Ann., 60 (1904), 177 - 180.

[59] K. Kunen. *Set Theory, An Introduction to Independence Proofs.* Studies in Logic and the Foundations of Mathematics, 102, North-Holland Publishing Company, 1980.

[60] K. Kunugui. Contributions à la théorie des ensembles boreliens et analytiques II and III. Jour. Fac. Sci. Hokkaido Univ., 8 (1939 - 40), 79 - 108.

[61] K. Kuratowski. Sur une généralisation de la notion d'homéomorphie. Fund. Math., 22 (1934), 206 - 220.

[62] K. Kuratowski and A. Mostowski. *Set Theory, With an Introduction to Descriptive Set Theory.* (Second Edition), Studies in Logic and the Foundations of Mathematics, 86, North-Holland Publishing Company, 1976.

[63] K. Kuratowski and C. Ryll-Nardzewski. A general theorem on selectors. Bull. Polish Acad. des Sciences, 13 (1965), 397 - 403.

[64] D. Lascar. Why some people are excited by Vaught's conjecture. J. Symb. Logic, 50 (1985), 973 - 982.

[65] H. Lebesgue. Sur les fonctions représentables analytiquement. Journal de Math., 6e serie 1 (1905), 139 - 216.

[66] A. Louveau. A separation theorem for Σ^1_1 sets. Trans. Amer. Math. Soc., 260 (1980), 363 - 378.

[67] N. Lusin. Sur la classification de M. Baire. C. R. Acad. des Sci., Paris, 164 (1917), 91 - 94.

[68] N. Lusin. Sur le problème de M. Emile Borel et les ensembles projectifs de M. Henri Lebesgue: les ensembles analytiqes. C. R. Acad. des Sci., Paris, 180 (1925), 1318 - 1320.

[69] N. Lusin. Sur les ensembles projectifs de M. Henri Lebesgue. C. R. Acad. des Sci., Paris, 180 (1925), 1572 - 1574.

[70] N. Lusin. Les proprieties des sensembles projectifs. C. R. Acad. des Sci., Paris, 180 (1925), 1817 - 1819.

[71] N. Lusin. Sur le problème de M. J. Hadamard d'uniformisation des ensembles. C. R. Acad. des Sci., Paris, 190 (1930), 349 - 351.

[72] G. W. Mackey. Borel structures in groups and their duals. Trans. Amer. Math. Soc., 85 (1957), 134 - 165.

[73] G. W. Mackey. *The Theory of Unitary Group Representations.* Univ. of Chicago Press, Chicago, 1976.

[74] A. Maitra. Discounted dynamic programming on compact metric spaces. Sankhyā A, 30 (1968), 211 - 216.

[75] A. Maitra. Selectors for Borel sets with large sections. Proc. Amer. Math. Soc., 89 (1983), 705 - 708.

[76] A. Maitra and C. Ryll-Nardzewski. On the Existence of Two Analytic Non-Borel Sets Which Are Not Isomorphic. Bull. Polish Acad. des Sciences, 28 (1970), 177 - 178.

[77] A. Maitra and B. V. Rao. Generalizations of Castaing's theorem on selectors. Coll. Math., 42 (1979), 295 - 300.

[78] A. Maitra and W. Sudderth. *Discrete Gambling and Stochastic Games.* Applications in Mathematics 32, Springer-Verlag, New York, 1996.

[79] R. Mansfield and G. Weitkamp. *Recursive Aspects of Descriptive Set Theory.* Oxford Logic Group, Oxford University Press, New York, Oxford, 1985.

[80] D. A. Martin. Borel determinacy. Annals of Mathematics, 102 (1975), 363 - 371.

[81] R. D. Mauldin. Bimeasurable Functions. Proc. Amer. Math. Soc., 29 (1980), 161 - 165.

[82] E. Michael. Continuous selections I. Annals of Mathematics, 63 (1956), 361 - 382.

[83] A. W. Miller. *Descriptive Set Theory and Forcing: How to prove theorems about Borel sets in a hard way.* Lecture notes in Logic 4, Springer-Verlag, New York, 1995.

[84] D. E. Miller. On the measurability of orbits in Borel actions. Proc. Amer. Math. Soc., 63 (1977), 165 - 170.

[85] D. E. Miller. A selector for equivalence relations with G_δ orbits. Proc. Amer. Math. Soc., 72 (1978), 365 - 369.

[86] G. Mokobodzki. Démonstration élémentaire d'un théorème de Novikov. Sém. de Prob. X, Lecture Notes in Math. 511, Springer-Vrlag, Heidelberg, 1976, 539 - 543.

[87] C. C. Moore. Appendix to"Polarization and unitary representations of solvable Lie groups". Invent. Math., 14 (1971), 351 - 354.

[88] Y. N. Moschovakis. *Descriptive Set Theory.* Studies in Logic and the Foundations of Mathematics, 100, North-Holland Publishing Company, 1980.

[89] M. G. Nadkarni. *Basic Ergodic Theory.* Texts and Readings in Mathematics 7, Hindustan Book Agency, New Delhi, 1995.

[90] P. Novikoff. Sur les fonctions implicites mesurables B. Fund. Math., 17(1931), 8-25.

[91] C. Olech. Existence theorems for optimal problems with vector-valued cost function. Trans. Amer. Math. Soc., 136 (1969), 159 - 180.

[92] D. Preiss. The convex generation of convex Borel sets in Banach spaces. Mathematika, 20 (1973), 1 - 3.

[93] R. Purves. On Bimeasurable Functions. Fund. Math., 58 (1966), 149 - 157.

[94] B. V. Rao. On discrete Borel spaces and projective sets. Bull. Amer. Math. Soc., 75 (1969), 614 - 617.

[95] B. V. Rao. Remarks on analytic sets. Fund. Math., 66 (1970), 237 - 239.

[96] B. V. Rao and S. M. Srivastava. An Elementary Proof of the Borel Isomorphism Theorem. Real Analysis Exchange, 20 (1), 1994/95 1 - 3.

[97] J. Saint Raymond. Boréliens à coupes K_σ. Bull. Soc. Math. France, (2) 100 (1978), 141 - 147.

[98] R. Sami. Polish group actions and the Vaught conjecture. Trans. Amer. Math. Soc., 341 (1994), 335 - 353.

[99] H. Sarbadhikari. A note on some properties of A-functions. Proc. Amer. Math. Soc., 56 (1976), 321 - 324.

[100] H. Sarbadhikari. Some uniformization results. Fund. Math., 97 (1977), 209 - 214.

[101] H. Sarbadhikari and S. M. Srivastava. Parametrizations of G_δ-valued multifunctions. Trans. Amer. Math. Soc., 258 (1980), 165 - 178.

[102] H. Sarbadhikari and S. M. Srivastava. Random theorems in topology. Fund. Math., 136 (1990), 65 - 72.

[103] M. Schäl. A selection theorem for optimization problems. Arch. Math., 25 (1974), 219 – 224.

[104] M. Schäl. Conditions for optimality in dynamic programming and for the limit of n-state optimal policies to be optimal. Z. Wahrscheinlichkeitstheorie and verw. Gebiete, 32 (1975), 179 – 196.

[105] W. Sierpiński. Sur une classe d'ensembles. Fund. Math., 7 (1925), 237 – 243.

[106] J. H. Silver. Counting the number of equivalence classes of Borel and coanalytic equivalence relations. Ann. Math. Logic, 18 (1980), 1 – 28.

[107] M. Sion. On capacitability and measurability. Ann. Inst. Fourier, Grenoble, 13 (1963), 88 – 99.

[108] S. Solecki. Equivalence relations induced by actions of Polish groups. Trans. Amer. Math. Soc., 347 (1995), 4765 – 4777.

[109] S. Solecki and S. M. Srivastava. Automatic continuity of group operations. Topology and its applications, 77 (1997), 65 – 75.

[110] R. M. Solovay. A model of set theory in which every set of reals is Lebesgue measurable. Annals of Mathematics, 92 (1970), 1 - 56.

[111] M. Souslin. Sur une definition des ensembles B sans nombres tranfinis. C. R. Acad. Sciences, Paris, 164 (1917), 88 – 91.

[112] J. R. Steel. On Vaught's conjecture. Cabal Seminar 1976 – 1977, Lecture Notes in Mathematics, 689, Springer-Verlag, Berlin, 1978.

[113] E. Szpilrajn-Marczewski. O mierzalności i warunku Baire'a. C. R. du I congrès des Math. des Pays Slaves, Varsovie 1929, p. 209.

[114] S. M. Srivastava. Selection theorems for G_δ-valued multifunctions. Trans. Amer. Math. Soc., 254 (1979), 283 – 293.

[115] S. M. Srivastava. A representation theorem for closed valued multifunctions. Bull. Polish Acad. des Sciences, 27 (1979), 511 – 514.

[116] S. M. Srivastava. A representation theorem for G_δ-valued multifunctions. American J. Math., 102 (1980), 165 – 178.

[117] S. M. Srivastava. Transfinite Numbers. Resonance, 2 (3), 1997, 58 – 68.

[118] J. Stern. Effective partitions of the real line into Borel sets of bounded rank. Ann. Math. Logic, 18 (1980), 29 – 60.

[119] J. Stern. On Lusin's restricted continuum hypothesis. Annals of Mathematics, 120 (1984), 7 – 37.

[120] A. H. Stone. Non-separable Borel sets. Dissertationes Mathematicae (Rozprawy Matematyczne), 28 (1962).

[121] S. M. Ulam. *A collection of mathematical problems.* Interscience, New York 1960.

[122] R. L. Vaught. Denumerable models of complete theories. Infinitistic Methods: Proceedings of the symposium on foundations of mathematics, PWN, Warsaw, 1961, 303 – 321.

[123] R. L. Vaught. Invariant sets in topology and logic. Fund. Math., 82 (1974), 269 – 293.

[124] J. Von Neumann. On rings of operators: Reduction Theory. Annals of Mathematics, 50 (1949), 401 – 485.

Glossary

$\equiv$ 1

$\mathbb{N}$ 2

$\mathbb{Q}$, X^k, $X^{<\mathbb{N}}$, e, $\mathbb{K}$, $\mathcal{P}(X)$, X^Y 3

$\leq_c$, $<_c$, 4

χ_A 5

$\mathbb{R}$ 6

$\succeq$, $\preceq$, $g|A$ 9

$|X|$, $\aleph_0$, $\mathfrak{c}$ 13

$\lambda + \mu$, $\lambda \cdot \mu$, λ^μ, $\lambda \leq \mu$, $\lambda < \mu$, $\Pi_i \lambda_i$, $\sum_i \lambda_i$ 14

ω_0 15

$\sim$, w^-, $W(w)$ 16

$\sum_{\alpha \in W} W_\alpha$, $W_a + W_b$, $W_1 \times W_2$ 17

$\prec$, $\preceq$ 20

$t(W)$ 21

ON, $\alpha < \beta$, $\alpha \leq \beta$, $\alpha + \beta$, $\alpha \cdot \beta$ 22

ω_1, $\aleph_1$, ω_α, $\aleph_\alpha$ 24

κ^+, $\beth_\alpha$ 25

$A^{<\mathbb{N}}$, $|s|$, a^n, $s|m$, $s \perp t$, $s\hat{\ }t$, $s\hat{\ }a$, $s \prec \alpha$ 26

$[T]$, T_u 27

$T_{[n]}$, $\bigvee_n T_n$ 28

$\leq_{KB}$ 29

$\rho(T)$ 30

$T[\alpha]$ 31

$\mathcal{F}_\sigma$, $\mathcal{F}_\delta$, $\mathcal{F}_s$, $\mathcal{F}_d$, $\neg\mathcal{F}$ 32

$\mathcal{A}_A$, $\mathcal{A}$, $\mathcal{A}_2$, $\mathcal{A}_A(\mathcal{F})$, $\Sigma(s)$ 33

$\mathrm{cl}(A)$, $\mathrm{int}(A)$, $x_n \to x$, $\lim x_n$ 42

A' 43

$d(x, A)$, G_δ, F_σ 44

$\hat{d}$, $\hat{X}$, $\mathbb{H}$ 50

$\bigoplus_n X_n$ 51

$\mathcal{C}$	52
O_f	54
K_σ, $C(X,Y)$	63
$irr(n)$	64
$\mathbb{T}$, $\mathbb{R}_\times$, $GL(n,\mathbb{F})$, $SO(n,\mathbb{R})$, S_∞	65
$K(X)$, $[U_0;U_1,\ldots,U_n]$	66
δ_H	67
$K_f(X)$, $K_p(X)$	69
K^α	72
$\rho(K)$	73
$\sigma(\mathcal{G})$	82
$\mathcal{D}\|Y$, $\mathcal{B}_X$	83
$\mathcal{F}_+$	84
$\bigotimes_i \mathcal{A}_i$	87
λ, $\mu\times\nu$	102
$\prod_n \mu_n$	103
$\mu^{\mathbf{N}}$, λ, $\overline{\mathcal{A}}^\mu$, μ^*	103
$\overline{\mathcal{B}_{\mathbb{R}}}^\lambda$	104
$E+x=\{y+x : y\in E\}$	105
BP	107
$A^{\Delta U}$, A^{*U}	112
$A^{\Delta U}$, A^{*U}, W^Δ, W^*	113
$\mathbf{\Sigma}^0_1$, $\mathbf{\Pi}^0_1$, $\mathbf{\Sigma}^0_\alpha$, $\mathbf{\Pi}^0_\alpha$, $\mathbf{\Delta}^0_\alpha$	115
$\exists^Y\mathbf{\Gamma}$, $\forall^Y\mathbf{\Gamma}$	127
$\mathbf{\Sigma}^1_1$, $\mathbf{\Pi}^1_1$, $\mathbf{\Delta}^1_1$	128
$\mathbf{\Sigma}^1_n$, $\mathbf{\Pi}^1_n$, $\mathbf{\Delta}^1_n$	130
Tr, WF	136
LO, WO, IF^*	137
$DIFF$	139
G_x	161
$\leq^{\mathbf{\Pi}^1_1}_\varphi$, $\leq^{\mathbf{\Sigma}^1_1}_\varphi$	165
$\leq^*_\varphi$, $<^*_\varphi$, $m <_x n$, $\leq^{\mathbf{\Sigma}^1_1}_{\|\cdot\|}$, $<^{\mathbf{\Sigma}^1_1}_{\|\cdot\|}$	166
$R(\alpha,K)$, $S(\alpha,K)$	167
$\leq_\rho$, $<_\rho$	168
Z_f	178
$G^{-1}(A)$, $\mathrm{gr}(G)$	184
A^*	186
$X/\mathbf{\Pi}$	188
$irr(n)$, $irr(n)/\sim$	196
E_G	211
$irr(A,H_n)$, $irr(A)$, $\bigoplus_n irr(A,H_n)$	214
$irr(A)/\sim$, $Prim(A)$, $\hat{\kappa}$	215
$\mathcal{C}\bigvee\mathcal{D}$	218
$D^\alpha(A)$, $\|A\|_D$, $D^\infty(A)$, Ω_D	222
X_L, $\mathcal{A}_x$, $L_{\omega_1\omega}$, $\bigvee_n\phi_n$, $\mathcal{A}_\sigma$	228
E_a	232
$<_{\mathrm{lex}}$	237

Index

A-function, 134
accumulation point, 42
action of a group, 113, 162
    Borel
        orbit, 197
    logic, *see* logic action
J. W. Addison, xiii
alephs, 24
P. Alexandrov, xii, 32, 52
algebra of sets, 81
    σ, *see* σ-algebra
algebraic number, 3
analytic determinacy, 147, 169, 236
analytic set, xi, 128, 130, 132, 141, 145
    σ-compact sections
        separation, 219
    closed sections
        separation, 157
    comeager sections, 226
    convex sections, 217
    countable sections, 216, 217
    first separation theorem, 147
        invariant version, 149
        proof using capacity, 175
    generalized first separation theorem, 155
    meager sections, 226
    second separation theorem, 175
antilexicographical order, 17
W. J. Arsenin, xii, 219
L. Auslander, 215
automatic continuity of group operations, 163
axiom of choice (**AC**), 7
axiom of constructibility, 169

R. Baire, xi
Baire
    σ-algebra, 108
    function, xi, 90
    measurable function, 110
    property, 107, 108
    space, 108, 109
Baire category theorem, 70
Banach category theorem, 73
Banach space, 151

T. Bartoszyński, xiii
D. Basu, 218
H. Becker, 229
Bernstein set, 94, 104, 108
S. Bhattacharya, 192
bimeasurable function, 94
  characterization, 181
binary relation
  antilexicographical, *see* antilexicographical orderrder17
  linear order, *see* linear order
  partial order, *see* partial order
  reflexive, 8
  transitive, 8
  well-order, *see* well-order
Blackwell – Mackey theorem, 154, 199
D. Blackwell, xii, xiv, 185, 208
E. Borei, xi
Borel code, 171
Borel conjecture, 145
Borel function, 86, 150
  countable-to-one, 180
  extension, 92
  of class α, 118
  one-to-one, 150, 153, 154
Borel homomorphism, 110, 144
Borel isomorphism, 94, 95
  extension, 100
Borel isomorphism theorem, 99, 134
Borel measure, 105
  finite, 105, 106
Borel on Borel, 206
Borel reducible, 136
Borel set, 83, 108
  G_δ sections, 220
  σ-compact sections, 220
    uniformization, 219
  closed sections, 158, 160
  compact sections, 159
    uniformization, 204
  convex, 149
  convex sections, 159
  countable sections, 180, 210
    uniformization, 205
  large sections
    uniformization, 206
  nonmeager sections
    uniformization, 207
  of additive class, 116
  of ambiguous class, 116, 123
  of multiplicative class, 116
  open sections, 157
  singleton sections, 180
  Solovay's code, 151, *see* Solovay's code of Borel sets
Borel σ- algebra, 83, 84
Borel space
  quotient, *see* quotient Borel space
boundedness theorem, 168
J. P. Burgess, 194, 201, 216, 217, 230

C^*-algebra, 64, 214
  GCR, 214
  locally type I, 215
  type I, 214
G. Cantor, xii, 3, 4, 25, 26, 58
Cantor – Bendixson derivative, 73
Cantor – Bendixson theorem, 76
Cantor intersection theorem, 49
Cantor ternary set, 58
capacitability, 174
  universal, 174
capacity, 172
  separation, 172
cardinal number
  as an initial ordinal, 25
  successor, *see* successor of a cardinal
cardinal number of a set, 13
cardinalities
  of σ-algebras, 86
  of analytic sets, 142

of Borel σ-algebras, 117
of Borel sets, 93
of coanalytic sets, 147
of Polish spaces, 76
Castaing's theorems, 192
characteristic function, 5
choice function, 7
G. Choquet, xiv, 175
Choquet capacitability theorem, 175
classification
problem, 196
space, 196, 197
clopen set, 52
closed set, 42
as the body of a tree, 56
closure, 42
coanalytic set, 128, 134
constituent, 146
reduction principle, 170
weak reduction principle, 156
P. J. Cohen, xiii, 26
comeager set, 70, 109
compact space, 57
complement of a sub σ-algebra, 218
weak, 218
minimal, 218
complete set
$\mathbf{\Pi}_1^1$-, 136
$\mathbf{\Sigma}_1^1$-, 135
completely metrizable group, 71
completely metrizable space, 55
conditional distribution, 209
proper, 209
continuum hypothesis (**CH**), 25, 88
generalized (**GCH**), 25
convergence
pointwise, 44
uniformly, 44
cross section theorem
Burgess, 201
Effros – Mackey, 196
Miller, 213
Srivastava, 213

R. Dedekind, 5
dense-in-itself set, 43
derivative of a set, 43
derivative operator on a Polish space, 221
derived set, *see* derivative of a set
diameter, 49
P. G. L. Dirichlet, xi
distribution function, 106
L. E. Dubins, 194

E. G. Effros, xiv, 195, 196
Effros Borel space, 97, 191, 221
embedding, 44
equinumerous sets, 1
equivalence relation
F_σ, 77
analytic, 217, 230, 235
number of equivalence classes, 230
Borel
equivalence classes countable, 212
closed, 76, 187
induced by a group action, 211
meager
number of equivalence classes, 231

J. Feldman, 212
Fell topology, 69
finite intersection property, 58
first order language, 227
formula
valid, *see* valid formula
function
Baire measurable, *see* Baire measurable function
bijection, 1
bimeasurable, *see* bimeasurable function

characteristic, *see* characteristic function
choice, *see* choice function
continuous, 43
at a point, 43
extension, 54
nowhere differentiable, 71
uniformly, 44
lower-semicontinuous, 47
measurable, *see* measurable function
Borel, *see* Borel function
oscillation at a point, *see* oscillation
simple, 89
upper-semicontinuous, 47, 61, 62

K. Gödel, xiii, 26, 147, 169
J. Glimm, 214
E. Grzegorek, 218

Hamel basis, 13
Hausdorff metric, 67
hereditary family, 221
hierarchy
Kalmar, 57
of Borel sets, 116
of projective sets, 131
Hilbert cube, 50
Hilbert space, 214
G. Hillard, 226
G. Hjorth, 164
homeomorphism, 44
hull – kernel topology, 215

ideal of sets, 103
idempotence of the Souslin operation, 35
immediate predecessor, 16
induction
on natural numbers
definition by, 18
proof by, 18
on well-founded trees
definition by, 30
proof by, 30
transfinite
definition by, 19
proof by, 19
initial ordinal, 25
initial segment, 16
interior, 42
irreducible
*-representation, 214
matrices, 64
isolated points, 43
isometry, 44
isomorphism theorem for measure spaces, 107

Jacobson density theorem, 64
J. E. Jayne, 123, 177
T. Jech, xiii
H. Judah, xiii

König's infinity lemma, 28
König's theorem on cardinal numbers, 15
R. R. Kallman, 215
I. Kaplanski, 214
A. S. Kechris, xii, xiii, 206, 207, 210, 229
Kleene – Brouwer ordering, 29
A. N. Kolmogorov, xiv
Kondô's theorem, 236
M. Kondô, xvi, 199, 236
D. König, 28
J. König, 13
B. Konstant, 215
K. Kunen, 230
K. Kunugui, xii, 157, 219
K. Kuratowski, 98, 170, 175, 189
Kuratowski – Ulam theorem, 112

M. Lavrentiev, 55
H. Lebesgue, xi
Lebesgue – Hausdorff theorem, 91, 124
Lebesgue measurable set, 104

Lebesgue measure, 102–105
Lebesgue number, 59
limit ordinal, 22
linear order, 8
logic action, 228
Lopez-Escobar, 228
A. Louveau, xiii, xvi, 227
N. Lusin, xii, xiii, 180, 205, 210, 211, 216, 217
Lusin set, 145

G. W. Mackey, xiv, 196
A. Maitra, xiv, 123, 169, 177, 191, 194
E. Marczewski, 114
D. A. Martin, xvi
matrix
- adjoint of, 64
- unitarily equivalent, 196
- unitary, 196

R. D. Mauldin, 182, 215
S. Mazurkiewicz, 139
meager set, 70, 109
measurable function, 86
- extension, 95
- of two variables, 89
- universally, 198

measurable set, 81
measurable space, 81
- isomorphism, 94

measure, 100
- σ-finite, 101
- Borel, *see* Borel measure
- continuous, 106
- counting, 101
- Dirac, 101
- extension, 102
- finite, 101, 102
- Lebesgue, *see* Lebesgue measure
- outer, *see* outer measure
- probability, 101
- product, *see* product measure
- regular, 104
- uniform, 101

measure space, 101
- σ-finite, 101
- complete, 103
- completion of a, 103
- finite, 101
- probability, 101

metric, 39
- complete, 49
- discrete, 40
- equivalent metrics, 41
- Hausdorff, *see* Hausdorff metric
- product, 40
- topology induced by a, 41
- totally bounded, 59
- usual, 40

metric space, 39
- compact, 78, 79
- complete, 49
- completion of a, 50
- second countable, 48
- separable, 48
- subspace, 40
- zero-dimensional, 55, 56, 59

metrizable space, *see* metric space
D. E. Miller, 160, 161, 197, 213
G. Mokobodzki, 155
monotone class, 85
monotone class theorem, 85
C. C. Moore, 212, 215
Y. N. Moschovakis, xiii, 165
μ-measurable set, 103
multifunction, 184
- closed-valued representation, 192
- compact-valued, 185, 191
- graph, 184
- lower-semicontinuous, 184
- measurable, 184
- section, 185
- selection, 184
- strongly measurable, 184
  - closed-valued, 192

upper-semicontinuous, 184

neighborhood of a point, 42
ϵ-net, 59
non-Lebesgue measurable set, 104, 105
nonmeager set, 109
norm, 165
Π^1_1-, 165
canonical, 167
P. Novikov, xii, 155, 157, 159, 204
nowhere dense, 69
null set, 103

open set, 41
order dense, 9
order isomorphic, 9
ordinal number, 22
even, 23
initial, *see* initial ordinal
limit, *see* limit ordinal
odd, 23
successor, *see* successor ordinal
outer measure, 103, 172
induced by a measure, 104

partial order, 8
partially ordered set, 8
chain, 9
maximal element, 9
partition, 186
closed, 186
countably separated, 187, 188
into σ-compact sets, 226
cross section, 186
into G_δ sets, 187
into closed sets
semicontinuous, 195
into countable sets, 205
lower-semicontinuous, 186
measurable, 186
into G_δ sets, 213
into closed sets, 196
saturation, 186
section, 186
semicontinuous
into G_δ sets, 213
upper-semicontinuous, 186
perfect set, 69
permutation, 65
Pettis theorem, 110
Π^1_1 on Π^1_1, 216
pointclass, 115
Borel
additive, 116
ambiguous, 116
invariance of, 193
multiplicative, 116
projective, 130
Polish group, 65, 71, 162, 163
closed subgroup, 160, 197
Polish space, 52, 76, 77, 79
characterization, 55
zero-dimensional, 56
poset, *see* partially ordered set
power set, 3
D. Preiss, 149
prewellordering, *see* norm
product measure, 102, 103
product σ-algebra, 87, 88
discrete, 88
projective set, xii, 130

quotient Borel space, 188

Ramsey - Mackey theorem, 100
rank function, 30
B. V. Rao, 88, 98, 141, 191
K. P. S. B. Rao, 218
reduction theorem
failure for analytic sets, 170
for additive Borel classes, 120
for coanalytic sets, 170
reflection theorem, 216
retract, 78
retraction, 78

B. Riemann, xi
C. A. Rogers, 123, 177
B. Russell, 7
C. Ryll-Nardzewski, 169, 189, 208

Saint Pierre, 217
J. Saint Raymond, 157, 219
R. Sami, 232
H. Sarbadhikari, xii, 134, 158, 207, 218, 226
L. J. Savage, 194
scale, 236
  κ-, 236
  Π^1_1-, 237
  very good, 237
    Π^1_1-, 237
M. Schäl, 194
Schröder – Bernstein Theorem, 5
Schröder – Bernstein theorem
  for measurable spaces, 96
selection theorem
  Blackwell and Ryll-Nardzewski, 208
  Dubins – Savage, 194
  for group actions, 201
  Kuratowski and Ryll-Nardzewski, 189
  Von Neumann, 198, 200
semidirect product of groups, 164
separation theorem
  failure for coanalytic sets, 170
  for multiplicative Borel classes, 121
sequence
  Cauchy, 48
  limit, 42
  of sets
    convergent, 122
set
  cardinal number, *see* cardinal number of a set
  countable, 2
  finite, 2
  infinite, 2
  partially ordered, *see* partially ordered set
  uncountable, 2
set function
  countably additive, 101
  countably subadditive, 101
  going up, 172
  monotone, 172
  right-continuous over compacta, 172
set of first category, *see* meager set
set of second category, *see* nonmeager set
W. Sierpiński, xii, 51, 84
σ-algebra, 81
  atom of a, 85
  atomic, 85
  closure under Souslin operation, 114
  countable-cocountable, 82, 86
  countably generated, 82, 86
  discrete, 82
  generator, 82
  generator of a, 82, 83
  indiscrete, 82
  Marczewski complete, 114
  trace, 83
σ-algebras
  Borel isomorphic, 94
σ-ideal, 103
J. H. Silver, xvi, 231
Silver's theorem, 231, 232, 234
S. Simpson, 143
S. Solecki, 163, 164, 232
R. M. Solovay, 105, 108, 164
solvable group, 215
M. Souslin, xi, xii, 31, 128, 148
Souslin operation, 31, 33, 34
  closure properties, 33

idempotence, *see* idempotence of the Souslin operation
The Souslin operation, 133
space of
rationals, 56
compact sets, 66, 68, 69
continuous functions, 63, 76
everywhere differentiable functions, 139
irreducible matrices, 64
nowhere differentiable continuous functions, 71
S. M. Srivastava, xiv, 98, 159, 163, 164, 192, 213, 215
stabilizer, 161
standard Borel space, 96
*-ideal, 215
J. Steel, 169
J. Stern, 230, 235, 236
A. H. Stone, 144
strong measure zero set, 144, 145
subcover, 57
successor, 16
successor of a cardinal, 25
successor ordinal, 22
system of sets, 32
associated map, 75
Cantor scheme, 75
Lusin scheme, 74
regular, 33
Souslin scheme, 74

Tietze extension theorem, 45
random version, 158
topological group, 65
Baire, 109
topological space, 41
Baire, *see* Baire space
completely metrizable, 52, 71
dense subset, 42
locally compact, 62, 63
normal, 45
second countable, 42
separable, 42
sequentially compact, 59
zero-dimensional, 52
topological sum, 51
topology, 41
base, 42
Borel-generated, 91
discrete, 41
Fell, *see* Fell topology
hull – kernel, *see* hull – kernel topology
of uniform convergence, 63
product, 48
subbase, 42
usual, 41
Vietoris, *see* Vietoris topology
transcendental number, 4
transition probability, 107
tree, 26
body, 27
body of a, 134
finitely splitting, 27
ill-founded, 27
node, 27
rank, *see* rank function
section, 31
terminal node, 27
well-founded, 27
the triangle inequality, 39
trichotomy theorem for well-ordered sets, 20
type of a well-ordered set, 21

S. Ulam, 88, 141
uniformization, xii, 185
unitarily equivalent, 214
universal pair, 122, 133
universal set, 119, 120, 129, 131, 141, 169
Urysohn lemma, 44, 171
random version, 158

valid formula, 228

R. L. Vaught, xvi, 229
Vaught conjecture (**VC**), 229
  topological (**TVC**), 230
    for abelian group, 232
    for locally compact Polish group, 230
    weak (**WTVC**), 227
Vaught transforms, 113, 229
vector space, 10
  basis, 10
Vietoris topology, 66, 67
Vitali partition, 105, 214

J. Von Neumann, xiii, 186, 198, 200

well-founded tree, 136
well-order, 15
well-ordered set, 15
  type, *see* well-ordered set
Well-Ordering Principle (**WOP**), 18

Zorn's Lemma, 9